EARTH
IS A
NUCLEAR PLANET

EARTH
IS A
NUCLEAR PLANET
THE ENVIRONMENTAL CASE
FOR NUCLEAR POWER

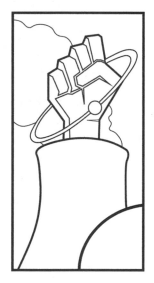

MIKE CONLEY & TIM MALONEY

Senior Science Advisor:
STEPHEN A. BOYD, Ph.D.

Earth Is a Nuclear Planet: The Environmental Case for Nuclear Power

First edition 2024

To find out more about Open Universe and Carus Books, visit our website at www.carusboooks.com.

Published by arrangement with Generation Atomic, GenerationAtomic.org, ENP@generationatomic.org

ISBN: 978-1-63770-059-4

This book is also available as an e-book (978-1-63770-060-0).

Library of Congress Control Number: 2023947379

Printed and bound in the United States of America. Printed on acid-free paper.

Cover art Aaftab Sheikh, www.Instagram.com/aaft.designs

Interior design and layout: 1106 Design

Illustrator: Erwan Sulistyo, erwan.syah21@gmail.com, https://www.fiverr.com/erwand

Photo of Catherine Sikorski on Author's Note page and photo of Junior in Chapter 17 © Michael Sean Conley

THANKS!

While writing this book, a small army of people have given us a wealth of notes, comments, and advice, for which we are deeply grateful. And though they did what they could, it is we the authors who remain solely responsible for any factual errors.

Science editors and advisors: Stephen A. Boyd, Ph.D.; Alex Cannara, Ph.D., MS (physics, engineering, statistics); Jaro Franta, P.Eng., retired AECL (deep geologic storage); Gregory Meyerson, Ph.D. (philosophy, education); William Pappert, Dr.Ph. (ABD) (epidemiology); John Schoonover, Ph.D. (nuclear physics); Bill Sacks Ph.D., MD (diagnostic radiology); and Robert Warshawski, MD, FRCP (radiology, nuclear medicine).

Editors and proofreaders: Matt Snider, John Flanagan, Lisa Sikorski, Eric Meyer, and John Slasky.

Special thanks to: Meredith Angwin, Sara Bancroft, Greg Barton, Mathijs Beckers, Ike Bottema, Tom Blees, Canon Bryan, Ed Calabrese, Jack Devanney, Cliff Gold, Robert Hargraves, Gabriel Ignetti, Lars Jorgensen, Jim Kennedy, Chris Keefer, John Kutsch, DJ LeClear, Gordon McDowell, Scott Medwid, Gene Nelson, Ed Norris, Ken Policard, Iida Rushalme, and Ed Sanders.

AUTHORS' NOTE

Over the last ten years, we have conversed and corresponded with dozens of nuclear scientists, physicists, engineers, and reactor operators, and we've read numerous books and papers on the subject.

Even so, don't take our word for it. We're the writers; they're the scientists. That's why all the endnotes, with plenty of extra links for the extra-curious reader. We're just explaining, as best we can, what we've been able to piece together. Throughout this project, we've been guided by a bit of sage advice, long attributed to Albert Einstein:

"You do not really understand something
unless you can explain it to your grandmother."

This book is dedicated to grandmothers everywhere
and to all the children they love.

Mike's grandmother Catherine Sikorski at
age seventeen (1920). With only a third-
grade education, she taught him how to
read soon after his fourth birthday.

ON JANUARY 7, 2021, a New Jersey high school student brought a small chip of pottery to science lab and left it there. It was about the size of a quarter, a fragment from an antique orange-glazed dinner plate, zipped in a baggie so no one would touch it.

Early next morning, while he was at home on the school's rotating Covid schedule, someone realized what was in the baggie. In "an abundance of caution" the entire school was evacuated and the hazmat team was called. In less than thirty minutes, they had removed the offending sample and cleared the school for re-entry. Then they paid a call on the student.

The budding young scientist was surprised when he opened his door. He thought it was common knowledge that anyone can buy old Fiestaware, online or in pawn shops and flea markets, all across the country. It's been a collector's item since the Seventies, when the company stopped using uranium-tinted glazes. He even put the chip in a baggie . . .

And that's how a little chunk of Fiestaware, with a whisper of a faint trace of radiation, cleared out an entire high school—even the science teachers—and made the national news.

It's nonsense like this that inspires us to write.

Contents

Acronyms and Abbreviations

ACGT – Adenine, Cytosine, Guanine, and Thymine

AEC – Atomic Energy Commission

AGW – Anthropogenic Global Warming

ALARA – As Low As Reasonably Achievable

AP-1000 – Advanced Pressurized (reactor) 1,000 megawatt-electric

APR-1400 – Advanced Pressurized Reactor 1400 megawatt-electric

BEAR – Biological Effects of Atomic Radiation

BED – Banana Equivalent Dose

BEIR – Biologic Effects of Ionizing Radiation

BMRC – British Medical Research Council

Bq – Becquerel

BS – bad science

BTU – British Thermal Unit

CANDU – Canadian Deuterium (heavy water) Uranium (reactor)

CAP-1000 – Chinese Advanced Pressurized (reactor) 1,000 megawatt-electric

CDC – Centers for Disease Control and Prevention

CF – Capacity Factor

CSP – Concentrated Solar Power

CT scan – Computerized Tomography
DGR – Deep Geologic Repository
DHS – Department of Homeland Security
DNA – Deoxyribonucleic Acid
DoE – Department of Energy
DU – Depleted Uranium
EBR-II – Experimental Breeder Reactor (version two)
EIA – Energy Information Administration
EPA – Environmental Protection Agency
EROEI – Energy Returned On Energy Invested
EROI – Energy Returned On Investment
eV – electron Volt
FP – Fission Product
FUD – Fear, Uncertainty, and Doubt
Gen-III – Generation Three
Gen-III+ – Generation Three Plus
Gen-IV – Generation Four
GND – Green New Deal
GW – Gigawatt
Gy – Gray
HALEU – High Assay Low-Enriched Uranium
HIV – Human Immunodeficiency Virus
HVDC – High Voltage Direct Current
IAEA – International Atomic Energy Agency
ICRP – International Council on Radiation Protection
IEA – International Energy Agency
IMSR – Integral Molten-Salt Reactor
INL – Idaho National Laboratory
IPCC – Intergovernmental Panel on Climate Change
ISL – In-Situ Leaching
kW – kilowatt

kWh – kilowatt hour

LD50 – Lethal Dose in 50% of cases

LEU – Low-Enriched Uranium

LNG – Liquefied Natural Gas

LNT – Linear No-Threshold

LSS – Life-Span Study

LTO – Long-Term Operation

LWR – Light-Water Reactor

MCFR – Molten Chloride Fast (neutron) Reactor

MeV – Megaelectron Volt

MIT – Massachusetts Institute of Technology

MOX – Mixed Oxides

MSR – Molten-Salt Reactor

MSRE – Molten-Salt Reactor Experiment

mSv – millisievert

MW – Megawatt

NAS – National Academy of Sciences (now NASEM)

NASA – National Aeronautics and Space Administration

NASEM – National Academies of Sciences, Engineering, and Medicine

NCRP – National Committee on Radiation Protection and Measurement

NERL – National Energy Research Laboratory

NIH – National Institutes of Health

NOAA – National Oceanic and Atmospheric Administration

NOEL – No Observable Effect Level

NRC – Nuclear Regulatory Commission

NSAID – Non-Steroidal Anti-Inflammatory Drug

NSWS – Nuclear Shipyard Worker Study

O&M – Operations and Maintenance

ORNL – Oak Ridge National Laboratory

PG&E – Pacific Gas and Electric

PV – Photovoltaic

PWR – Pressurized Water Reactor

rad – radiation absorbed dose

RBMK – *reaktor bolshoy moshchnosti kanalnyy* (high-power channel-type reactor)

RE – Renewable Energy

REE – Rare Earth Element

RERF – Radiation Effects Research Foundation

RMR – Rock to Metal Ratio

RPV – Reactor Pressure Vessel

SI – Système International

SNF – Spent Nuclear Fuel

SONGS – San Onofre Nuclear Generation Station

Sv – Sievert

SWAG – Scientific Wild-Ass Guess

TEPCO – Tokyo Electric Power Company

TMI – Three Mile Island

TRU – Transuranic

TW – Terawatt

TWh – Terawatt hour

UNSCEAR – United Nations Scientific Committee on the Effects of Atomic Radiation

UTES – Underground Thermal Energy Storage

WHO – World Health Organization

Wt% – Weight percent

WWS – Wind Water Sunshine

µSv – microsievert

Foreword

OUR NUCLEAR PLANET WAS BORN in a nuclear universe.

Mind-boggling cosmic explosions forged heavy elements like uranium in the chaotic dance of neutron stars. When these dense remnants collided, their physics-defying explosions scattered radioactive matter across the cosmos. Some found their way to our solar system. This is the special ingredient a planet needs to develop advanced forms of life.

Amazingly, the radioactive decay in our planet's core drives the convective heat of molten, magnetically-charged iron that generates our planet's magnetic field. This shield protects our atmosphere from the solar wind. Without our magnetosphere, the air we breathe and the water we drink would not exist. Our planet would look a lot like Mars, and you wouldn't be reading this sentence.

The natural nuclear reactors discovered in western Africa at Oklo, Gabon provide further evidence that this is a nuclear planet. Around two billion years ago, groundwater transported dissolved uranium ore to the site. Pools of rainwater allowed the neutrons to slow down and fission, creating the right conditions for a sustained nuclear chain reaction.

These natural reactors generated about 100 kilowatts of heat, enough to continuously power 10,000 modern 10-watt LED light bulbs or 40 electric ovens—*for hundreds of thousands of years*. Like the fictional "vibranium" of *Black Panther's* Wakanda, there was a special metal in Africa that could release an incredible amount of energy: uranium. Nuclear energy has been part of our planet's story from the very beginning, and the authors do a fantastic job of telling that story in an engaging and accessible way.

Our nuclear capabilities give us immense power to harm, or to provide abundant clean energy. This book is a firehose of common-sense, plain-language information that will help you better understand the origins of that power and the possibilities that lie ahead. It will expand your knowledge and give you hope for the future. Discuss it with friends, suggest it for your book club, and join communities to explore these ideas. Our future on this nuclear planet depends on curious, informed citizens—like you!

Stay positively radiant. Together, we'll keep writing nuclear's unfolding story—a tale as endless and surprising as the cosmos itself.

Eric G. Meyer
Executive Director, Generation Atomic

It May Surprise You to Learn That . . .

MORE AMERICANS HAVE BEEN SHOT by their own dog than have ever been harmed by nuclear power or nuclear waste. It sounds preposterous, but it's true—about once a year, someone puts their gun down or stuffs it in a bag or purse, and their dog steps on the trigger. In contrast, the last sixty-plus years of American nuclear power have caused only sixteen deaths (ten from construction) and one severe injury (see Chapter 4). Some other things you should know:

- More people die each year working on wind turbines and rooftop solar than have ever died from American nuclear power. (Chapter 8)

- The average dose received by anyone downwind from Three Mile Island was less than a chest X-ray. (Chapter 8)

- If no one had evacuated Fukushima, the maximum downwind dose would have equaled one CT scan. (Chapter 1)

- The false estimate of 4,000 future deaths from Chernobyl was based on a 1927 experiment, conducted before low levels of radiation could even be accurately measured and delivered. (Chapter 9)

- The bad science that came from this early work—the idea that there is no safe dose of radiation, and that all doses are cumulative—has impeded the buildout of nuclear power. (Chapter 10)

This book is a challenge to our fellow environmentalists to rethink their position on nuclear power. Too many of us seem to believe that it's a fate worse than global warming, and that we must devise a national clean-energy solution based entirely, or mostly, on renewables instead.

Ideas are being offered and debates are being waged on the best way to get the world to Carbon Zero. The goal is to undo the damage from two centuries of burning fossil fuel, and do what we can to minimize the consequences. But even if the world stopped burning all fossil fuel today, humanity has already put enough CO_2 into the atmosphere to risk a warmer and chaotic climate for the next several hundred years.

The world will require a stupendous amount of clean energy to achieve Carbon Zero. Then we'll need even more energy to go Carbon Negative, to restore the world's pre-industrial climate. This energy will have to be produced above and beyond the energy we'll need to run the machinery of civilization. And all of it must be carbon free. The trick, of course, is to do this as fast as *humanely* possible.

Passions are running high, because the decisions we make today will affect the future of civilization as we know it. That's usually just a figure of speech, but in this case it's all too real. Complicating matters is the fact that climate science has become a political chew toy, especially in the United States.

But this issue is beyond politics, or should be. The human-generated excess CO_2 that is warming the planet and acidifying the oceans is a real and worsening crisis, regardless of anyone's political persuasion. So if all this climate stuff sounds concerning, that's good. If you're not concerned, you haven't been paying attention.

The science is the science, and most of it comes down to basic high school physics and chemistry. How to respond is the only productive point of debate. As we see it, the best response with the least disruption to society would include a rapid buildout of nuclear power. France and the province of Ontario, where one in three Canadians live, have already demonstrated that this is an entirely feasible solution.

Credit: Gerhard Mester, translated by Lommes (CC-BY-SA-4.0)

Renewables are fine for what they can accomplish, but they're not enough, all by themselves, to reliably power the planet. The staggering amount of clean, on-demand energy we'll need in the years ahead defies comprehension. If people are serious about building a carbon-free world by midcentury, or even by 2100, nuclear power will have to be an integral part of the solution.

Earth Is a Nuclear Planet is the first book in our series on clean energy. In this book we address radioactivity, nuclear fear, spent fuel, and nuclear waste.

The LNT Report (www.TheLNTReport.com) is a short companion book to this volume, a guided tour through the long, strange tale of the linear no-threshold (LNT) model of radiation risk, devised nearly a century ago by Hermann Muller and now being critically reexamined, notably by Dr. Ed Calabrese at the University of Massachusetts, Amherst. We hope our synopsis of his work will help it gain the attention it deserves.

Roadmap to Nowhere (www.RoadmaptoNowhere.com) is the third book, our nuts-and-bolts comparison of a proposed 100% national renewables grid with a hypothetical 100% nuclear grid.

Power to the Planet (www.PowertothePlanetBook.com) will be the fourth book, where we explore nuclear physics, nuclear power, reactor design, thorium, liquid fuel vs. solid fuel, and the numerous ways nuclear power can help humanity in the years ahead.

The first edition of *Roadmap to Nowhere* has been available online since December 2017. We're revising the book in light of new information, but our thesis and conclusions remain the same. In fact, they're now more compelling than ever.

Though we wrote *Roadmap to Nowhere* first, this book is actually the first one to read if you have any serious concerns about radiation, nuclear energy, spent fuel, waste, proliferation, nuclear medicine, or all the above.

Over the last half-century, nuclear power has attracted more than its fair share of scare stories, cherry-picked facts, and overwrought concerns. These misunderstandings have impeded efforts to solve the world's energy crisis. In our view, the biggest obstacle to finding common ground on clean energy can be summed up in two words: *nuclear fear.*

So first things first.

CHAPTER ONE

The Only Thing We Have to Fear

"The only thing we have to fear
is fear itself."
—FRANKLIN D. ROOSEVELT

WE ALL KNOW THE FEELING—that clutch of anxiety when boarding a plane. We remind ourselves that air travel is the safest mode of transportation, and that driving to the airport is far riskier than flying. But most of us don't get scared behind the wheel because we're familiar with automobiles and highways, and we're the ones in control. We know it's a false sense of security, and yet we feel safe.

We also know that our fear of flying is not entirely rational. But we feel it anyway, especially as we're being herded through a wobbly jet bridge to the door of the plane, where the crew greets us with reassuring smiles. And then, like millions of people every day around the world, we swallow our fear, smile back, and step over the threshold.

Like nuclear power, flying can be scary, but it's not dangerous.

The risk / benefit calculations we make in situations like this are shaped by the fact that the goods and services we have come to expect in modern life are rarely risk free. So we take the best precautions we can, keep calm, and carry on.

Consider these facts:

- Since 1970, commercial airline crashes have caused over 80,000 deaths. Yet global air traffic has increased year after year. [1]

- In August 1975, the Banqiao Dam in China failed, causing an estimated 170,000 to 230,000 deaths and displacing more than ten million people. [2] The dam was rebuilt with improved safety measures, and the flood plain has since been repopulated. Despite the deadly risks of dam failure, some renewable energy advocates have been pushing for even more hydroelectric power. [3]

- In December 1984, an isocyanate gas leak from a pesticide plant in Bhopal, India, injured more than 550,000 downwind residents, with more than 3,500 immediate deaths and 15,000 additional deaths in the weeks that followed. Nearly forty years later, the cleanup continues. The company (Union Carbide) is still in business, and the same pesticides are still in widespread use, even in our own back yards.

- In September 2010, a natural gas pipe exploded in San Bruno, south of San Francisco. [4] Fifty homes and eight lives were lost in the inferno. The pipeline was repaired, the utility was fined, the neighborhood was rebuilt, and most of the residents rebuilt and moved back in. To this day, gas explosions, fires, and casualties from natural gas continue to mount, all around the world. [5]

Now consider the public reaction to these three incidents:

- In March 1979, a reactor at Three Mile Island in Pennsylvania suffered a partial meltdown, releasing a small amount of radioactive steam. The average dose of radiation received by the public in the vicinity of the plant was equal to less than one chest X-ray. [6]

But sensational media reports, and celebrity-led protests that linked nuclear power to nuclear weapons, stirred an understandably nervous public into unwarranted panic. The only serious fallout from the event was that the popularity of nuclear power suffered a near-fatal blow, and we expanded our use of coal. The direct result was approximately 100,000 premature deaths from increased air pollution. [7]

- In April 1986, a Soviet reactor at the Chernobyl plant in Ukraine suffered a major meltdown, with multiple explosions and fires. [8] With no containment structure, the fire raged for ten days, sending radioactive fallout thousands of miles downwind. [9] Twenty-eight first responders died from radiation in the months after the accident, [10] along with fifteen local children who succumbed to thyroid cancer. [11]

  Twenty additional first responders died in the two decades after the accident, though from indeterminate causes. [12] The infamous false estimate of 4,000 future casualties was based on a deeply flawed radiation risk model called LNT (linear no-threshold), which has since been discredited [13] (see Chapters 9 through 14). Yet some still believe, with no evidence, that the death toll was actually 1,000,000.

  Neither of these scare stories have proven to be true, but one verifiable calamity did occur: Between 100,000 and 200,000 abortions of wanted pregnancies were reluctantly chosen by women who lived downwind, fearing they would give birth to deformed or sickly children. [14]

  Another provable calamity was the decline in life expectancy for the downwinders, brought on by alcoholism and depression aggravated by nuclear fear. Several studies have shown that these consequences were actually triggered by the fear of radiation, rather than radiation itself. [15]

The follow-on effects of that fear caused yet another calamity: the permanent and unnecessary relocation of over 300,000 people. Many suffered from the fatalism that descends when people are told they're the victims of something incurable and beyond their control.

- In March 2011, a tsunami following a major earthquake caused three reactors in Fukushima, Japan, to melt down and suffer hydrogen gas explosions. No one was killed or even harmed by radiation, and UNSCEAR (United Nations Scientific Committee on the Effects of Atomic Radiation) concluded that it is unlikely anyone will be harmed in the years ahead. [16]

  But nuclear fear killed more than 2,000 Japanese citizens. Some 1,600 of them died during the panicked and largely unnecessary evacuations in the cold March weather. Others suffered stress-related deaths and some were driven to suicide, triggered by long-term dislocation from their homes, land, and businesses. [17] And even though 20,000 died from the earthquake and tsunami, and none from radiation, nuclear fear drove the news cycle, and still does.

1.1 FUKUSHIMA

Since the March 2011 accident is still fresh in our collective memory, let's look closer:

On March 11, 2011, at 2:46 pm local time, a 9.0 earthquake struck off the northeast coast of Honshu, Japan's main island. Every nuclear reactor in the country automatically shut down within seconds, with no significant damage. Fifty minutes later, a 15-meter

(50-foot) tsunami inundated the coast, killing thousands of people and devastating the local infrastructure. With their backup generators drowned, the reactors at the Fukushima plant had no way to circulate the water needed to keep themselves cool, leading to three meltdowns.

Of the 1,700 evacuees who were closest to the power plant, only ten people received a dose of more than 10 mSv, [18] the equivalent of one CT scan. [19] (We'll explain mSv, or milliSieverts, in the next chapter). Even so, rumors of a fire in Unit 4's spent-fuel cooling pool prompted the Japanese government to order a much wider evacuation than they originally called for. [20]

The gist of the rumor was that the spent-fuel pool of the Unit 4 reactor was on fire because its water may have leaked out. With no way to stay cool, the decay heat of the spent fuel rods could have indeed started a fire.

But none of this actually happened. In fact, there were no troubles with Unit 4 until four days later. There had been a brief fire in the building, but the source was flammable hydrogen gas that entered the building from neighboring Unit 3, through their common gas-management system. The hydrogen had come from the superheated water in Unit 3's melted core, but neither reactor's spent fuel pool was damaged. [21]

Nevertheless, the rumor was given credence by Gregory Jaczko, the anti-nuclear head of the United States NRC (Nuclear Regulatory Commission) at the time, who had sent an advisory team to Japan. Jaczko's hand waving about a fuel pool fire that never happened was largely responsible for the long-term dislocation of tens of thousands of people. [22] His nuclear fear carried into his testimony to the US Congress on the unfolding situation. The performance, duly recorded on C-SPAN, went viral. [23]

Thanks to Jaczko, people were worried about spent fuel pools catching fire, even though the chances of that happening were—and are—vanishingly small, even after a record earthquake and tsunami. [24]

The reactors were safely in shutdown mode and just needed to be kept cool, but the tsunami swamped the emergency diesel generators so the water pumps didn't work. Those same pumps also kept the spent fuel pools topped up. The concern was that if the water level dropped, the exposed fuel would overheat and its zirconium "cladding"—the long, hollow rods that hold the fuel pellets—would react with steam to produce flammable hydrogen gas.

The entire cascade of events at Fukushima can be traced back to the glaringly obvious blunder of placing the reactors' backup generators *in a basement on the beach.* (True story; see Fig. 1.) TEPCO (Tokyo Electric Power Company) installed them where they did with the approval of the Japanese government, in spite of clear warnings from our (pre-Jaczko) NRC at the time. And sure enough, the emergency diesel generators were drowned by the tsunami of 2011. The resulting fiasco supposedly proved that nuclear power is dangerous, when in fact the Fukushima meltdowns proved precisely the opposite:

Despite the arrogance of those who built where they did, and the corporate conformity of those who signed off on it, and in spite of a 9.0 earthquake, followed by a 15-meter tsunami that triggered three major meltdowns with hydrogen explosions, no one was harmed or killed by radiation.

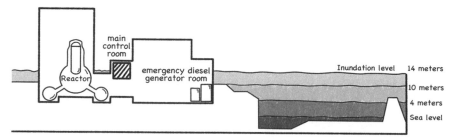

Fig. 1: Schematic of Fukushima Dai-Ichi at Mean and Storm-surge Tides, and Tsunami Inundation Level

Source: https://www.nirs.org/wp-content/uploads/fukushima/naiic_report.pdf (see frame 14)

Imagine if an airliner crashed, and no one died or was even injured. It would be a spectacular confirmation of aircraft safety. But when it comes to reactors, the public is encouraged to panic over the same result.

As it turned out, the quake was not the proximate cause of the accident. Every reactor in Japan, including the ones at Fukushima, survived the earthquake intact and shut themselves down within seconds, exactly as designed. The non-lethal industrial accident at Fukushima was due to a blatant construction management error and a disastrous lack of regulatory oversight, not nuclear technology. Indeed, the official report by the Japanese government on the accident put it like this:

> "What must be admitted—very painfully—is that this was a disaster 'Made in Japan.' Its fundamental causes are to be found in the ingrained conventions of Japanese culture: our reflexive obedience; our reluctance to question authority; our devotion to 'sticking with the program'; our groupism; and our insularity." [25]

The seaside parcel of land on which Fukushima was built originally featured a high natural bluff, some thirty-five meters (115 feet) above sea level. The Fukushima inundation, one of the largest tsunamis in recorded history, reached less than half that height. But since it's easier (and cheaper) to bring in construction materials by sea, TEPCO excavated the site down to just ten meters above sea level. [26] They apparently ignored the hand-chiseled "tsunami stones" stones that dot the coast, left behind by their ancestors to warn them of such foolishness. [27]

Loosely translated, the messages boil down to this cautionary haiku:

Don't build on the beach
in the same country that coined
the word tsunami.

7

Tsunami Stone
Credit: T. Kishimoto (CC-BY-SA-4.0)

Eighty miles up the coast, north of the town of Sendai, the Onagawa nuclear power plant was closer to the epicenter than any other nuclear plant in Japan; if any reactor melted down it should have been at Onagawa. But all four Onagawa reactors came through the disaster unscathed, providing power and shelter to the local population. [28] This is because Onagawa put their backup generators where they belong—on a seaside bluff, higher than any tsunami has ever reached.

The best automobile ever made will fail without a functioning cooling system. And just like most cars, a feature of Generation-II and Gen-III reactors is an active water-cooling system that continues to run after shutdown. You may have noticed this with a big car on a hot day, when the radiator fan keeps running after the engine has been shut off. This strategy, and its limitations, are the same for

Fig. 2: Honshu Island (Japan)

water-cooled reactors. Think of them as the big-block V-8s of the nuclear industry.

To qualify as a Generation-IV design, a reactor must be able to shut itself down and cool itself off with no outside help by humans, machines, or computers. Plant personnel could literally walk away and have a cup of coffee. A Gen-III+ reactor like the Westinghouse AP-1000 or the Chinese CAP-1000 comes close to this ideal—you could walk away for seventy-two hours before operator action is required. [29] With a Gen-IV reactor, you could walk away forever.

The Fukushima evacuation radius set by the Japanese government on March 12, one day after the disaster, was initially set at twenty kilometers (12 miles). But in response to the nonexistent fuel-pool fire, our NRC director Jaczko recommended a 50-mile radius for any US citizens in the area. In the midst of such an unprecedented emergency, Japan deferred to his purported expertise and imposed his recommendation on their own citizens. [30]

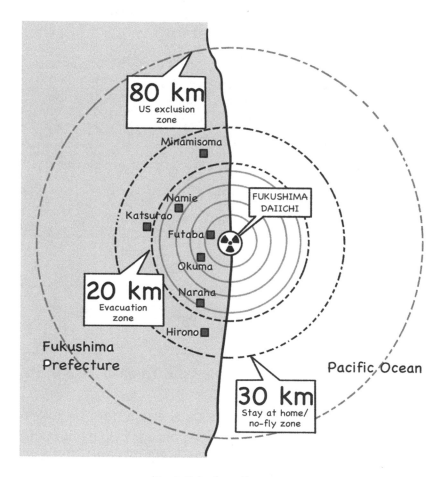

Fig. 3: Fukushima Evacuation Map
Source: https://sites.suffolk.edu/fmadkour/2012/01/30
/a-lesson-learned-through-fukushima-daiichi/

By acceding to the head of our NRC, the Japanese government's authority on nuclear matters diminished in the eyes of their people. [31] In fact, Jaczko's fearmongering was so effective that, even now, most evacuees still won't return. This is in spite of encouragement to do so by their own government, which originally ordered them to flee. After years of unnecessarily frightening their citizens, the government has now resorted to paying them nearly $20,000 to return. [32]

This ongoing tragedy of errors was caused, in no small part, by Gregory Jaczko's bad advice. With a doctorate in particle physics but no particular experience in nuclear technology, he worked as an aide to the staunchly anti-nuclear congressman (now senator) Ed Markey (D-MA), who later co-authored the 2019 Green New Deal. [33] Jaczko then served as a science policy advisor to Senate Majority Leader Harry Reid (D-NV), whose state hosts the Yucca Mountain spent fuel repository, a project Reid had long opposed. Two years before Fukushima, Reid persuaded President Obama to appoint Jaczko as chairman of the NRC, and things at the agency went downhill from there. [34]

Motivated by nuclear fear, the misplaced panic in Japan has inflated the country's cleanup and relocation costs to an outlandish degree, and this somehow "proves" that nuclear is expensive. [35] Similar to Chernobyl a quarter-century before, nuclear fear has magnified the long-term social and psychological consequences—and the expense—of an already disastrous earthquake and tsunami.

Instead of evacuating, if everyone downwind had just taken the usual precautions for fallout (shelter in place for a few days, seal openings to minimize the intrusion of fallout dust, etc.), no one in the evacuation zone would have received a dose anywhere close to 100 mSv, which is a safe yearly dose for any adult. This is something the French Academy of Sciences has known for years:

"Epidemiologic studies have been unable to detect in humans a significant increase of cancer incidence for doses below about 100 mSv." [36]

Our planet's average natural background radiation dose is around 3 mSv per year, but the scientific literature shows that receiving up to

100 mSv per year, even all at once, is no cause for alarm. [37] Indeed, life on Earth evolved billions of years ago by adapting to much higher background radiation levels than we have today. A dose of 100 mSv is not a cause for wholesale evacuation, or for removing millions of tonnes of topsoil from Fukushima farmland as a precaution.

> **NERD NOTE:** A "tonne" is a standard metric measurement of mass (often referred to as weight), equal to 1,000 kilograms, or 2,204.6 lbs; also known as a metric ton.
>
> Other terms you may encounter (but not in this book): A "short ton" is a measurement commonly used in the United States, equal to 2,000 lbs. A "long ton," still used in the UK, is equal to 2,240 lbs.

Nonetheless, the Japanese government initially set the repopulation threshold at a ridiculously low 20 mSv per year. [38] They've since relaxed the threshold to 28 mSv, even though the difference between these two "safety standards" is essentially meaningless. As the Health Physics Society points out:

> "Below levels of about 100 mSv [greater than] background from all sources combined, the observed radiation effects in people are *not statistically different from zero.*" [39]

The Japanese government's overwrought caution has been playing out for more than ten years now, though several populated areas around the world have considerably higher background levels than 28 mSv per year. And these regions have normal rates of cancer.

You may have heard that the Japanese Health, Labor, and Welfare Ministry awarded compensation to a Fukushima cleanup worker who developed leukemia. The claim was paid even though the man received a cumulative dose of just 15.7 mSv. [40] In another instance, the family of a man who received 74 mSv and died of lung cancer was awarded a death benefit. Even so, this does not mean the melt-downs were the cause: in Japan, workers' comp pays out regardless of whether the injury is job-related. The claim was awarded despite the fact that lung cancer takes decades to manifest, and had no apparent link to the accident. [41] This point was lost on the global public, who assumed that since the award was paid, meltdown radiation must have been the cause. It wasn't.

Another common argument against nuclear power is that higher rates of thyroid tumors were found post-Fukushima. This new statistic, however, is the result of the Japanese government testing every downwind child. [42] Usually, doctors only test if symptoms are present, since most thyroid nodules are asymptomatic and benign. If it requires medical attention, thyroid cancer will announce itself, and when it does the success rate for treatment is over 90%. [43] That's why testing every child after Fukushima showed an apparent increase in cases. Even so, no excess thyroid cancer deaths have been reported since the accident.

1.2 "LET US NOT TALK FALSELY NOW, THE HOUR IS GETTING LATE." — BOB DYLAN

Perceptions of danger can be strongly influenced by disinformation and fear. The misinformed response to Fukushima, for example, has turned a casualty-free industrial accident into a traumatic global event. Nuclear fear has made the recovery from Fukushima far worse than the accident itself.

While researching these books, we have become more sensitive to the concerns that some people have about nuclear power. The more we examined the facts, the more empathy we've come to have for the misinformed among us, because no one can reach a correct conclusion with incorrect information. Our ire is reserved for those who continue to misinform—and in some cases deliberately disinform—the public with talking points that have repeatedly been debunked. It's tiresome, and the world no longer has the luxury of rehashing the same bogus issues, over and over again. So we're losing patience. Scaring the public away from a safe and well-proven technology that can do the job that needs to be done, in the time we need to do it, is an outrage.

So please bear with us if we fly off the handle every now and then, or drop the occasional snide remark. It's a sign of the times.

CHAPTER TWO

Zoomies! We're Doomed!

(Heads up: This chapter is a deep dive into the basics of radiation. You can skip it, skim it, or refer to it later, but reading it now will serve you well in the chapters ahead.)

BE NOT AFRAID. Unless you were born and raised in a salt mine, you've been bombarded with radiation your entire life. Welcome to planet Earth.

Cosmic rays are but one small example. These are mostly high-velocity protons from our sun, along with some high-energy gamma rays that come zooming down to us from beyond the solar system. Most swerve away upon entering the magnetosphere, but some still reach the planet surface. And if you're in the way, they'll zoom right through your body. [1]

About 13% of the background radiation we encounter can be attributed to cosmic rays, and the less atmosphere there is above us the more intense the effect. People living at high elevations, people at cruising altitude, and astronauts get zapped by zoomies worse than anyone. No one is left untouched.

(Note: Just to be clear, "zoomie" is US Navy submariner slang for radiation, and does not refer to dogs spinning in circles or running crazy around the house.)

We'll be focusing on the zoomies that originate right here on Earth. Categorized as alpha, beta, or gamma radiation, they come from inside an atom's nucleus. That's why zoomies are called sub-atomic particles—they're much smaller than any atom.

15

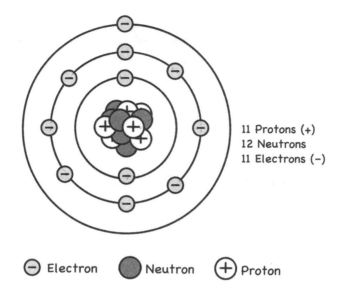

11 Protons (+)
12 Neutrons
11 Electrons (−)

⊖ Electron   ⬤ Neutron   ⊕ Proton

Fig. 4: The Sodium Atom

(Note: To keep our sketches simple, we don't depict every proton and neutron. Just pretend the rest of them are on the far side of the nucleus.)

If you really want to understand the actual hazards of radiation, and the lack thereof, it's important to get this zoomie stuff right, so put your thinking cap on and embrace your inner nerd. We'll walk you through it.

The enormous nuclei (centers) of uranium and thorium atoms are two examples of zoomie emitters. These heavy-metal elements, and the isotopes into which they decay, are found all over the world in rock, soil, and water. Only slightly radioactive, these atoms will throw off the occasional zoomie as they slowly transmute (change) into completely different elements, getting smaller and weaker over time. One of these elements is radium, which decays into the radon lurking in your basement (see below), and in morning dew. [2]

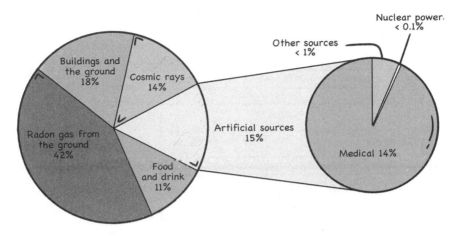

Fig. 5: Background Radiation

Establishing sensible safety standards requires a thorough understanding of the effects of radiation on the human body. But just as important, it also requires that we update our understanding—and our regulations—as science and medicine evolve.

Zoomies are everywhere, and there is no escape. They're in sunlight; they're even in bananas, loaded with nutritious potassium. Most potassium (K-39) is radiologically inert. But a tiny percentage (0.012%) of any batch of potassium is radioactive K-40. This is why the average banana contains about 0.1 microSievert of radiation.

As a matter of fact, bananas are so radioactive that a truckload can set off alarms at a border crossing; so can avocados. The average avocado has a bit less K-40, at about 0.08 microSievert. [3] The Department of Homeland Security (DHS) actually had to recalibrate their testing apparatus for incoming shipments.

"Banana Equivalent Dose" (BED) is a jocular way of assessing low doses of radiation, by comparing the doses we commonly receive in our daily lives with the consumption of ridiculous amounts of bananas. The point being is that we all need to dial back on our nuclear fears and keep low-dose radiation in perspective. While high doses can indeed be dangerous, low doses are not the dire threat that some claim them to be.

Number of bananas	Equivalent exposure
100,000,000	Fatal dose (death in 15 days)
20,000,000	Single session of radiotherapy (targeted dose)
70,000	Computerized tomography (CT) scan of chest
20,000	Single mammogram
200 – 1000	Chest X-ray
700	One year's residence in stone, brick, or concrete structure
400	Typical transatlantic flight (UK to US)
100	Average daily background dose
50	Dental X-ray
1 – 100	One year's residence near a nuclear power plant

Bananas contain a trace amount of Potassium-40, a radioactive isotope.

Eating one banana = 1 BED = 0.1 μSv = 0.01 mrem

Fig. 6: Banana Equivalency Table
Source: https://livinglfs.org/mutants-vs-radiation-trying-to-understand-radiation-in-lfs/

NERD NOTE:

- Sievert (Sv) is the standard unit for quantifying the *biological effect* of a dose of ionizing radiation on a lifeform, called the effective dose.
- milliSievert (mSv) = one thousandth of a Sievert.
- microSievert (µSv) = one millionth of a Sievert.

We get even more zoomies from man-made sources like X-rays and CT scans. Zoomies are also in beer, and every time you down a frosty brew they zap you from within. (It's from the potassium in the grains.) And then there's all that guacamole.

Any granite has traces of radioactive material—this includes your kitchen countertop. Even clean seawater has a "becquerel count" of about 12,000 Bqs per cubic meter. Most of this comes from potassium-40 and rubidium-87, along with a smattering of uranium. To put this in perspective, adult humans have a continuous internal becquerel count of about 5,000, nearly all of it from the potassium in their cells—about 80% in muscle cells, and 20% in bone cells. [4] (See Chapter 15.)

> **NERD NOTE:** A becquerel (Bq) is a standard measurement of basic radioactivity. One Bq simply means one decay emission per second, regardless of what is being radiated or how energetic the emission happens to be. Whenever an atom spits out a zoomie, that's one becquerel.

Whatever subatomic particles an atom happens to emit—protons, neutrons, electrons, or gamma rays—the atom is classified as radioactive because it's actively radiating zoomies. Even if it only ejects one weak zoomie every hundred years, it's still considered radioactive.

A popular bogeyman, radon-222 contributes to nearly half the average background radiation we humans typically receive on this nuclear planet. The popular linear no-threshold (LNT) model, which claims there is no safe dose of radiation, is flatly contradicted by the real-world data on radon.

The graph below summarizes a 1995 study of 1,729 US counties, comprising 90% of the American population at that time. [5] Increasing radon levels are shown on the *x*-axis (horizontal), and increasing cancer rates are shown on the *y*-axis (vertical). When the data is examined, it's clear that the low-dose radon levels typically found in our homes do not correlate with increased rates of cancer. In fact, quite the opposite: Lung cancer mortalities are actually *lower* in households with higher-than-average radon levels.

In certain places, such as underground mines, radon can be much higher than average, and if it poses a risk it should of course be mitigated. But just like the mitigation of any potentially hazardous substance, the question is: *To what degree?* Science, not science fiction, should be our guide.

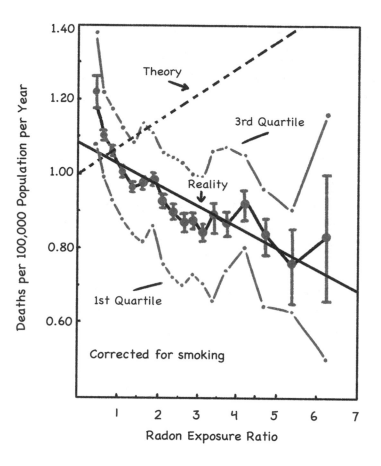

Fig. 7: Radon and Lung Cancer
Source: http://www.phyast.pitt.edu/~blc/LNT-1995.PDF
The x-axis shows radon exposure, and the y-axis shows lung cancer mortality. The dashed "Theory" line shows the predicted increase of radon-induced cancer per the LNT model. The solid line shows real-world cancer rates trending *lower* with *increasing* levels of radon.

2.1 SOME THINGS ARE MORE ZOOMACEOUS THAN OTHERS

On modern-day planet Earth, the average background dose is about 3 mSv (milliSieverts) per year. But not to worry—our skin is a protective outer organ that shields our innards from most of the radiation we encounter. This is why the primary concern with nuclear material is inhaling or ingesting the stuff, rather than simply being around it. [6] This also explains why we should seal any openings to avoid inhaling fallout particles after a nuclear incident.

The overall biologic effect depends on the type of radiation, and the circumstances of contact:

- What is the collective energy of the zoomies that strike?

- Is the radiation source inside the body, or near the body?

- Is it partially blocked, slowed, or dispersed, or is it penetrating?

- If it penetrates, does it come in like a bowling ball, a BB, or a bullet (alpha, beta, or gamma ray)?

- What organ or organs are being struck?

- What is the intensity and duration of the zoomie storm (the dose-rate)?

All these variables must be considered when assessing a zoomie's *biologic effect*, which is properly expressed in Sieverts. That's because the focus of radiation safety is on the biologic effect of a received dose, typically called the *effective* dose. This is why Sievert is the preferred unit of measurement, rather than Gray, which measures the *absorbed* dose.

NERD NOTE: To be precise, one Sv represents the biologic effect that one joule of radiative energy has on one kilogram of living matter.

Below 100 mSv, the distinction between Grays and Sieverts is often disregarded, and the terms are used interchangeably. (See Fig. 11 below. Also see our supplement "Sieverts and Grays." [7])

Sometimes, however, these terms are not interchangeable. A good example is the Sievert measurement of an alpha particle's effect. Because of the alpha's +2 formal charge (which makes it very electron-hungry), and because of its size, the alpha particle is "weighted" on the Sievert scale as having twenty times the biologic effect of a stream of beta particles with the same collective kinetic energy.

Billions of years ago, when life on Earth was just getting started, background radiation levels were much higher than they are today. But as the eons passed, radioactive elements decayed, reducing background radiation. This happened because any radioactive atom is a decaying atom, throwing off bits and pieces as it transmutes into a series of other elements (a decay chain), until it finally stabilizes.

Most uranium eventually decays into lead. And not radioactive lead, just plain old boring lead, which is the freaky thing about radioactive material—it changes into *entirely different stuff*. Iron, for example, is always iron, no matter what: Rust is iron plus oxygen, but the iron is still there. But lead is not uranium plus anything. Iron doesn't reduce to anything less than iron, while radionuclides (radioactive isotopes) will decay into completely different and lighter elements.

The alchemists were onto something; they were just fiddling with the wrong stuff. Radioactive decay, in which zoomies are released, is part of the ongoing process of transmutation, in which one substance, slowly or quickly, becomes an entirely different substance. It's a simple sci-fi plot twist from there to conjure scare stories of mad scientists wielding death rays—the same pulp-fiction / nuclear-fear button from a century ago that is still being pushed today.

Some decay chains are super-fast, lasting less than a second, and others can take forever. The uranium, thorium, and potassium in the earth's crust, for example, have been decaying since the earth formed, more than four billion years ago.

In fact, more than half the heat in the earth's core comes from the naturally-occurring radioactive decay of thorium, uranium, and potassium-40, along with the heat of friction from the circulating molten core, and a bit of leftover heat from the formation of the planet itself. [8] The thermal energy from this geodynamo keeps the earth's outer core in a molten, or liquid, state, and radiation is responsible for most of it. That's why Earth is a nuclear planet. [9]

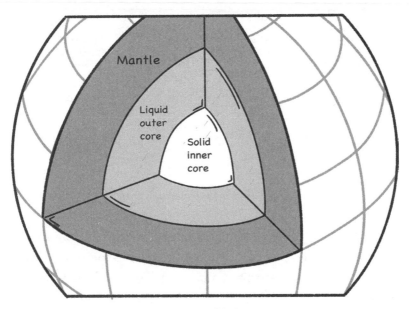

Fig. 8: Earth's Core

The constant motion of the free charge within this molten metal, which is mostly iron, produces our planet's magnetic shield. The magnetic shield is what deflects the charged particles of the solar wind (mostly high-energy protons), which would otherwise strip away Earth's atmosphere.

Good planets are hard to find. NASA has recently discovered that over eons of time, the thin mantle of Mars allowed enough heat to escape, to where its core ceased being molten and thus stopped circulating. As a result, the red planet has no magnetic shield, and though the solar wind is much weaker that far from the sun, the rate of solar flux is still strong enough to strip away atmospheric gases. While some think it's possible to create a Martian atmosphere for human colonization, others contend that without a magnetic shield, it would be an endless and Sisyphean effort to exceed the strip rate. [10]

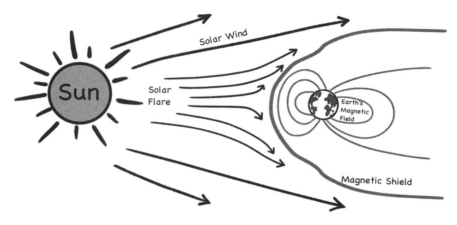

Fig. 9: Solar wind and magnetic shield

Whatever misgivings you may have about nuclear material, realize that this is indeed a nuclear planet—there wouldn't be any life on Earth if it wasn't. To take care of our home planet, we need to understand what makes it tick. And rather than fearing the very thing that makes life on Earth possible, we should use it to our best advantage. Everyone should have a healthy respect for radiation, but nuclear fear doesn't help anyone at all (except the fossil fuel industry).

2.2 WHEN ZOOMIES STRIKE!

Toss a bullet at a wall and it bounces right off. Fire it from a gun and note the difference. In much the same way, a zoomie's biologic effect is determined by its mass, speed, size, and electric charge. The predominating feature will depend upon the type of zoomies involved—protons, neutrons, and electrons, as well as the bizarre, positively-charged "anti-electrons" that decaying atoms sometimes emit, called positrons. (Gammas are different; we'll explain below.)

An alpha particle can be stopped by a piece of paper or human skin, which has several protective layers of dead cells that can stop alphas cold. So can a mere quarter-inch of air. But if a large dose of alpha particles is inhaled, their presence beside living lung tissue can pose a problem. This is

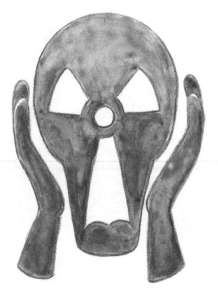

Credit: Scott E. Sutton

why fallout dust can be a concern. However, *ingested* alpha emitters pass through the GI tract rather quickly.

An alpha particle is essentially a helium nucleus—a tight bundle of two protons and two neutrons—but without helium's two orbiting electrons. Since neutrons are electrically neutral (hence the term), the alpha's two protons give it a formal charge of +2. This makes alpha particles electron-hungry.

The sheer size of an alpha particle—a cluster of four bowling balls in a world of penny-sized zoomies—is another reason why radiation's biological effects are measured in Sieverts rather than Grays. Measuring radiation in Grays focuses solely on the amount

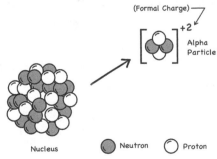

Fig. 10: Alpha Decay

of absorbed energy, and not on the type of particle, the size of the particle, or the biological effect of its impact.

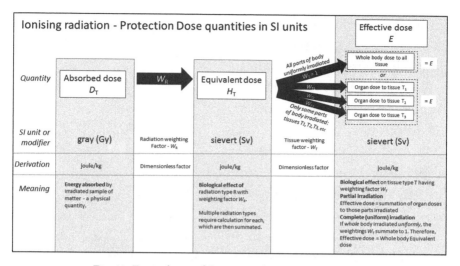

Fig. 11: Equivalency of Sieverts and Grays at Low Doses
Credit: By Doug Sim. Own work (CC-BY-SA-3.0)

Aside from its size and speed, the alpha's biological effect is mainly due to its charge imbalance. After smashing through a cell membrane, the electron-hungry alpha will steal two electrons from the other atoms present, ionizing those atoms in the process. Hence the term ionizing radiation.

The alpha decay of radioactive material in the earth's crust (i.e., the high-speed ejection of a helium-nucleus equivalent) is the world's major source of helium: These alpha zoomies steal two electrons from whatever they slam into, and settle down to become helium atoms.

Elemental helium is a byproduct of the fossil fuel industry, a lightweight gas that comes to the surface along with heavier natural gas, crude oil, and coal. As the world weans itself from fossil fuel, our helium supplies will dwindle as well. And we've been squandering this limited resource filling up party balloons.

A beta particle comes in two flavors: negatively-charged electrons and their positively-charged equivalents called anti-electrons, or positrons. Betas are about 8,000 times smaller than alphas—just one tiny electron, compared to a massive cluster of two neutrons and two protons, which are nearly identical in size and mass.

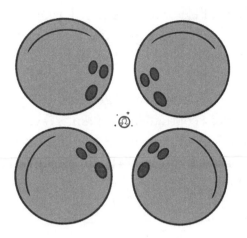

Fig. 12: Alpha Particle vs. Beta Particle
Like a penny vs. four bowling balls

While alphas can be stopped by a piece of paper, the tiny beta particle can be far more penetrating, depending of course on its energy level. When you're that small, the solid wall of atoms that forms a piece of paper can be seen for what it actually is—mostly empty space. Nevertheless, betas can be stopped by a few millimeters of wood or aluminum, as they eventually hit something and fizzle out.

Gamma rays are high-frequency electromagnetic waves, but they can also act like "massless particles." Due to the short wavelengths of their electric and magnetic fields, gammas are the most difficult type of radiation to block, and may have a substantial biologic effect depending on their energy level. These waves/packets of light can be stopped, but it requires a foot of lead, six feet of concrete, or about fifteen feet of water.

A gamma ray is a highly penetrating form of electromagnetic radiation, and can be thought of as a packet of pure energy—a photon, or discrete quantity of light. Depending on their total energy, gamma rays can impart a greater biologic effect than alphas or betas.

Like any form of electromagnetic radiation, light is very weird stuff—sometimes it acts like a particle, and sometimes it acts like a wave. This wave/particle duality is a fundamental aspect of the theory of quantum mechanics. [11]

At any energy level, gammas are simply "particles of light," which can seem like a contradiction in terms if you're not a quantum mechanic. [12] While photons have no "rest mass," they do have "mass equivalence" as a result of their velocity, commonly denoted as C, or the speed of light. This is the equivalence of mass and energy that Einstein referenced in his formula $E = mc^2$.

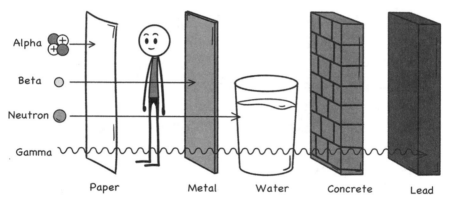

Fig. 13: Penetration of Various Zoomies

A zoomie can act like a cotton ball, a bowling ball, or a bullet, depending on its size, charge, and kinetic energy. At lower energies, a zoomie may only excite an atom, causing one of its electrons to temporarily jump to a higher orbital.

If a zoomie is energetic enough, it can knock an electron away from its home atom for good, turning the bereft atom, shorn of its electron, into an ion of its former self. Because of their charge imbalance, these ions (positive or negative) can either be more chemically stable or more chemically reactive than the atoms from which they originate.

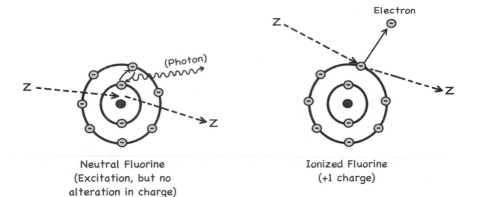

Fig. 14: Excitation vs. Ionization

So what's the big deal? It's just one atom, right? Maybe yes, and maybe no. Changing an atom's ionic state can change the function of the molecule that atom is a part of. And if the molecule is in a strand of DNA, it must be repaired.

2.3 NO WORRIES

Living cells can perform up to ten thousand repairs per day, per cell. This is an underreported and pretty damn remarkable feat that we humans (and other living things) routinely accomplish every day of our lives. That deserves repeating:

> *Living cells can repair their own*
> *radiation and chemical damage,*
> *up to 10,000 times a day.*

Science didn't understand this in 1946, when the erroneous idea that "all radiation is harmful" gained an undeserved reputation as settled science. This false no-threshold claim persists to this day, inspiring a fresh outbreak of nuclear fear whenever a zoomie is found in the wild.

Research on cell repair began back in the 1930s. In 1951 Rosalind Franklin, a British chemist and X-ray crystallographer, discovered key aspects of the DNA molecule, recording her work in a series of X-ray photographs. [13] Without her knowledge, one of her unpublished photos was shown to Watson and Crick by Franklin's lab partner. From there, the three men detailed the structure of DNA's double helix configuration, for which they, but not Ms. Franklin, received the Nobel Prize.

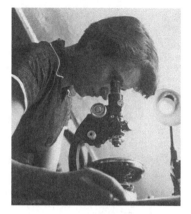

Rosalind Franklin
Courtesy of MRC Laboratory of Molecular Biology (CC-BY-SA-4.0)

Understanding DNA structure greatly improved our ability to examine the hypothesis of cell repair. At the time, however, the very idea of cells repairing their own damaged DNA was dismissed by the biological sciences community at large. But research in the field doggedly progressed, so that "by the end of the 1970s, it was evident that cells have evolved *multiple diverse mechanisms* [of DNA repair]." [14]

Evidence continued to accumulate, and in 2015 the Nobel Prize in Chemistry was awarded to three of the principal pioneers in the now-established field of cell repair. [15] We humans host

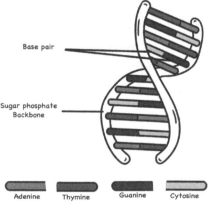

Fig. 15: Basic DNA Structure

about ten trillion cells each, doing their regular jobs while also seeing to their own daily repairs. No wonder we need a good night's sleep.

2.4 WAY, WAY BACK IN THE DAY

Uranium is element 92 on the Periodic Table. Most of the uranium found in nature is the U-238 isotope, with 92 protons and 146 neutrons. In any atom, the number of protons determines the element, while the number of neutrons determines the isotope. In the Periodic Table of Elements, the atomic number is the proton count, while the bottom number is the element's average mass. Since many elements have multiple isotopes, the mass number is usually a weighted average.

Starting with sodium, element 11 on the Periodic Table, larger atoms tend to have more and more neutrons than protons. Down along the bottom of the Table, the actinides have more than half again as many neutrons as protons. The reason why is a bit complicated, but in simple-ish terms:

NERD NOTE: Since protons are positively charged, they naturally push each other away. This is the electrostatic force we're all familiar with: Like charges repel and opposite charges attract.

However, there is another force working at the subatomic level that most people are not familiar with: the strong nuclear force. This potent force makes positively-charged protons override their natural repulsion and attract each other instead, but the strong force is only effective across a *very* short distance.

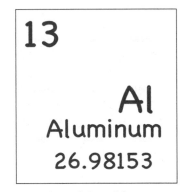

Fig. 16: Detail from Periodic Table

Imagine forcing two magnets together so firmly that they're just about to touch, and then they suddenly bind together. (Sub-atomic physics is weird; just roll with it.) Even though large clumps of protons will bind like this, each proton in the bunch will also repulse the protons they're not directly bound to. Imagine a mosh pit of ravers in a group hug, even as they elbow the others around them—a tribal dance for Tasmanian Devils.

This is where neutrons come into play: As an uncharged, neutral version of a proton, neutrons can schmooze equally well with protons and each other. They serve as chaperones, the social diplomats that hold the mosh pit together and keep the party going, but it's a trick that works less and less well in larger nuclei.

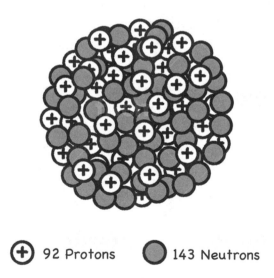

⊕ 92 Protons   ⬤ 143 Neutrons

Fig. 17: Neutrons Help Hold the Nucleus Together

If we smack into one of these big, cobbled-together nuclei with a free-flying neutron zooming along at just the right speed, we can fission (split) the atom. This releases some of the fantastically compact and potent binding energy that's been (barely) holding the nucleus together all these eons, along with the release of one, two, or sometimes three zooming neutrons. These fly off and split other fuel atoms, and thus we have chain reactions.

This is entirely unprecedented in the field of energy production. Prior to fission, energy typically came from burning various materials to release their chemical bonds or form new ones. At the atomic level, combustion rearranges fuel molecules into different molecules, but that's about it. A lot of electrons get shuffled around, throwing off some energy in the process, but the nuclei of the atoms involved remain serenely undisturbed. Chemistry is basically electron management.

Splitting the atom is an entirely different process. And of all the atoms found in the natural world, the easiest one to split is uranium-235. Billions of years ago, natural uranium contained more than 5% U-235—a lighter, less stable, and thus more radioactive isotope of uranium. With the requisite 92 protons (that's what makes it uranium) but a lower neutron count of 143, U-235 is like a stripped-down hot rod of the far more sedate uranium-238, the full-size atomic sedan.

With fewer neutrons to hold it together, a U-235 nucleus is easier to split with a free-flying neutron, which its neighboring U-235 atoms will sometimes spontaneously eject. This being the case, most of Earth's unstable U-235 has long since decayed away or fissioned, leaving a concentration of just 0.7% in today's natural uranium.

Way back in the day, the more radioactive U-235 isotope was everywhere, to one degree or another, in every Precambrian rock and puddle, and life had to adapt. Along with cosmic rays, uranium and some of the isotopes into which it decays were among the first evolutionary incentives, by stimulating the cell's ability to repair its own radiation damage. This mechanism became an integral part of every living cell's tool kit. If it didn't, we wouldn't even be here. That's worth repeating:

The ability to repair radiation damage
is part of every living cell's tool kit.

The ability to self-repair radiation damage is the same mechanism our cells employ to handle the effects of chemical damage, such as oxidative stress. We take antioxidants to aid our metabolism, but most of us aren't aware that our cells' ability to repair oxidation damage is actually an adaptation of the earlier mechanism of radiation repair.

Similarly, most people aren't aware that atmospheric oxygen (O_2) was an environmental latecomer, a corrosive substance that early life adapted to and learned to utilize. [16] Cells evolved to use the oxygen molecule, repairing the damage it caused by replicating their earlier success with radiation repair. The fact is, living cells had successfully adapted to radiation long before complex lifeforms drew their first collective breath. [17]

Even now, billions of years later, and despite how highly evolved we are, the oxygen we need to survive is *still* corrosive to our cells. As Reuter, *et al.*, explain, "observations to date suggest that oxidative stress, chronic inflammation, and cancer are closely linked." [18]

2.5 WELL, MAYBE SOME WORRIES

As rugged as cells have become in adapting to life on this nuclear planet, they can only do so much. If their rate of repair is overwhelmed by a zoomie storm, there will be trouble unless there is enough quiet time for repairs. But sometimes the zoomies come so hard and fast, and the storm lasts so long, that cells simply can't recover. And when they die, flushing their carcasses out of your carcass can make you sick—hence the term "radiation sickness." [19]

If the over-irradiated cells don't die, things can become even more problematic—their DNA may have been damaged to where they start making mutated copies of themselves. In concert with other factors, cell mutation can prompt a beneficial evolutionary adaptation, or it can lead to cancer and other maladies. But except for very high doses in the range of several Sieverts (an extremely rare occurrence), ionizing radiation is a weak cancer agent compared to most other natural hazards, such as the air we breathe and the food we eat. [20]

As scary as radiation has been made out to be—a sort of "sub-atomic Ebola"—the descendants of the Hiroshima and Nagasaki victims have shown no signs of inherited mutations, even though their forebears were irradiated more than any population in human history. [21] That's also worth repeating:

> *The descendants of*
> *Hiroshima and Nagasaki victims*
> *have shown no signs*
> *of inherited mutations.*

Contrary to popular belief, the cancer rates among the *hibakusha* (the Japanese bomb survivors) are nearly identical to the wider Japanese population, differing by less than 1%. [22] This obviously does not justify the use of nuclear weapons, nor does it minimize their horrific effects. But of all the cancer risks we face, it does demonstrate just how ineffective radiation actually is.

It doesn't matter where zoomies come from—kitty litter, guaca-mole, radon gas, bananas or beer, an X-ray or a CT scan, a nuclear reactor or outer space. The origin is irrelevant; the paramount concern is their health effect on humans and other living things. There are only three important details about any form of radiation:

- The dose, or cumulative energy, these subatomic particles and waves of photons impart as they strike, pass through, or pass close by a target (such as yourself)

- The rate, or the amount of time in which the dose is received

- The size of the affected area.

Zoomies are zoomies, whether they come from beyond the stars or a humble rock in the desert. If a zoomie plows through the periphery of a cell, it may not be a big deal. Even if it strikes a DNA strand, it might not be a big deal, either—unless that cell is in the process of dividing. This is because cells can't always fully repair themselves during cell division (mitosis) or reproduction (conception). They can, however, repair themselves afterwards, but the process isn't always foolproof.

This is why some researchers consider embryos and fetuses to be more susceptible to radiation damage—they're constantly growing, meaning more cell divisions per unit of time. [23] However, since infants have stronger immune systems, others consider the net effect to be a wash.

This is also why radiation therapy can kill a patient's hair, along with their targeted cancer. Since both hair and cancer cells have relatively rapid growth rates, they divide more often than other cells. This means there is a greater chance of zoomies hitting them during cell division, when they can't repair the hit as easily. Thus, the faster cells grow, the more susceptible they are to radiation damage, with more opportunities to flub the "proofreading" they use to self-repair any damaged DNA. [24]

This also explains why radiation therapy includes a recovery period between sessions. A radiologist will take care to divide a dose into multiple beams converging on the target tissue. That way, the cells surrounding the target will receive a lower dose. Even so, your body needs time to repair the zoomie damage and flush out the dead cells.

2.6 A SPECIAL KIND OF AWFUL

Nuclear skeptics seem to think that radiation from the nuclear industry is somehow more harmful than natural background radiation, as if it possessed a special kind of awful, like an engineered toxin that doesn't appear in nature.

Not true. Once again: *Zoomies are zoomies.* Like any other projectile, zoomies can be no big deal, or kind of a big deal, or a very big deal, indeed: cotton balls, BBs, bullets, or photon torpedoes. It all depends, because the issue is not just about dose, but *dose-rate*: how energetic the zoomies are (dose), and how many of them strike over what period of time (rate). Getting hit with a thousand cotton balls all at once would probably be unpleasant. But that doesn't mean that getting hit with one per second, until you accumulate a total dose of one thousand hits, is unsafe. Annoying perhaps, but not unsafe.

So the next time you read an alarming article about "37 Trillion Becquerels!" see if the reporter mentions the two most important details of any radiation event—the dose (cumulative energy) and the rate (period of exposure), expressed together as the dose-rate. If they don't mention these important factoids, they're probably just trying to scare you with big numbers, or they themselves are scared and are spreading their nuclear fear like a virus. You can mask up by knowing the facts.

Background radiation is everywhere on this nuclear planet. But as things turned out, living cells (including ours) learned how to repair themselves, and we can all breathe easy because of this singular evolutionary development. Once again, we wouldn't even exist if our cells lacked the ability to self-repair radiation damage. Even so, radiation is something that far too many people fear, regardless of the dose.

Like any kind of fear, nuclear fear can be highly contagious. The remedy, as always, is a generous serving of facts, seasoned with a bit of perspective. [25] But even more important than getting the facts straight is a person's willingness to change their mind in light of new information. Few people have said it better than Bertrand Russell:

> *"The essence of the liberal outlook lies not in what opinions are held, but in how they are held: instead of being held dogmatically, they are held tentatively, and with a consciousness that new evidence may at any moment lead to their abandonment."*

CHAPTER THREE

A Half-life Well Lived

YOU'VE PROBABLY HEARD OF RADIOACTIVE HALF-LIVES, which can be anywhere from a nanosecond to several billion years. The half-life of a radioactive material has elapsed when one half of any given batch of the stuff changes into a different isotope, or a completely different element.

Some of the hazardous materials in science and industry are toxic forever, like the cadmium in coal ash, which is also found in solar panels. One good thing about radioactive material is that it's in a constant state of decay—a continual process of becoming benign.

The original material decays by changing into one radioactive isotope after another, taking any number of zigzag paths down the Periodic Table (Fig. 18) until it finally becomes a stable, non-radioactive element. Uranium and plutonium, for example, usually become lead ("Pb" in the decay chain below, from the Latin word *plumbum*, from which we get the word plumbing).

Of course, the larger the original batch, the more total radioactivity there is. But regardless of the size of the batch, at the end of one half-life, half the radioactivity will be gone. As a general rule of thumb, ten half-lives will render a batch benign: The radioactive half that remains after the first period will be reduced by half in the second period, and so on, until the original batch becomes a lump of radioactively benign stuff and virtually all the zoomies are gone.

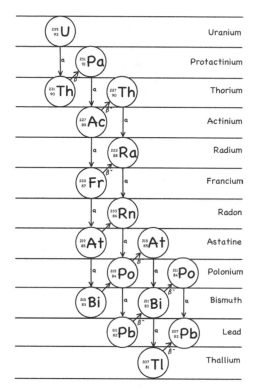

Fig. 18: Decay Chain of Uranium-238 Becoming Lead
Source: https://en.wikipedia.org/wiki/Decay_chain

Check out the graph below. After four half-lives, the material is only 6.25% as radioactive as it was when starting out. Total radio-activity will of course depend on the size of the original batch, but if you extrapolate the math, ten half-lives result in less than 0.1% (one one-thousandth) of the original amount of radiation.

For example, the half-life of plutonium-239 is 24,000 years. This means it will be considered inert after it ages for about 240,000 years, long enough for virtually the entire batch to transmute into stable, non-radioactive lead. Of course, a better solution than burying it for a quarter-million years might be to use it as fuel in a fast-neutron reactor. This will break it down into much shorter-lived fission products (FPs), and we'll get loads of clean energy for our good deed.

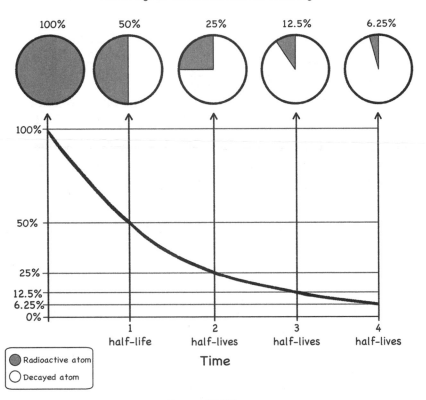

Fig. 19: Half-lives
Source: SW Infographics / Knowable

3.1 RECYCLING FOR FUN AND PROFIT

If you're a fan of the environment, you'll be happy to know that recycling used reactor fuel would do far more to reduce society's waste stream than our ongoing efforts to recycle plastic. Petroleum industry hype to the contrary, our efforts over the last thirty years to recycle their plastic waste have largely been greenwashing.

In 1989, the Exxon Valdez spilled 257,000 barrels of crude oil into Alaska's Prince William Sound. [1] The disaster caused more environmental damage than any spill in history, including the larger-volume Deepwater Horizon blowout of 2010. The Valdez incident sparked global outrage and a determined pushback against petroleum and petrochemicals, including plastic.

After Chernobyl in 1986 and the Exxon Valdez in 1989, interest in renewable energy understandably soared. Sensing a long-term sea change in public opinion, Big Oil began promoting municipal recycling programs. [2] They figured if people could start going green by cleaning up plastic waste, maybe we would all be hating on petroleum a little bit less, because the plastic made from it is kinda / sorta recyclable. Many of us bought into the idea, chauffeuring the kids to school in gas-guzzling SUVs and diligently recycling the trash at home.

Thankfully, there is something that would actually help: By using the energy stored in used reactor fuel, municipalities could power the carbon-free recycling of *all* their waste, including plastic. A plasma-arc smokeless furnace (Fig. 20) operating at 1,800° C will cleanly reduce any waste down to its constituent metals, mixed vitrified solids (glass), ethanol and methanol, both of which are useful as carbon-neutral fuels. This is not incineration; this is the molecular-level breakdown of any and all municipal trash—medical waste, hazardous material, food waste, even Mom's old couch—without smoke, chemical residue, or other pollution. [3]

The drawback of plasma-furnace waste remediation has always been the enormous amount of electricity required. A dedicated fleet of fast-neutron reactors, running on "spent" nuclear fuel (SNF), could resolve this dilemma, powering a 21st-century solution to a 20th-century problem. The strategy is elegantly simple:

> *Use nuclear waste*
> *to power the cleanup*
> *of all our other waste.*

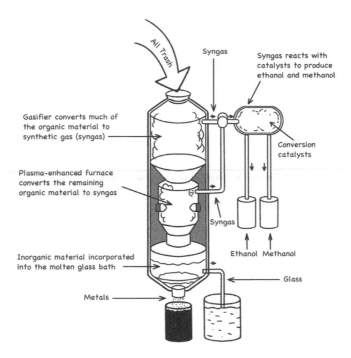

Fig. 20: Plasma-Arc Smokeless Trash Incinerator
Source: https://energy.mit.edu/news/turning-waste-into-clean-fuels/

3.2 TURNING BIG ONES INTO LITTLE ONES

Fissioning (splitting) a big fuel atom to release the energy locked in its nucleus results in two (sometimes three) smaller atoms. Though these fission products (FPs) are unstable and therefore radioactive, they have much shorter half-lives than the fuel atom from whence they came. As the philosopher Lao Tzu famously observed: "The flame that burns twice as bright burns half as long."

Some of these short-lived fission products can shed copious amounts of thermal energy. This is a good thing, because they can be chemically extracted from spent fuel and used to produce thermoelectric energy—the direct conversion of heat to electricity. For example, Russia has been using slugs of strontium-90 reactor "waste" to electrify their remote lighthouses. The amount required is small enough to fit in a tall Red Bull can. Think of it: Enough energy to power a lighthouse for decades, in the palm of your hand. So where's the waste? [4]

Used reactor fuel can also be harvested for the isotopes used in medicine, science, and industry. Examples include technetium-99m for monitoring blood flow, chromium-51 for blood cell tagging, cobalt-60 for sterilizing harmful bacteria, and strontium-90 for precision thickness gauges.

When it comes to used fuel, or any kind of nuclear "waste," always keep one thing firmly in mind:

It's only waste if we waste it.

Plutonium-238, made from the Neptunium-237 extracted from used reactor fuel, powers the Mars rovers and all deep-space probes. Voyager II, for example, was launched nearly a half century ago and has finally left the solar system. After 44 years, the probe is still operating at 70% power, and has recently been reprogrammed to explore interstellar space. [5] You can't boldly go with solar panels.

Since a radioactive atom, by definition, actively radiates (ejects) pieces of itself, it follows that any radioactive source can only spit out a finite number of zoomies before it stops being radioactive. It can't just keep creating new zoomies out of thin air. Which means that the longer the isotope's half-life, the less radioactive it is.

That may seem counterintuitive at first, but upon reflection it soon becomes clear. And yet, half-lives are one of the most widely misunderstood aspects of nuclear science, and are routinely used to scare uninformed people: "OMG! Thorium is radioactive for fourteen billion years!" (*Zoomies! We're doomed!*)

Take a breath, good citizen. A half-life of fourteen billion years means that thorium is so weakly radioactive, it will take that long for one half of any given batch of the stuff to eject the occasional zoomie as it ever so *slooooooowly* turns into lead. The same billion-year scenario applies to potassium and natural uranium.

3.3 LEAVE NO FUEL UNBURNED

EROI (pronounced E-roy) means Energy Returned On Investment. The term refers to the total amount of energy invested in building, fueling, operating, and decommissioning an energy production device, relative to the total amount of useful energy derived from said device over its lifecycle. [6]

You'll also see it spelled EROEI, Energy Returned On Energy Invested, which some feel is a better phrasing, but the original EROI acronym stuck. Charles Hall, the noted ecologist who coined the term, explains EROI with this simple formula:

$$EROI = \frac{\text{Energy returned to society}}{\text{Energy required to get that energy}}$$

As shown in the graph, the EROI for a fast-neutron reactor (see Chapter 5) can be a whopping 2,000 to one. The skinny detail on the right shows the actual scale—the energy density of a fast reactor is literally off the chart. In stark contrast, the EROIs of solar, wind, and biomass are in the single digits.

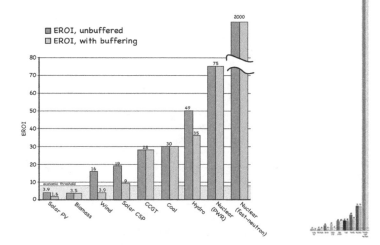

Fig. 21: EROI Energy Returned on Investment
Source: https://festkoerper-kernphysik.de/Weissbach_EROI_preprint.pdf

Notice that the EROI numbers for renewables are even worse when "buffering" is factored in. This term refers to the backup provided by energy storage or other power plants. The even-lower EROI values for buffered renewables are due to the energy it takes to build and maintain the buffering device.

NERD NOTE: The buffering of intermittent energy (by storing it in a battery, for example) is like the buffering in a video player, and for much the same reason: Stuff arrives in bits and pieces and is allowed to build up in storage, so it can then be smoothly streamed without interruption.

EROI analysis has been criticized because as technology improves and prices drop, EROIs improve as well. Instead of 19 for wind and 3.9 for solar, as shown in the 2014 graph above, their estimated EROIs in 2020 ranged as high as 25:1 (Fig. 21-A). But this misses the larger point: As wind and solar become more than 20% of our energy mix, they'll need more buffering than our conventional power plants now provide for free. And as they supply more and more of their own buffering, their EROIs will drop off a cliff.

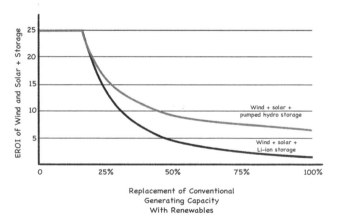

Fig. 21-A: The EROI Cliff
Source: Energy Storage and Civilization (Palmer and Floyd, 2020)
https://link.springer.com/book/10.1007/978-3-030-33093-4

Adequate buffering for renewable systems is a substantial expense, but is often dismissed as an externalized cost. Indeed, renewables can seem quite affordable until the buffering required to stabilize their erratic performance (grid-scale batteries, gas power plants, etc.) is added to the sticker price. Most fans of renewable energy haven't quite absorbed this sticker shock, and what it portends for the future. (We explore the details in *Roadmap to Nowhere*.)

The EROI threshold for a viable power source in a developed society is an energy investment ratio of at least 7:1. [7] This means that over its service life, an energy source must produce at least seven times the energy that was invested in building, operating, and decommissioning the thing. If it can't do that, then it's not fit to serve as a utility power plant in a modern economy (more on this in Chapter 7).

Charles Hall points out that with an EROI of 3:1 or even 5:1, you couldn't do much more than power a civilization like England had in the 16th century. [8] At that time, up to 50% of the economy revolved around fuel. While some wood was available for space heating, most of it was prioritized for forging iron tools and weapons. Civilization grew at a snail's pace, with little energy available for anything more than basic subsistence.

The introduction of coal in the 1750s made the steam engine practical. With the wide availability of coke (refined charcoal, purified by heating coal without burning it), the high temperatures for steel production became feasible. In the market economy that coal and steam created, fuel's share fell to about 25%.

With the expansion of petroleum around the turn of the 20th century, fuel's portion of the modern market economy gradually shrank to around 5%–10%, even as the world became electrified and mobilized. That's what energy density is all about.

For a developing society, or a low-industrial society, an EROI below 7:1 may suffice. But if a society intends to advance beyond basic necessities, they must grow their capacity for energy production as a crucial first step. Energy is what drives economic progress and the advance of civilization. The steam engine came before the Industrial Revolution, not the other way around. Energy comes first.

CHAPTER FOUR

A Fate Worse than Global Warming

Sᴏᴍᴇ ᴇɴᴠɪʀᴏɴᴍᴇɴᴛᴀʟɪsᴛs ʜᴀᴠᴇ such a pronounced distaste for nuclear energy that they categorically dismiss the subject without further discussion. This is unhelpful to say the least, especially given nuclear's excellent record of performance and safety.

The accidents at Fukushima and Three Mile Island killed no one, and no one will ever build a reactor like Chernobyl again. And yet, some people seem almost stubbornly afraid of nuclear power. It's like writing off every automobile that's ever been built, or ever will be, because some Pinto gas tanks blew up a half-century ago. Chernobyl was the Pinto of nuclear reactors.

1970 Ford Pinto, With Faulty Gas Tank
Credit: John Lloyd (CC-BY-SA-2.0)

As renowned green journalist George Monbiot wrote in *The Guardian* newspaper, just days after Fukushima:

"While nuclear causes calamities when it goes wrong, coal causes calamities when it goes right, and coal goes right a lot more often than nuclear goes wrong." [1]

This perspective doesn't seem to matter to some people, but it should. Because the actual harm caused by the production of nuclear power is nearly nonexistent.

After sixty-five years, with about 100 reactors operating in the US and another 100 or so at sea, only sixteen deaths have been attributed to the production of American nuclear power: Three operators perished from the partial meltdown of an experimental reactor at Idaho National Laboratory in 1961. [2] Ten additional deaths came from construction accidents at various commercial power plants, and there have been three deaths and one severe injury in the US fuel-enrichment and processing industry. [3] After generating about 27 trillion kilowatt-hours of clean electric energy over the last several decades [4] (equal to 29 years of electricity for all US single-family homes), not a single member of the American public has been injured or killed by the production of nuclear power, or by "spent" nuclear fuel (SNF).

By health and safety statistics, an operating nuclear power plant is one of the safest work environments in the country. [5] And as it turns out, the elevated leukemia rates found near some nuclear plants, formerly blamed on radiation, have now been linked to other industries, both past and present, in the same industrial zone. [6]

The fact is, nuclear power is so safe that a person in the United States actually has a greater statistical chance of being shot by their own dog than being harmed by either nuclear power or nuclear "waste." [7] Amazing, but true: About one American each year gets shot by their own dog. Allowing for population growth and recent gun proliferation, we can guesstimate that there have been roughly thirty-five dog-involved shootings in the US since the advent of commercial nuclear power in 1957. Compare this to the casualties cited above.

But even if we do accept nuclear's stellar safety record, what about the mining?

4.1 DIGGING DEEP

There was a substantial number of cancers and mortalities in the early days of US uranium mining and milling, most of which took place in the Navajo Nation with local labor. [8] At the time, cigarette smoking was common, and safety regulations were either lax or non-existent. As it turned out, inhalation of mining dust and high levels of radon gas (which decays into the same radioactive polonium found in cigarette smoke) were contributing factors at any type of underground metal mine, and not just uranium digs.

The arms race against the Soviets and the systemic neglect of Indigenous American health and safety made the situation even worse. Wind-blown dust from abandoned mine tailings and waste rock may well have caused cancers and shortened lives, and leaching chemicals from in-situ mining were not always properly monitored, which polluted local aquifers (note the monitor wells in Fig. 23). As regulations and oversight improved, ventilation and respirators were introduced to deep-shaft uranium mining. Casualty rates soon reduced to the same levels found in other metal mining operations. [9]

Canada mines far more uranium than the US, and a major study has shown that the health of their miners is on par with their American counterparts. And just how healthy are their miners? *"There has been no known case of illness caused by radiation among uranium miners in Australia or Canada."* [10]

Another Canadian study shows that the CO_2 released by uranium mining and milling amounts to just 1.1 grams per kilowatt-hour of electricity produced from the resulting fuel. [11] When power plant construction, O&M (operations and maintenance), and decommissioning are factored in, nuclear power's total lifecycle carbon emissions are only 6 g /kWh. [12] Compare this to the whopping 380 grams per kWh emitted by heavy coal-burning countries like Germany. [13]

And this was before Russia invaded Ukraine. Eight months later, in November 2022, anti-nuclear Germany was up to 459 grams of CO_2-equivalent emissions and facing the onset of winter. As shown in Fig. 22 below, nuclear has the lowest greenhouse gas (GHG) emissions of any clean-energy technology.

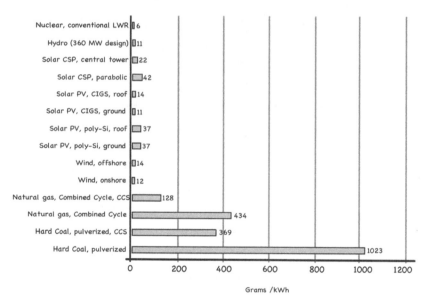

Fig. 22: Lifetime CO_2-equivalent Emissions
Source: Life Cycle Assessment of Electricity Generation Options | UNECE

Nuclear's meager emissions can be further reduced to nearly zero with electrically powered equipment for mining, milling, processing, and construction, energized by a nuclear grid or an on-site microreactor.

Even so, the total environmental impact of an energy system comes from more than its carbon emissions. The resources consumed in building, fueling, operating, maintaining, and decommissioning the system must also be taken into account, along with the mining

Caterpillar Model 793, Now Available with Electric Drive
Credit: Rob Simmons (CC-BY-SA-4.0)

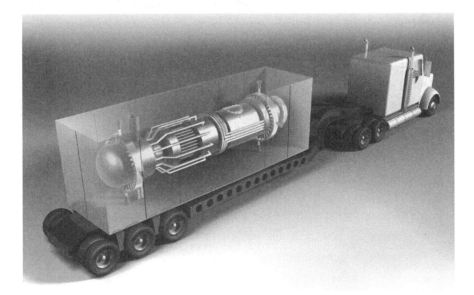

Mobile Micro Reactor for Off-Grid Power
Source: inl.gov/

to procure said resources. The essential question is this: What is the total footprint of a given energy system, compared to the total energy derived from that system?

"Throughput" is engineering speak for the resource consumption of a system. Conventional power plants consume fuel as they operate, but they also "consume" a variety of other material for extended periods of time. Construction components like steel and concrete are typically amortized over the productive life of the plant.

The steel throughput of a power plant, for example, is the quantity of steel used relative to the energy produced by the plant. If 70 tonnes of steel are used to build a nuclear plant with a yearly output of 1 TWh (one terawatt-hour) and an estimated lifespan of 60 years, the plant has a steel throughput of about 1.17 tonnes per TWh (70 ÷ 60 = 1.166).

But caution is strongly advised—relying on this popular metric can sometimes be awfully misleading, because any throughput calculation is only a snapshot. As technologies evolve, they often reduce their material throughput, sometimes dramatically. For example, engineers now have the technology to make a reactor last for eighty years or more, doubling the original forty-year estimated lifespan. [14]

This doubling cuts the steel throughput of a nuclear plant in half, which dramatically alters any quantitative analysis. And since doubling the life of a reactor effectively doubles the life of the entire plant, the throughput calculations for concrete, copper, glass, aluminum, and other materials used in building the plant will reduce as well.

NERD NOTE: The metal walls of a Reactor Pressure Vessel (RPV) become embrittled over the years, as the zooming neutrons in the core bury themselves in the metal walls of the RPV. Through physics wizardry, some metals will slowly transform into wall-weakening elements such as helium-4, iron-56, and other pesky isotopes.

Over the years, these insults eventually degrade the metal's crystalline structure, making the vessel susceptible to microscopic cracks. Since this is a well-known issue, operators conduct embrittlement inspections during each refueling cycle.

Luckily, a reactor vessel's metal crystalline structure can be restored by "annealing" the metal. [15] The emptied vessel is heated at 550º C for about 200 hours, then slowly allowed to cool. This repairs the crystalline structure of the metal, even after years of neutron bombardment. Large RPVs can be annealed in place with induction heaters—the Russians are already doing this. [16] The RPVs of small modular reactors will be taken back to the factory for annealing, refueling, and other maintenance.

Since high-temperature Generation-IV systems like the molten-salt reactor (MSR) will operate well above 550º C, the tantalizing possibility exists that MSRs will be continuously self-annealing, enabling service lives of well more than a century.

(They should have fusion figured out by then. ☺)

Energy generation requires mining for two different things: the material for building the plant, and the material to make the fuel. In contrast, passive energy-harvesting technologies like wind and solar do not require fuel. (Wind and sunshine are not fuels; see Chapter 7). Nuclear power takes the extra step of excavating deep geologic repositories (DGRs) for the long-term storage of used reactor fuel (see Chapter 18). This mining must be considered as well. [17]

Currently, nuclear must dig for its fuel, while renewables obviously do not. Their "fuel" comes to them free of charge and with no mining karma attached, which is nice work if you can get it. About 90% of the uranium for US reactors is mined and processed outside the country, where ore concentrations are typically higher. [18]

Nowadays, most uranium miners don't even go underground. This partly stems from advances in automation, and partly because 57% of the world's uranium, and virtually all uranium mined in the US, is now being extracted by in-situ leaching (ISL). [19] This simple process injects a leaching solution into an underground uranium deposit, either a sodium carbonate mixture (basically, washing soda) or a dilute, vinegar-strength sulfuric acid, depending on the chemistry of the ore body. Pumps draw up the dissolved uranium, which is then piped to an on-site mill where the slurry is dried and filtered.

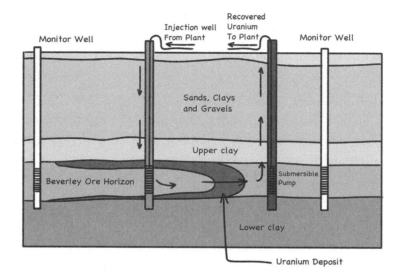

Fig. 23: In-Situ Leaching
Source: https://world-nuclear.org/information-library/nuclear-fuel-cycle/mining-of-uranium/in-situ-leach-mining-of-uranium.aspx

One nice thing about in-situ leaching is that it doesn't generate waste rock and leaves almost no "tailings," the final waste material of any dirt-mining process. Some copper mines also use ISL, but most copper is extracted through either deep-shaft or open-pit mining, and renewables use a *lot* of copper (see Fig. 24 below).

But whether a mineral-bearing ore is scraped from the surface, or brought up to the surface, or left where it is (in-situ) and leached of its desired mineral, the same chemicals are used to leach uranium and copper from their respective ore. Whether it's a uranium mine or a copper mine, the mineral-bearing ore is ground into rubble and milled into a sand or powder, then the material is piled in heaps and sprinkled with an alkaline or acid wash, depending on the ore chemistry. The "pregnant liquor" is collected from below, and processed to remove the leached-out metal. Finally, the depleted liquor is sent to a tailings pond, where it dries to a solid.

4.2 DIRT MINING

The 43% of mined global uranium that is not recovered by in-situ leaching is mined by either underground or open-pit methods. We estimate that the combined global average waste-to-ore ratio for uranium dirt mining is about 3.9:1. This means that for every tonne of uranium-bearing ore dug from the earth, about 3.9 tonnes of waste rock will be displaced as well. (See Part One of our mining supplement: [20])

In open-pit mines, this is called the "strip ratio." The only real difference between open-pit and underground mining is whether the ore body is close to the surface or deep underground. But once the ore is accessed, the process is the same: The ore-bearing raw material is excavated, waste rock and all, and sent to the mill for processing.

Once the mill extracts the mineral from the ore, the depleted ore is considered waste rock as well. Although, when prices rise, depleted ore tailings are a top candidate for re-leaching. Since the leaching process is imperfect, there will always be a bit more uranium that can still be extracted from the rubble and dust of an evaporated waste pond—for the right market price.

Ore concentrations will vary from mine to mine. Some of the world's best uranium deposits have ore concentrations of 0.5% or higher, but most mines have significantly lower values. We show in our mining supplement that the world's average uranium ore concentration (often called the "ore grade") at today's mines is a paltry 0.078%.

You don't need to be a uranium prospector to know this means some mighty slim pickings: For every kilogram of uranium-bearing ore dug out of the earth, chances are that only about 0.078% of each kilo (78 grams) will be uranium oxide.

A comprehensive assessment of any system's environmental footprint would have to consider the total material dug out of the earth in order to build, operate, and maintain the system. This includes the waste rock, the mineral-bearing ore within the waste rock, and the finished material (refined mineral) derived from the ore (steel, copper, etc.). The recycling of finished material must also be taken into account, as this can significantly reduce the mining and milling footprint.

Only 43% of global uranium is obtained from dirt mining, while 57% comes from in-situ leaching. While the finished material throughput for nuclear is tiny compared to other energy technologies, nuclear's raw material throughput can be substantial, due to the volume of waste rock dug up by 43% of global uranium production.

Aside from digging into the numbers on uranium, we also considered the raw material (displaced earth) behind the other finished materials needed to build and maintain nuclear, wind, and solar energy systems. (See Part Two of our supplement: [21]) For some perspective on this: The global rock-to-metal ratio (RMR) of elemental iron, from which steel is made, is 9:1. That is, for every tonne of iron-bearing ore, nine tonnes of waste rock are typically exhumed as well.

More fun mining facts:

- Rock-to-cement is about 1.6:1

- Rock-to-cement-to-concrete is about 1.1:1

- Rock-to-aluminum is about 7:1

- Rock-to-chromium is about 18:1

- Rock-to-zinc is about 71:1

- Rock-to-copper is about 513:1

- Rock-to-uranium is about 3,000:1

- Rock-to-fuel pellets is about 21,700:1

These last two numbers clearly suck, especially for a massive and long-term buildout of nuclear power fueled by freshly-mined uranium. But as karma would have it, there are several alternative sources for reactor fuel which can greatly reduce—and potentially eliminate—any future need to displace more raw material to make even more fresh fuel pellets. We'll explore these game-changing options in the next two chapters.

To be entirely candid, it will be difficult for existing nuclear technology to build out big enough and fast enough, and do the job it can plainly do, so long as it continues to rely upon business-as-usual uranium mining for its fuel supply. The first automobiles were built by hand, but things had to change to put the world behind the wheel. Sources and methods must change as well for the buildout of nuclear power.

For an in-depth discussion of where uranium mining sits at present, we invite you to dig through our mining supplement (bring a calculator). But as you explore the numbers, keep in mind that they are only a snapshot of existing conditions. Methods can and will change bigly in the years to come.

The legacy of uranium dirt mining leaves little to be proud of, but things have changed. Stringent regulations now ensure that existing and future uranium digs must set aside sufficient funds for land rehabilitation. The former Mary Kathleen Mine in Australia, for example, is now an operating cattle station with unrestricted access. [22]

In Kazakhstan, where the world gets nearly 40% of its uranium (Fig. 24), virtually all U mining is now done by in-situ leaching. The same holds true for every U mine in the United States, most of those in Uzbekistan, and some in Russia as well. Most of the uranium dirt mining that is still being done in this world is taking place in Canada and Australia, where mine rehabilitation is strictly enforced. Much of the rest is happening in Namibia and Niger, where China has been scooping up underperforming mines.

Digging for fuel doesn't have to leave a scarred landscape. Unlike coal mining, in which half or more of the exhumed material is carted away and burned, uranium mines have plenty of waste rock to fill whatever hole they dig. That's because the recovered uranium will often average just 0.078% of the total displaced earth—less than one-tenth of one percent.

At a properly managed uranium mine, *displaced earth becomes replaced earth.* Once the mine is played out, the hole is filled and the land is rehabilitated for future use. This approach even passes muster in anti-nuclear Australia, where they're grazing cattle atop a rehabilitated uranium mine. Ironically, nuclear fear helps to ensure this happy outcome. Unlike iron and copper mines, where rehabilitation may not be strictly enforced, the onus on uranium has inspired both locals and environmentalists around the world to pay very close attention to the rehabilitation of any uranium dig—even though the rubble at a granite quarry can be as radioactive as uranium mine tailings. [23]

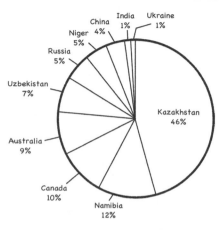

Uranium Production

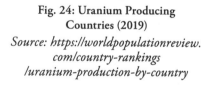
Fig. 24: Uranium Producing Countries (2019)
Source: https://worldpopulationreview. com/country-rankings /uranium-production-by-country

The good news is that the mining footprint of nuclear fuel can be reduced to virtually zero with existing and emerging technology. This will come in handy, because as we show in *Roadmap to Nowhere*, the US may need about 1,800 GW of nuclear power to supply all "primary energy" by 2050, meaning all forms of energy, not just electricity. The country now has about 95 GW up and running, so a national nuclear grid would amount to a twenty-fold increase. That means twenty times the fuel, but it does not necessarily mean twenty times the mining.

With the alternative sources and methods discussed in the next few chapters, both a national and global buildout of nuclear power would be entirely feasible—and sustainable—with an environmental footprint a mere fraction of what renewables could ever hope to achieve, even with dirt-cheap grid-scale batteries. By midcentury, the throughput for nuclear power could look something like this (see our mining supplement for calculations):

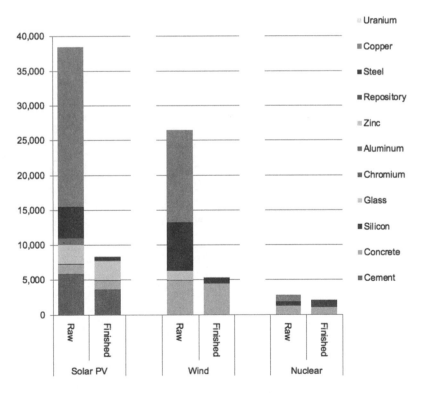

Fig. 25: Raw and Finished Material Throughput (Tonnes/TWh)
with Recycling (and no uranium mining)
Credit: By the authors

CHAPTER FIVE

"Waste Not, Want Not."

– BENJAMIN FRANKLIN, ELECTRICITY PIONEER

Nuclear power's primary mission is to ensure a reliable and sustainable supply of clean energy for the nation and the world by replacing fossil fuel with the smallest and cleanest possible footprint. A key part of this process is to wean the world off coal by producing energy cheaper and cleaner than coal. Nuclear is already doing this; it just needs to be doing a lot more of it, and in a lot more places. [1]

For nuclear to be a genuine long-term solution, its sustainability model must include the reduction or (much better) the elimination of uranium dirt mining, coupled with the freedom to recycle "spent" fuel rather than sending all of it to long-term storage (see Chapter 18).

The good news is that uranium mining can be significantly reduced, or even eliminated, through any combination of these five technologies:

- **Exploit** our depleted uranium as fuel for fast-neutron reactors.

- **Recycle** "spent" nuclear fuel in fast-neutron reactors.

- **Breed** new fuel from feedstocks of natural uranium and thorium.

- **Reprocess** used fuel for another round in light-water reactors.

- **Extract** uranium from seawater to fuel existing and future reactors.

Reprocessing used fuel for a second run in light-water reactors, like France has been doing for the last forty years, is the low-hanging fruit that could cut uranium mining waste nearly in half. After its second run, the twice-used fuel could then be processed and used again in fast-neutron reactors for complete burnup. The alternatives listed above could all be in widespread use by midcentury.

Before we dig any deeper into the subject of energy derived from fuel, we should take a moment to refresh those cherished high school memories of ions and isotopes, and the difference between chemistry and nuclear physics. But this time, let's do it with lots of pictures and (almost) no math.

First off, you can think of *isotopes* like cars—identical models of the same sedan, but with different engines. Hydrogen is a simple example: Most hydrogen atoms have nothing but a single proton in the nucleus, and that's it. When people refer to hydrogen, this is the isotope they're thinking of: protium, or hydrogen-1. Think of it as the Protium, hydrogen's base-model ride.

Some hydrogen atoms have a neutron in the nucleus to keep the proton company—that's the deuterium model. Tritium, or hydrogen-3, is a rare, limited-edition hydrogen isotope sporting twin neutrons under the hood. But so long as it just has one proton, it's only a hydrogen sedan. (Your neutrons may vary.)

Chemistry is all about the electrons, and does not involve the nucleus of any atom. All chemistry does is rearrange the electrons orbiting the atoms involved in the chemical process. Electrons are negatively charged, and the protons in the nucleus they orbit are positively charged.

An *ion* is a custom version of an atom's basic model. The ionization process offers a variety of trim packages that give the atom a greater or lesser number of electrons whizzing around its nucleus in a cloudy blur. Changing the atom's electron count will give it a formal minus-charge, or a formal plus-charge, depending on whether electrons were gained or lost in the process.

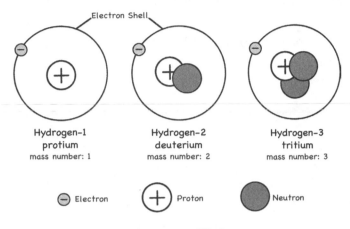

Fig. 26: Isotopes of Hydrogen

Hydrogen, for example, typically has one electron orbiting a solitary proton; this protium atom becomes a free-flying proton when its electron is stripped away. That's how the sun, our first operating fusion reactor, showers us with cosmic rays.

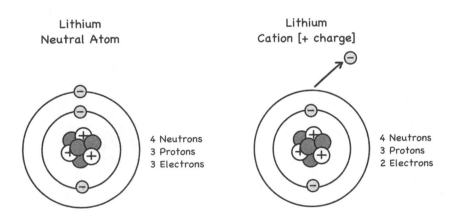

Fig. 27: Positive Ions ("CAT-eye-ons")

Metals (such as lithium) tend to lose electrons and become cations, and some elements like fluorine tend to gain electrons and become anions. A chemical reaction can alter an atom's net electric charge by removing or adding electrons, turning the atom into a positive or negative ion of its former self. This can make it attract or repel another atom, depending on the other atom's charge.

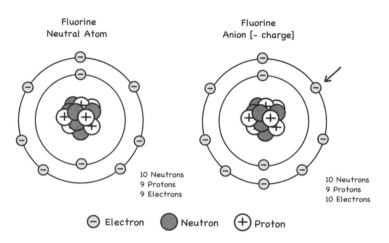

Fluorine
Neutral Atom

Fluorine
Anion [- charge]

10 Neutrons
9 Protons
9 Electrons

10 Neutrons
9 Protons
10 Electrons

⊖ Electron   ⬤ Neutron   ⊕ Proton

Fig. 28: Negative Ions ("AN-eye-ons")

Think of ions as two servings of the same ice cream, but for some reason you're attracted to one and repulsed by the other.

(We're sorry you had to see that.)

Chemical energy comes from breaking and forming electron bonds between the atoms involved in a chemical reaction. The process either gives off energy (exothermic), or absorbs energy (endothermic), but either way the field of chemistry is all about the electrons.

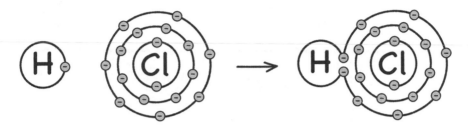

Fig. 29: Formation of Hydrogen Chloride

Nuclear energy, as the name implies, is all about the nucleus, and is typically produced by splitting a uranium or plutonium nucleus with a free-flying neutron zooming along at just the right speed. Because it has no charge, a neutron can zoom undeterred through a thick cloud of electrons and strike a target nucleus.

The neutron could be absorbed, but if it splits the nucleus instead, some of the binding energy that held the nucleus together is suddenly released. Some fast-flying neutrons will zoom away from this high-energy event as two new smaller atoms are formed (sometimes three), with most of the original electrons settling into orbitals around the new atoms. [2]

Zooming neutrons, whether fast, slow, or in between, are the projectiles that crack open the nuclei of heavy atoms—which, conveniently enough, eject one or more zooming neutrons as they split into smaller atoms. And thus we have controlled chain reactions. But not to worry: It is impossible for a reactor to have a runaway chain reaction like a bomb. Fuel enrichment levels are far below the required concentration, at 4%–19% versus 85% or higher for nuclear weapons.

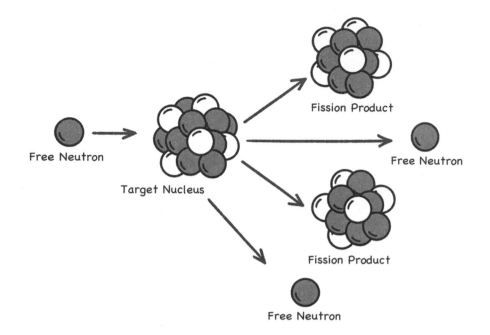

Fig. 30: Nuclear Fission

The energy we get from combustion (wood, coal, diesel, etc.) comes from the sudden rearrangement of electrons when carbon fuel atoms and oxygen atoms react (ignite, combust, explode, etc.). The energy derived from fission comes from splitting some of the largest atoms in the universe. Think of it like this:

The electrons of a large fissionable atom are like a swarm of TIE fighters protecting the Death Star. Chemistry is when the TIE fighters engage in a dogfight with the X-Wing fighters of the Rebel Alliance fleet hovering nearby. Nuclear is when Luke obliterates the Death Star with a proton torpedo down the thermal exhaust vent.

Except nuclear fission uses neutron torpedoes, and the nucleus doesn't get blown to smithereens. It just breaks into a pair of smaller atoms (sometimes three), and one or two neutrons (sometimes three) go zooming away as well, along with the release of a stupendous amount of energy. But you get the idea.

5.1 MEANWHILE, BACK ON PLANET EARTH

To make fuel pellets for a light-water reactor, uranium is chemically extracted from uranium-bearing ore. The resulting yellowcake powder is purified natural uranium, a metallic powder that's 99.3% U-238 oxide and 0.7% U-235 oxide.

Yellowcake
Credit: US Nuclear Regulatory Commission (CC-BY-SA-2.0)

Fuel Pellets
Credit: US Nuclear Regulatory Commission (CA-BY-SA-2.0)

While uranium-238 will not fission, U-235 will, and this is best accomplished with a slow-moving or "moderated" neutron. Any neutron is fast and furious when it zooms away from a fission event, but maintaining its speed (and thus its energy) depends on what it bumps into. Conveniently, water molecules do an excellent job of slowing down zooming neutrons to the proper speed for splitting uranium-235.

The idea of fissioning U-235 in a water-moderated reactor, to make steam to spin a turbine to make electricity, was a logical step in early reactor design. The result was the light-water reactor (LWR), a workhorse technology virtually impervious to climate or weather, and capable of generating on-demand energy for about two years straight before refueling.

Plutonium-239 can form when U-238 absorbs a neutron. This is because while U-238 is not fissile, it is "fertile" and can thus be "bred" into Pu-239, which is an excellent reactor fuel. (A bunch of 20th-century guys came up with these terms.)

Though it rarely happens, a bit of Pu-239 can also be formed when the U-235 in fuel pellets absorbs multiple neutrons instead of fissioning. But however it's made, the heavier plutonium-239 isotope typically requires a fast (unmoderated) neutron to fission.

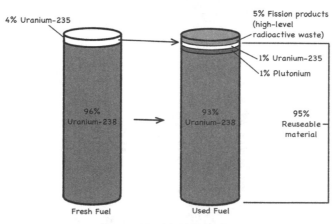

Fig. 31: Used Fuel Makeup

For a number of technical reasons we'll explore in *Power to the Planet*, the Pu-239 that forms in a fuel pellet has a 40% chance of fission. But even so, plutonium is such a terrific fuel (it "burns hot") that more than one-third of the energy produced in a light-water reactor comes from breeding and "burning" plutonium (1950s slang for fissioning nuclear fuel). This is known as the uranium-plutonium fuel cycle. [3]

The remaining 60% of the Pu-239 that forms in a reactor's fuel pellets will just sit there gobbling up more neutrons like immobilized Pac-Men, turning into larger and larger isotopes: Pu-240, -241, -242, and so on. Higher elements will also be produced, such as americium-241, the radioactive material used in smoke detectors.

Aside from plutonium-239, however, most self-respecting transuranic isotopes wouldn't even consider fissioning with anything less than a seriously fast neutron. And that's where complications arise.

5.2 TRANSURANICS HAVE NOTHING TO DO WITH EXPLORING URANUS

Transuranics (TRUs) are the man-made elements that lie beyond (trans-) uranium on the Periodic Table: neptunium, plutonium, americium, curium, berkelium, etc. They're down in the bottom row, separated from the main table (see Fig. 31 below).

NERD NOTE: The Periodic Table primarily organizes elements by their proton count. All isotopes of uranium, for example, have 92 protons—that's what makes them uranium atoms. But the Table also organizes elements according to how many electrons occupy their various "orbitals," arranged as shells within shells like a Russian doll.

Nature has some very clever ways of fitting electrons into these orbitals, but in simple terms, lanthanum and actinium begin their respective rows below the main Table because they are the first elements with f-electron orbitals. Since there can be up to fourteen f-electrons, placing these oddball elements in the Table would make it too unwieldly, so they're depicted in separate rows below.

Periodic Table of Elements

IA																	VIIIA
1 H Hydrogen 1.01	IIA											IIIA	IVA	VA	VIA	VIIA	2 He Helium 4.00
3 Li Lithium 6.94	4 Be Beryllium 9.01											5 B Boron 10.81	6 C Carbon 12.01	7 N Nitrogen 14.01	8 O Oxygen 16.00	9 F Fluorine 19.00	10 Ne Neon 20.18
11 Na Sodium 22.99	12 Mg Magnesium 24.31	IIIB	IVB	VB	VIB	VIIB	VIIIB	VIIIB	VIIIB	IB	IIB	13 Al Aluminum 26.98	14 Si Silicon 28.09	15 P Phosphorus 30.97	16 S Sulfur 32.06	17 Cl Chlorine 35.45	18 Ar Argon 39.95
19 K Potassium 39.10	20 Ca Calcium 40.08	21 Sc Scandium 44.96	22 Ti Titanium 47.87	23 V Vanadium 50.94	24 Cr Chromium 52.00	25 Mn Manganese 54.94	26 Fe Iron 55.85	27 Co Cobalt 58.93	28 Ni Nickel 58.69	29 Cu Copper 63.55	30 Zn Zinc 65.38	31 Ga Gallium 69.72	32 Ge Germanium 72.63	33 As Arsenic 74.92	34 Se Selenium 78.97	35 Br Bromine 79.90	36 Kr Krypton 83.80
37 Rb Rubidium 85.47	38 Sr Strontium 87.62	39 Y Yttrium 88.91	40 Zr Zirconium 91.22	41 Nb Niobium 92.91	42 Mo Molybdenum 95.95	43 Tc Technetium (98)	44 Ru Ruthenium 101.07	45 Rh Rhodium 102.91	46 Pd Palladium 106.42	47 Ag Silver 107.87	48 Cd Cadmium 112.41	49 In Indium 114.82	50 Sn Tin 118.71	51 Sb Antimony 121.76	52 Te Tellurium 127.60	53 I Iodine 126.90	54 Xe Xenon 131.29
55 Cs Cesium 132.91	56 Ba Barium 137.33	57 - 71 Lanthanides	72 Hf Hafnium 178.49	73 Ta Tantalum 180.95	74 W Tungsten 183.84	75 Re Rhenium 186.21	76 Os Osmium 190.23	77 Ir Iridium 192.22	78 Pt Platinum 195.08	79 Au Gold 196.97	80 Hg Mercury 200.59	81 Tl Thallium 204.38	82 Pb Lead 207.20	83 Bi Bismuth 208.98	84 Po Polonium (209)	85 At Astatine (210)	86 Rn Radon (222)
87 Fr Francium (223)	88 Ra Radium (226)	89 - 103 Actinides	104 Rf Rutherfordium (265)	105 Db Dubnium (268)	106 Sg Seaborgium (271)	107 Bh Bohrium (270)	108 Hs Hassium (277)	109 Mt Meitnerium (276)	110 Ds Darmstadtium (281)	111 Rg Roentgenium (280)	112 Cn Copernicium (285)	113 Nh Nihonium (284)	114 Fl Flerovium 289	115 Mc Moscovium (288)	116 Lv Livermorium (293)	117 Ts Tennessine (294)	118 Og Oganesson (294)

57 La Lanthanum 138.91	58 Ce Cerium 140.12	59 Pr Praseodymium 140.91	60 Nd Neodymium 144.24	61 Pm Promethium (145)	62 Sm Samarium 150.36	63 Eu Europium 151.96	64 Gd Gadolinium 157.25	65 Tb Terbium 158.93	66 Dy Dysprosium 162.50	67 Ho Holmium 164.93	68 Er Erbium 167.26	69 Tm Thulium 168.93	70 Yb Ytterbium 173.05	71 Lu Lutetium 174.97
89 Ac Actinium (227)	90 Th Thorium 232.04	91 Pa Protactinium 231.04	92 U Uranium 238.03	93 Np Neptunium (237)	94 Pu Plutonium (244)	95 Am Americium (243)	96 Cm Curium (247)	97 Bk Berkelium (247)	98 Cf Californium (251)	99 Es Einsteinium (252)	100 Fm Fermium (257)	101 Md Mendelevium (258)	102 No Nobelium (259)	103 Lr Lawrencium (262)

Fig. 32: Periodic Table of Elements
Credit: Emeka Udenze (CC-BY-SA-4.0)

Only a handful of actinide isotopes are long-lived. These troublesome TRUs come about when the plutonium-239 that forms in reactor fuel absorbs even more neutrons without splitting. It's these long-lived, man-made isotopes that make used fuel seem like such a vexing problem, even though it's actually not. For one thing, the longer the half-life, the lower the radioactivity and thus the lesser probability of harm.

Along with a smattering of these long-lived isotopes, used fuel also has a collection of short-lived fission products (FPs), newly-minted atoms that are too small to fission. They form when big fuel atoms split in two, and some of the resulting FPs are harshly radioactive.

Since fission product atoms are formed from much larger atoms, and since these humongous atoms are held together with an overabundance of neutrons, an FP will usually start life with several more neutrons than its elemental peers—radioactive cesium-137, for example, versus plain old cesium-133. Fission products shake off (radiate) this extra baggage so they can settle into their new life as smaller, normal atoms.

When fuel atoms split, the fission products that result tend to occupy two distinct regions of the Periodic Table. (Sorry, alchemists, but gold isn't on the list; that's not how the nuclear cookie crumbles.)

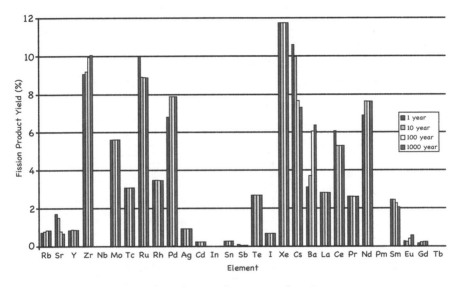

Fig. 33: Distribution of Fission Product Creation
*Source: https://en.wikipedia.org/wiki/Fission_product_yield#
/media/File:Fission_yield_volatile_2.png*

The problem with FPs and TRUs forming in fuel pellets is that they're pretty much evenly distributed throughout several tonnes of (very) mildly radioactive U-238 filler, which constitutes about 95% of the typical fuel pellet (Fig. 31). This is why all used fuel is treated as long-term radioactive waste. Although, considering the amount of energy produced per kilo of fissioned fuel, the volume of used fuel is downright minuscule, even from reactors that only burn 3% of their total fuel load (Figs. 31 and 34).

The good news is that both used fuel and depleted uranium can be recycled and thoroughly fissioned in fast-neutron reactors. This is different than reprocessing used fuel for another round in slow-spectrum LWRs, which for all their stellar performance are inefficient fuel burners. The unmoderated neutrons in a fast reactor can break down (think: "compost") the 100,000-year transuranic waste that forms in used fuel into 300-year fission-product waste, producing gigawatts of power in the process. [4]

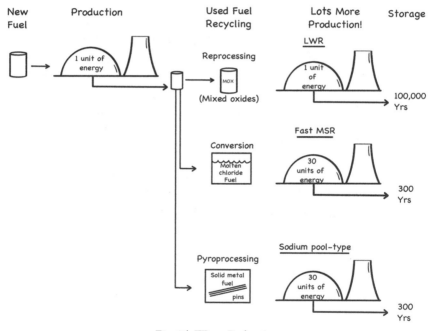

Fig. 34: Waste Reduction
Credit: By the authors

By using fast-neutron reactors, the US would have enough "spent" nuclear fuel for about 180 years of electricity at the nation's current rate of consumption. [5] However, we should note that processing and recycling used fuel for complete burnup in fast reactors would be considerably more complicated, and expensive, than using depleted uranium.

5.3 DEPLETED, BUT FAR FROM SPENT

Depleted uranium (DU) is the leftover material that results from the solid-fuel enrichment process for light-water reactors. Several batches of natural uranium yellowcake must be depleted of their uranium-235 to enrich a single batch, which is then made into fuel pellets. The depleted uranium from which the U-235 has been separated is currently regarded as waste, but our nation's stockpile of DU is actually a treasure trove of free fuel for fast reactors. However, there's just one hitch:

You need eight times as much fissile material to kick-start a new fast-neutron reactor, than you need for a new slow-neutron reactor. This is because the faster a neutron flies, the smaller every fissionable target nucleus seems to be. This means a fast reactor needs a *lot* more neutrons to get the party started.

NERD NOTE: From the viewpoint of a zooming neutron, the perceived size or "cross-section" of a target nucleus is largely determined by the speed of said neutron, and is measured in "barns." The term comes from the wry observation that if a neutron is going slow enough, every target nucleus will seem to be as big as a barn.

To a fast neutron, however, a target nucleus may seem to be only one barn wide, rather than its actual width of eight barns. This of course makes the target eight times smaller and thus eight times harder to hit.

The solution is to throw eight times more neutrons at the target. Thus, the initial fissile load for a fast reactor is about eight times the startup load of a slow-spectrum reactor like the LWR.

Another way of looking at it—you'll need eight bags of briquettes to fire up the grill. But once a fast reactor is up and running, it can gradually start breeding more and more of its own fuel from an increasing diet of DU or used LWR fuel, which is about 95% depleted uranium-238

(Fig. 31). With proper care and feeding, a fledgling fast reactor will mature into a self-sustaining compost bin for nuclear "waste."

Having to build a bonfire in order to have a weenie roast is why fast reactors should always have a slightly positive breeding ratio. That is, instead of being an "iso-breeder" that makes just enough fuel for its own use, and not a smidgen more, a fast breeder reactor would breed a little more fuel than it actually needs, and set it aside. Depending on the breeding ratio, there will eventually be enough set-aside fissile to kick-start another fast reactor—without mining, refining, and processing a humongous new 8x startup load. And not to worry, the stockpiled fissile material from a breeder will be a mix of "reactor-grade" plutonium and other actinides: Great for nuclear power but useless for nuclear weapons.

With fast breeders making extra startup fuel to power their own buildout, our depleted uranium "waste" could serve as a long-term supply of free fuel for a national fleet of fast reactors. We have hundreds of thousands of tonnes of the stuff—about five or six times more than all the light-water reactor fuel we have enriched since the 1950s. [6]

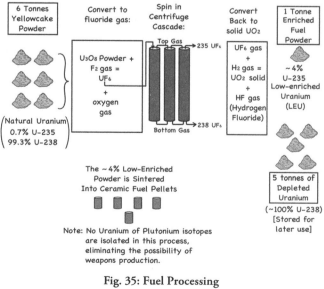

Fig. 35: Fuel Processing
Credit: By the authors

Unlike slow-neutron reactors, which tend to be picky eaters, fast reactors will eat everything they're fed, and the mining has already been done.

NERD NOTE: Natural uranium contains only 0.7% of the scarce U-235 isotope fissioned in existing slow-neutron reactors. To make a batch of low-enriched uranium (LEU) to fuel a nuclear power plant, several batches of yellowcake powder are converted to uranium hexafluoride gas (UF_6) and spun in a cascade (sequence) of high-speed centrifuges.

Centrifuge Cascade
Credit: US Nuclear Regulatory Commission (CC-BY-SA-2.0)

Spinning at 50,000 rpm, a centrifuge separates the uranium hexafluoride gas ever so slightly, with the heavier U-238 gas tending to drift toward the outside. The lighter U-235 gas tends to remain a bit closer to the center, and being a bit lighter, it will also tend to rise toward the top.

The "top gas" is drawn off and fed to the next centrifuge, and so on down the cascade, thousands of times over, with a tiny bit of separation (and thus enrichment) in each centrifuge.

When the desired enrichment level is finally reached, the U-235 top gas and U-238 bottom gas are chemically converted back into separate batches of solid uranium oxide (a mix of UO_2 and U_3O_8).

The enriched yellowcake powder made from the top gas is then "sintered" into fuel pellets—applying dry heat and pressure to form a dense, insoluble ceramic. The depleted uranium powder made from the bottom gas is set aside as waste.

Since DU is nearly pure uranium-238, it's only slightly radioactive and can be easily transported on public highways. However, regulations have made it nearly impossible to transport more-radioactive used fuel to either a recycling facility or to a deep geologic repository (DGR). This overabundance of caution is in spite of the long-demonstrated safety of transport casks, which can withstand the direct impact of a rocket-assisted locomotive. (See the videos: [7])

It should be noted that when used fuel is transported to a reprocessing facility, recycling this "spent" nuclear fuel into another round of reactor fuel requires sophisticated shielding and handling, both of which are expensive. So while either DU or SNF can be used to generate power in a fast reactor, depleted uranium is a lot less hassle all around. And like any country that enriches uranium, the US has a *lot* more DU than used fuel.

Indeed, after processing low-enriched reactor fuel for more than sixty years, the US has accumulated about 750,000 tonnes of depleted uranium. If this was fissioned in a fleet of fast reactors, it could power the nation for more than 1,500 years at our current rate of electricity consumption, or about 600 years when we transition to an all-electric society. (See our supplement on DU reserves: [8])

Don't take our word for it. Here's how climate scientist James Hansen puts it:

"With a fourth generation of nuclear power [Gen-IV reactors], you can have a technology that will burn more than 99 percent of the energy in the fuel. It would mean that you don't need to mine uranium for the next thousand years." [9]

In the long run, we think DU will prove to be more popular than SNF. As we see it, the widespread recycling of used LWR fuel will probably be done only if the public insists. And to be fair, the public just might insist, because thoroughly recycled used fuel would produce gigawatts of clean power, while also reducing its radioactive longevity from 100,000 years or more, to 300 years or less.

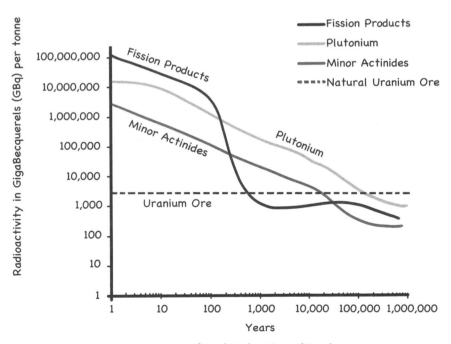

Fig. 36: Decay of Used Fuel to Natural Levels
*Source: Thorium: Energy Cheaper Than Coal (2012)
by Robert Hargraves, pg. 206*

In any case, the radioactivity in used fuel is not the looming, rectify-or-die threat to humanity it's often made out to be. It is true that the half-life of plutonium-239 is 24,000 years, and as we saw in Chapter 3 it takes ten half-lives to render a radioactive material harmless. This means the Pu-239 in used fuel is considered radioactive for 240,000 years.

At first glance, this might seem like a showstopper for nuclear power all by itself, but the hazard window for human health and safety is significantly shorter, despite any fearmongering to the contrary. In fact, after just 600 years, all penetrating gamma radiation is gone. From that point forward, you would have to literally grind the stuff up and inhale it or ingest it for the remaining alpha and beta radiation to do you any harm. [10] After 1,000 years, the radioactivity of used fuel will be no more intense than the mine from whence it came.

If scavengers in some far-flung future ever do go through the trouble of exhuming used fuel from an ancient repository, they would also need nuclear chemists and engineers to process and refine the stuff into any sort of significant threat. And if they had the technology to do that, they would *also* have the technology to make plutonium in a production reactor from a bit of yellowcake (see Chapter 19), and save themselves all the excavation and chemistry.

Like most scare stories about nuclear power, the "unsolvable problem of nuclear waste" isn't a problem at all—unless it's been made into one.

CHAPTER SIX

Back to the Future
(except for the fusion part)

FEW PEOPLE KNOW that the Experimental Breeder Reactor at Idaho National Lab ran on spent nuclear fuel for nearly thirty years. [1] Relying solely on recycled SNF that was "pyro-processed" at the lab, [2] the EBR-II produced most of the federal facility's electricity and space heating for almost three decades.

The EBR-II at Idaho National Lab
Source: https://en.wikipedia.org/wiki/Experimental_Breeder_Reactor_II

Equally important, the EBR-II also passed a landmark emergency shutdown test. When all power to the facility was cut, including the reactor's coolant pumps and emergency backup generators, the reactor shut itself down and cooled itself off with no human or mechanical assistance.

Dramatic live video of the 1986 test preserved this milestone in the early days of Generation-IV technology. [3] The control room engineers may look calm, but they're intently watching their gauges to see if the reactor would shut down, or melt down. As predicted, when the core temperature rose to a predetermined point above normal, the reactor shut itself down and began to cool.

Such a remarkable achievement should have been a global good news story—a reactor that automatically shuts itself down if it starts to overheat. Even better, this safety feature did not depend on clever feats of engineering, but on the laws of physics instead (which is genius engineering).

Just two weeks later, in April 1986, a poorly-designed Soviet reactor in Ukraine melted down, and the EBR-II's success story was lost to history. Then for reasons that remain unclear, the Clinton administration shuttered the successful EBR-II program in 1994. [4] But since that time, the climate has shifted favorably on nuclear power. This has been inspired by the sobering realities of air pollution, drought, global warming, and the lackluster performance of wind and solar—all of which have increased Europe's dependence on Russian natural gas. An abundance of nuclear power can effectively address these issues and more.

The EBR-II's fast-neutron reactor technology is being revived by Terrapower, Bill Gates's nuclear energy company. Their first Natrium reactor will be built at a retired coal plant in southwest Wyoming, in partnership with Warren Buffett and with the support of the governor.

The Natrium is a small, modular, and simplified update of the EBR-II, with the same physics-driven passive shutdown and walk-away safety. The 345-MW power plant will also feature 500 MW of molten salt thermal-energy storage. This isn't nuclear fuel salt, but rather the "solar salt" used at concentrated solar power plants to store thermal energy.

In 2022, the DOE released a report concluding that 80% of our medium and large coal plants are good candidates for nuclear power conversion. [5] Anticipating this opportunity, the Natrium was designed to replace the furnace of a coal plant, while utilizing the plant's existing turbine, generator, and power transmission equipment.

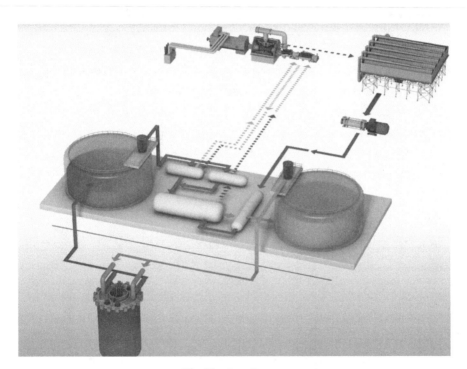

The Natrium Reactor
Source: https://www.terrapower.com/our-work/natriumpower/

As we saw, a fast-neutron reactor can burn anything that fissions, or turn it into something that will. And since actinides are metals, any combination of these metals can be made into solid fuel pins for a liquid-sodium-cooled fast reactor.

The fissioning fuel pins in the Natrium reactor heat an unpressurized pool of molten sodium, a soft metal with a low melting point and an extremely high boiling point. This liquid metallic sodium does double duty as the reactor's coolant and its heat transfer medium—the "working fluid" that transfers heat from the reactor to the steam generators, keeping the core at the right operating temperature.

NERD NOTE: Natrium is the Latin word for sodium. The Generation-IV Natrium reactor is a solid-fuel, molten-sodium-cooled, pool-type fast reactor modeled on the EBR-II at Idaho National Lab. Sounds complicated, but it's not:

Fuel assemblies with dozens of thin, solid-metal fuel "pins" are immersed in a sealed, unpressurized vessel (pool) of molten sodium, a liquid metal that does not slow neutrons.

Heated by the fissioning metallic fuel pins, the molten sodium working fluid is pumped out of the reactor and sent through a heat exchanger to make steam to power a turbine, before returning to the reactor to pick up more heat.

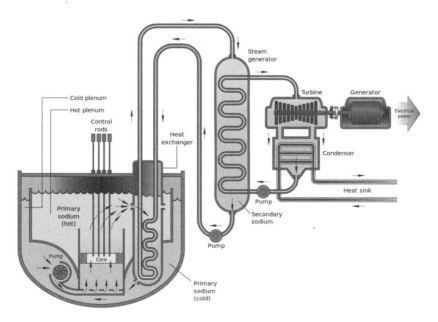

Fig. 37: Sodium Pool-Type Fast Reactor
Source: https://en.wikipedia.org/wiki/Sodium-cooled_fast_reactor

6.1 LIQUID FUEL CHANGES EVERYTHING

Another Terrapower design is in development with Southern Company, this one a molten chloride fast reactor (MCFR). [6] A true liquid-fuel molten-salt reactor, the MCFR will do an even better job of composting depleted uranium, used LWR fuel, and transuranics.

Like molten metallic sodium, molten chloride salt does not slow neutrons. But unlike the Natrium's *molten metal* working fluid, which is heated by the reactor's solid-metal fuel pins, the *molten salt* working fluid of the MCFR is self-heated by a salt form of nuclear fuel dissolved in the working fluid itself.

Combining these three functions—fuel, coolant, and working fluid—in an unpressurized system allows the fuel salt to expand and contract as it makes and gives off heat. The result is a self-heating, self-balancing, unpressurized liquid that never overheats, boils, or melts down. And if the molten salt should ever leak, it would just spread out, stop fissioning, and cool like candle wax.

Liquid fuel is such a radically different approach, with so many unique advantages and intrinsic safety features, that some nuclear wonks think of liquid-fuel MSRs as Generation-V reactors. (We'll explore the details in *Power to the Planet.*)

To recycle used solid fuel into molten salt fuel for a fast reactor,

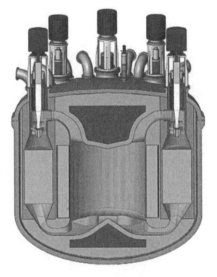

Fig. 38: Molten Chloride Fast Reactor (MCFR)

Source: https://www.terrapower.com /our-work/molten-chloride-fast -reactor-technology/

the uranium oxide pellets (UO_2) are chopped into pieces and dissolved in a hot bath of molten chloride salt (see the drawing below). A strong oxidizer like zirconium tetrachloride ($ZrCl_4$) is added, which strips the oxygen atoms from the uranium and replaces them with chlorine atoms. Since all actinides are metals, this can be done with any actinide found in used fuel, converting all of it into chloride salt fuel for a fast MSR. [7]

The result is molten hodge-podge of uranium chloride salt (UCl_3), with traces of plutonium salt, americium salt, neptunium salt, and a pinch of berkelium salt as well—the usual TRUs. Having done its job, the aggressive zirconium oxidizer becomes sedate zirconium oxide, which precipitates to a solid and is filtered from the mix. Now the salt is ready to use as fuel in a chloride-salt fast MSR. And notice that it's the same hopeless hodge-podge of isotopes that it was back in its fuel pellet days—useless for weapons but perfect for a fast reactor.

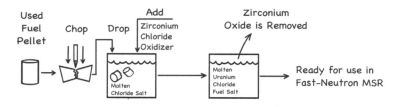

The pellets dissolve in the hot salt. When the zirconium oxidizer is added, the metal oxides from the dissolved fuel pellets exchange their oxygen atoms for the zirconium oxidizer's chlorine atoms.

This is a one-step chemical conversion process, from used LWR fuel to ready-to-use fast reactor MSR fuel.

Note: There is no isotopic separation in this process. All uranium isotopes remain in the mix at all times. Thus, weapons-grade material cannot be isolated.

Fig. 39: Fuel Conversion
Credit: By the authors

NERD NOTE: Actinides are metals and chlorine is a "halide." This means it's in the halogen column on the Periodic Table (second column from the right). And metal + halide = salt. Sodium chloride, or table salt, is a common example.

This means that any actinide (thorium, uranium, plutonium, etc.) can be made into molten-salt reactor fuel. The conversion process, however, does not pose a weapons proliferation threat, since it changes all the metal oxides in used fuel into a mix of chloride salts.

Since fuel conversion is a one-step chemical batch process, it cannot be used to separate one uranium isotope from another, which is required for weapons production.

At this point in time, there is no plan to run the Natrium reactor on recycled used fuel—fresh fuel is easier and cheaper to get in today's market. However, the Natrium is entirely capable of burning used fuel, should the need arise.

To recycle used LWR fuel for a solid-fuel fast reactor like the Natrium, the "spent" fuel pellets are dissolved in a molten chloride salt bath, just like the conversion process, but the actinide metals are electrolyzed out of the mix instead, and formed into solid metal fuel pins. This "pyroprocessing" method is how Idaho National Lab fueled their EBR-II, the Natrium's predecessor.

Alternatively, the used solid oxide fuel pellets from a light-water reactor can be dissolved in a nitric acid bath—the old-school, tried-and-true way of processing used fuel. The metals are electrolyzed from the bath and formed into solid metallic fuel pins—not metal oxides, but pure or "reduced" metals. (This costs money.)

Making liquid fuel for a molten-salt fast reactor is much easier than processing one form of solid fuel into another. Once the used fuel pellets are chopped up and dissolved in molten chloride salt, the zirconium chloride oxidizer does its chemistry trick, turns into zirconium oxide, and gets filtered out. And that's it. The result is a molten chloride fuel salt, ready to burn in a fast MSR, with no electrolysis or pin fabrication required. (This saves money.)

Liquid fuel saves even more money, because it burns more completely. The fissile material isn't stuck inside a pellet stacked in a fuel rod, but floats freely as ionic components of the molten salt, circulating in a loop from the reactor to the heat exchanger and back again. This gives every bit of fuel multiple chances to be in the center of the reactor's core—the sweet spot where the chances of fission are best. This is just one reason why MSRs have significantly better burn rates than solid-fuel reactors. With an equal opportunity for fission, no fuel is marginalized in an MSR.

When fuel atoms fission, some of the FPs they split into are gases, like xenon, krypton, and samarium. This poses a problem for solid fuel, just like any gas that forms inside a solid material. Gas buildup eventually cracks the pellet, distorting its shape and shortening its life. One of the many advantages of MSR design is that gaseous fission products simply bubble out of the molten salt, all on their own.

Since most of these gases love to absorb neutrons, their absence extends the productive life of the fuel load by improving the MSR's "neutron economy," as they say in the reactor biz. The fewer neutrons absorbed by FPs, the more neutrons are available to fission the fuel, and to make more fuel as well. Gaseous FPs can be cooled to a liquid state and trapped in activated carbon filters, or (much better) trapped in crystalline compounds called MOFs—metal-organic frameworks.

Being a solid-fuel crystal in a solid-fuel pellet, stuck inside a long, thin zirconium rod, is not unlike having an assigned seat at a banquet—you may not be anywhere near the scintillating conversation at the energetic center of the table.

An MSR is like a hot party in a crowded room, and everybody's circulating. They'll stream outside and cool off, but then they'll come back in and keep the party going. And all the boring gasbags just drift away, without being asked to leave.

With a one-step conversion process and no fuel pins (or fuel pin assemblies) to manufacture, a fast MSR is ideally suited to burn anything that fissions, or turn it into something that will. The technology should be in commercial use by the mid-2030s, if not before.

Until then, the used fuel from existing reactors can be stored in dry casks, or recycled into mixed-oxide (MOX) fuel pellets. [8] Burning MOX in an LWR does help reduce waste, but it's nowhere near as thorough as breaking it all down with fast neutrons. Even so, burning MOX is why France has only half the used fuel per unit of energy as any other nuclear country. (A fuel so nice, they use it twice. ☺)

With 75% of their grid powered by a nuclear fleet (which only took about fifteen years to build), France enjoys half the CO_2 emissions per unit of electricity produced compared to other European countries. In fact, French emissions per kilowatt-hour are *one-eighth* that of Germany, the continent's renewable-energy leader. (You can track the numbers in real-time here: [9]) Despite all the wind and solar deployed in Germany, residential electricity remains half the price in nuclear-powered France. And that was before Russia invaded Ukraine.

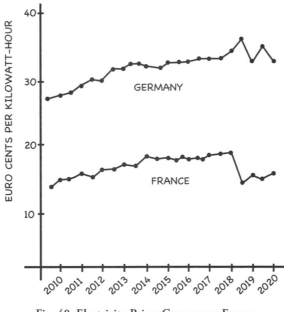

Fig. 40: Electricity Prices Germany vs France
Sources:
https://www.statista.com/statistics/418078/electricity-prices-for-households-in-germany/
https://www.statista.com/statistics/418087/electricity-prices-for-households-in-france/

6.2 THORIUM

Thorium-232 can be bred into fissile uranium-233, and the perfect place to do this is in the core of a molten-salt reactor (Fig. 41). The working fluid in this case would be a fluoride salt, not a chloride salt. Fluoride salt slows, or moderates, neutrons like water does, but it has such a fantastically high boiling point that the reactor can operate without the hassles of using highly pressurized water as the "working fluid." This is the liquid material that removes the heat produced in the reactor's core, by flowing in a continuous "primary loop" that runs through the reactor to the heat exchanger and back again.

In what should have been a Sputnik Moment for the US nuclear industry, China has built their first air-cooled slow-spectrum MSR in the unforgiving heat of the Gobi Desert. [10] Their design is based on the original experimental MSR, co-invented by Alvin Weinberg and Eugene Wigner and developed at Oak Ridge National Lab (ORNL) in the 1960s and '70s. [11] Their pioneering work later became the declassified, open-source science on molten-salt reactors that China used to launch their own MSR research and development. They got the technology fair and square from our own national lab, in cooperation with the Department of Energy. (We'll explore the China/Oak Ridge connection in *Power to the Planet*.)

Estimates vary, but the world apparently has enough proven reserves of thorium to supply all global primary energy for more than 300 years. [12] This does not include the recoverable thorium (and uranium) "waste" in the tailings of other industries, such as rare earth, sulfide, and phosphate mines. [13]

Like any actinide, thorium is a metal, about four times more abundant than land-sourced natural uranium. And there's plenty more on the moon and Mars, for whatever we care to do up there—Jeff and Elon, take note. [14] Barely radioactive, and unable to fission in its native state, fertile thorium-232 can absorb a slow neutron and become highly fissile uranium-233.

This is known as the thorium-uranium fuel cycle. Unlike the uranium-plutonium fuel cycle in fast reactors, a slow-spectrum MSR does not need eight years' worth of fissile fuel for an initial startup load. Like any slow-spectrum reactor, one year's worth of kick-start fuel will suffice to fire up a new fluoride-salt MSR. Once it's up and running, a bit of thorium salt is added each day, and it'll run that way for years.

It's actually a slight misnomer to call thorium a fuel. Like uranium-238, thorium-232 is a fertile feedstock that can be bred into U-233 fuel and burned inside a reactor's core. As we saw, some Pu-239 can also be made from U-238 and burned in slow spectrum, but it's not the most productive environment for breeding and fissioning plutonium fuel.

In simple terms: A fast-neutron reactor can breed and burn plutonium from uranium, and a slow-neutron reactor can breed and burn uranium from thorium. This makes thorium an excellent addition to either new or recycled solid fuel for light-water reactors, and for heavy-water CANDU reactors as well. [15] (We'll explain CANDU technology in *Power to the Planet*.)

Even so, like any nuclear material, whether fertile or fissile, thorium would perform best in a liquid-fuel environment. To that end, the original MSR experiment (MSRE) at Oak Ridge was a fluoride-salt reactor designed to take full advantage of the thorium-uranium fuel cycle. Anything that likes slow neutrons was grist for the MSRE mill.

Another important advantage of the Th–U fuel cycle over the U–Pu fuel cycle is that U-233 is the smallest and lightest fissionable isotope there is. Because this is so, and because its rate of fission is so high in slow spectrum, U-233 will produce less than 10% of the long-lived transuranics that come from burning the same amount of U-235 (we break down the numbers here: [16]).

Thorium is easily mined; it's even found in beach sand, and does not have to be enriched—virtually all the thorium found in nature is Th-232. It's good to go right out of the ground, and can be harvested in abundance from the tailings of our domestic phosphate mines. [17]

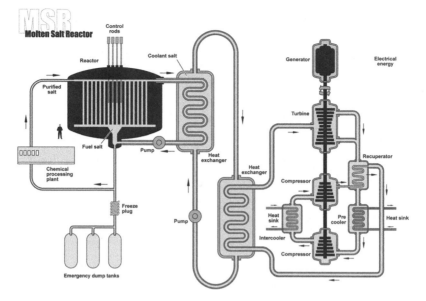

Fig. 41: The Molten-Salt Reactor
Source: https://en.wikipedia.org/wiki/Molten_salt_reactor

In contrast, the U-235 fuel for an LWR must be enriched to about six times the concentration found in nature—from 0.7% to about 4%–5%. Using thorium as a feedstock to make and burn U-233 would require much less mining and refining than enriching and burning U-235 in a light-water reactor.

And yet, for all its advantages, thorium is not a self-starter. Being fertile, not fissile, thorium must always be paired with fissile material to start a new reactor. But once the neutrons start flying, it's all good: Mild-mannered Th-232 changes into highly fissionable U-233 "superfuel." [18]

There are tonnes of thorium in the tailings of China's massive rare-earth mines. Over the last three decades, these mines have been the primary source for the neodymium magnets in wind turbines, to the tune of about one-half tonne of refined neodymium per large turbine. [19] Here's hoping that China's buildout of molten-salt reactors will take advantage of this already-mined fuel.

6.3 SEAWATER-SOURCED URANIUM

This promising alternative to mined uranium also offers one of the best guarantees for the long-term sustainability of nuclear power, with energy security, energy independence, and the smallest footprint all rolled into one.

About 4.5 billion tonnes of uranium oxide are dissolved in the world's oceans, continuously replenished by natural processes. This is an inexhaustible resource that can be harvested with cheap, eco-friendly and reusable artificial sponges. [20] Even better, an extraction method recently developed in India can remove more than 90% of the uranium in a given volume of seawater in just two hours, rather than the current sponge-harvesting cycle of ninety days. [21] And, it's renewable:

> "The uranium in seawater is controlled by steady-state chemical reactions between the water and rocks that contain uranium, such that whenever uranium is extracted from seawater, the same amount is leached from the rocks to replace it." [22]

Yellowcake made from seawater-sourced uranium is exactly like the dirt-mined or in-situ leached variety, ready to be enriched and fissioned in any reactor. The difference is that seawater uranium can be a virtually endless supply of carbon-free fuel, with no scarcity, mining, or access issues. This ensures a sustainable future of true energy security and independence for any country that can enrich and burn its own uranium, with extraction methods that are already being optimized.

Should things pan out for seawater uranium—and there is every reason to believe they will—we won't have to wait for Gen-IV technology to build a sustainable national nuclear fleet. The key to any nuclear buildout is having enough fuel, and for light-water reactors this means plenty of fresh uranium-235. Problem solved.

Another thing to consider is this: Nearly all advanced reactors in the first wave of a nuclear buildout will need fresh U-235. Even the Natrium's fuel pins will be made with fresh uranium, enriched to just under 20%. This category of reactor fuel is called HALEU (high-assay low-enriched uranium, pronounced "HAY-loo") because 20% is the high end of the "low-enriched" category for commercial reactor fuel, which is typically enriched to about 4%. You'll be seeing a lot more of this term in the years ahead.

The molten-salt reactor will be an exception to this rule of thumb. Terrestrial Energy, a Canadian company with a US branch, will be using a blend of low-enriched uranium (LEU) and thorium to power their Integral Molten Salt Reactor (IMSR), a liquid-fuel MSR designed to cogenerate high-temperature process heat and electric power for industrial applications. [23] Using technology closely related to the original work at Oak Ridge, Terrestrial intends to start marketing and building IMSRs by the mid-2020s.

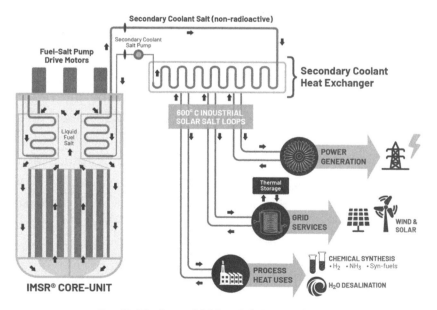

Fig. 42: The Integral Molten Salt Reactor (IMSR)
Credit: Terrestrial Energy (CC-BY-SA-4.0)

Aside from liquid-fuel MSRs, most upcoming Gen-IV reactors will be using HALEU fuel. The drawback of using HALEU is that five times the enrichment means five times the mining, along with increased market competition for five time the quantity of this scarce and increasingly in-demand resource. As we saw in Chapter 4, nearly 40% of today's uranium comes from Kazakhstan, a vassal state of Russia, and China is investing heavily in Africa's major uranium digs. [24]

For a comprehensive solution, the nuclear industry should advance to Gen-IV technology so the world can start burning its stockpiles of DU and spent fuel. But even as these advances are being made, seawater uranium could eliminate the *necessity* to advance to Gen-IV in order to expand nuclear power and sustain an effective buildout. Our graph at the end of Chapter 4, compiled from our mining supplement, is worth another look:

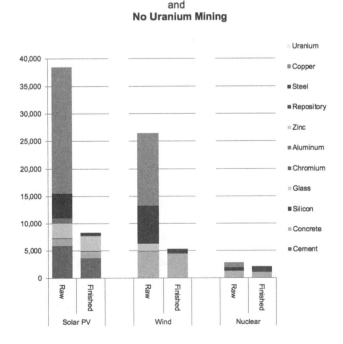

Material Throughput
(tonnes per terawatt-hour)
With Recycling
and
No Uranium Mining

With seawater uranium, existing reactors can enjoy this same minuscule footprint without waiting for Gen-IV to mature before we can fill the holes and put down the shovels. The transition to sustainable, mine-free, ocean-caught uranium can begin today—fresh from our reactor to you! The ads practically write themselves. For all intents and purposes:

Seawater uranium
is a virtually endless
and sustainable
source of carbon-free energy.

Ocean-sourced uranium can provide humanity with a whopping 240,000 years of primary energy at our current rate of consumption, give or take a few centuries (we break down the numbers here: [25]). This does not include the uranium that would leach from river silt and undersea rocks to replace whatever we harvest from the sea.

Unlike depleted uranium or used fuel, both of which can only be fully exploited in fast-neutron reactors, seawater-sourced yellowcake can be enriched in centrifuges and burned in any reactor. While recycling used fuel for a second round would cut uranium mining in half, seawater uranium could entirely eliminate mining from the front end of the nuclear fuel production cycle. At the back end of the cycle, fast reactors could eliminate the issue of long-lived "waste." (Remember, it's only waste if we waste it. ☺)

The near-future projected cost of ocean-harvested yellowcake is already quite reasonable, even without improved extraction methods. (We explore uranium harvesting prices here: [26]) This is great news, because by the year 2050 the world will need a *lot* more uranium—global primary energy consumption is predicted to increase by almost half. [27]

This sobering prediction does not include the energy required for any future efforts to remove excess CO_2 from Earth's atmosphere and oceans. If and when humanity does work to remove and sequester excess CO_2 (and clean up the rest of our trash with plasma-arc furnaces while we're at it), primary energy requirements will of course be much greater. Even so, the additional energy demand would hardly make a dent in the ocean's supply of uranium. Global stockpiles of used fuel and depleted uranium pale in comparison, and this "waste" alone could power the world for several centuries.

Fast reactors, depleted uranium, used fuel, thorium, and especially seawater uranium, can all reduce nuclear's footprint, both upstream from the power plant and down. As we saw in Chapter 4, detailed in our mining supplement and distilled in the previous graph, the material throughput of a nuclear power plant is downright minuscule, including the fuel. The only difference is the fuel *source*—whether it's dirt-mined, ISL'ed, recycled from used fuel, or extracted from seawater.

Wind and solar are passive harvesting technologies of diffuse and intermittent energy, requiring vast amounts of short-lived equipment. That being the case, it's hard to see how renewables will ever have a substantially smaller footprint than what they now require (see Chapter 18). On the other hand, nuclear power's footprint can be dramatically reduced with existing technology and near-future developments.

Nuclear is as clean as clean energy gets, and we have plenty of carbon-free fuel. All we have to do is use it, stepping lightly as we go.

CHAPTER SEVEN

To Be Perfectly Blunt

WHILE RENEWABLES CAN PLAY AN EFFECTIVE ROLE in clean-energy portfolios, they are not the universal solution that many believe them to be. It is in this context that we're being so hard on renewables. They're great for what they are, but they are clearly not enough to power our world, all by themselves. Insisting that renewables are a comprehensive solution—by themselves or even mostly by themselves—will only complicate the task before us.

There has been a lot of news lately about the rapid growth of renewable energy. This however must be kept in context, something the renewables industry seldom provides. The media will typically cite whatever numbers the industry gives them, and when it comes to renewables the numbers typically refer to installed capacity, rather than Capacity Factor (CF).

This key concept (sometimes called average capacity) refers to the output that can reasonably be expected each year, on average, from the particular equipment described when used at a given location. Installed capacity is a much different concept, referring to the equipment's peak performance under ideal conditions. With passive energy-harvesting systems, real-world averages will always be dramatically lower.

The rationale for citing installed capacity is that this number doesn't change while the capacity factor will, depending on the location. This, however, seems like a flimsy excuse, since the CF of any wind or solar farm would have already been estimated before the project was built. A more likely rationale is that by emphasizing installed capacity, the RE industry can cite much higher numbers than average capacity, often by a factor of four or more, which makes wind and solar seem much more productive than they actually are. [1]

> **NERD NOTE:** The Capacity Factor (CF) of a generating system is always a fractional number less than one. It's the amount of energy a system can actually produce over the course of a year, compared to the maximum amount of energy the system could produce under ideal conditions during that same period of time. The more casual term "average capacity" is often used as well, meaning what the power the plant will produce, on average, over the course of a year.
>
> For wind or solar, "ideal conditions" means the equipment is in perfect condition, the wind is blowing at the optimum speed, or the sun is overhead in a cool, clear sky. This, of course, doesn't happen all day long, every day of the year. Thus, the capacity factor of an RE farm will always be a fraction less than one, and is usually expressed as a percent. For example, a CF of 0.20 is a 20% capacity factor.

In a developing country or other low-energy scenario, an influx of even moderate amounts of renewable energy can transform lives. But to even attempt to function as a bona-fide utility for a developed society, an RE farm must be overbuilt to compensate for its lack of stable production: With a 20% capacity factor, a 100-MW solar farm must be overbuilt to about 500 MW to account for inconveniences like clouds and sunsets.

But that alone won't make up the difference. The farm must also be supported by adequate energy storage, which can be prohibitively expensive, or by on-demand ("dispatchable") backup

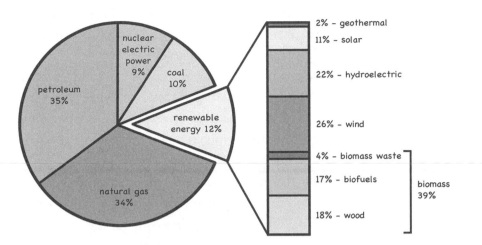

Fig. 43: US Primary Energy Consumption by Source
*Source: US Energy Information Administration "Monthly Energy Review"
Tables 1.3 and 10.1, April 2021 (preliminary data)*

generation from actual power plants, preferably both. Without these enhancements, renewables cannot be relied upon to do much more than augment fueled power plants and hydroelectric dams, much less replace them.

Indeed, a wind or solar farm backed by natural gas is actually a gas plant augmented by renewables. Why? Because a 500-MW installed capacity solar farm with a 20% capacity factor is a 100 MW-average solar farm with copious amounts of gas backup. Or it has batteries, or some other means of "firming" the farm's intermittent production—if passive energy harvesting can even qualify as "production."

Grid-scale wind and solar can, however, be a cheap way of reducing fossil fuel consumption, at least in the short term. Long-term costs are a different matter. Local wind and solar (rooftop, neighborhood, etc.) can also reduce demands on the overall grid, which is great. All of this will be helpful in a nuclear buildout to replace coal and gas plants, retiring hydroelectric dams, and older reactors. But modern society primarily depends upon "always there" services like water and power, police, fire, etc. Such services must be available more than 99% of the time to be considered "utility grade." Reactors can easily meet this standard; by themselves, renewables cannot.

Currently, grid-scale wind and solar can come in at a low wholesale price of about $40 per MWh, [2] but this is because they have externalized the cost of whatever backup systems are needed to firm their erratic performance. By themselves, renewables can only augment an already-functioning grid. Solar-plus-storage packages are a step in the right direction, but the four hours of storage typically included in these popular combo packs are not nearly enough to turn a passive energy-harvesting facility into a replacement for a power plant that runs on actual fuel. (Wind, falling water, and sunshine are not fuels. See section 7.2.)

Lithium batteries are now approaching $150 per kilowatt-hour of storage capacity, though the all-in cost can be much higher than that. [3] A renewables-heavy grid would require a practical and scalable storage solution about ten times cheaper than that, in the range of $10 to $20 /kWh. [4] Iron-air batteries look promising in this regard, and may eventually reach that price point in full-scale production. [5]

But until such time arrives, renewables will only add to the peak capacity of an electric grid, and not to its stability, reliability, or flexibility. Those responsibilities are now being shouldered by fueled plants and hydroelectric dams, whose stabilizing performance props up grid-scale wind and solar, making them seem like clean-energy heroes.

Unfortunately, most of our fueled power plants were not designed for such a cockamamie arrangement, and neither was the grid itself. Regardless of the technology used, electricity must be consumed or stored the moment it's made. Since energy storage is expensive, a system operator's job is to maintain a constant balance between supply and demand. The problem is, no one who built our national grid ever anticipated that the system would have to accommodate an expanding fleet of unreliable energy sources. But here we are.

7.1 BENDING OVER BACKWARD

There's been a lot of talk lately about flexible power production, and how the grid should have a "flexible base." The concept, however, applies to all the equipment on the grid that has to be flexible to accommodate renewables. It does not apply to renewables themselves, because renewable energy isn't flexible—it's erratic. Subject to the whims of Mother Nature, renewables are passive energy *harvesting* systems, not active energy *production* systems. That's worth repeating:

Renewables are
passive energy harvesting systems
not
active energy production systems.

Aside from large hydroelectric projects, renewables, by themselves, are inherently unsuitable for generating the dependable, grid-scale power that modern society requires. Unless, of course, they also have enough dirt-cheap energy storage to mimic actual utilities. (Batteries sold separately.)

As passive harvesters of ambient natural energy, wind and solar can mimic the performance of energy production systems only if they are buffered by adequate storage, or firmed by other power plants, and it's unfortunately just as simple as that.

"The notion that renewable energy and batteries alone will provide all needed energy is fantastical. Worse, tricking the public to accept the fantasy of 100% renewables means that, in reality, fossil fuels reign and climate change grows."
– JAMES HANSEN

As long as renewables keep coming to the party without their own buffering, they will always be a marginal source of energy—marginal in the sense that they can only add to what is already there by contributing bursts of intermittent energy to an otherwise stable grid of fueled power plants and hydro.

That may seem like a harsh assessment, but the foundation of any system cannot be erratic at heart. Adding fast and cheap renewables to a grid is like adding a second-floor addition without strengthening the existing walls and foundation. Adding even more floors may be just as cheap, but the result is even more instability.

Clouds and wind are fickle phenomena, with disruptions both large and small that can happen several times a day. To mandate a buildout of RE, without telling prospective RE farms to "Bring Your Own Buffering" (BYOB), obligates our existing power plants to adjust their production—at their own expense—to accommodate the new kids on the block.

Like any machine, large or small, a functional grid must demonstrate, as a core competency, reliable operation. The more stable the system is, the better it can absorb the contributions of marginal producers. But the core of any system cannot be unpredictable by nature, with its stability maintained by peripheral systems. Engineering, computing, business, and even life itself doesn't work that way.

But that's what happens when renewable energy is given top priority on a regional grid. Unless it comes to the party with its own backup and storage, a renewable system of any size cannot function as a free-standing utility; it must be supported by actual energy production, storage, or both.

Renewables have been touted as a disruptive technology, and indeed they are—but not in the way their advocates like to think. After producing reliable power for decades, many of our large workhorse reactors have now been slated for early retirement because they can't perform like contortionists whenever Mother Nature takes a break and renewables run short on "fuel."

The power for a developed society must be reliably generated by on-demand energy production systems, or obtained from enough dirt-cheap batteries that can mimic power plants. Wind and solar are virtually endless sources of free energy, harvested from nature's bounty, which, we grant you, is a tremendously appealing idea. But from an engineering perspective, renewables are sprawling, passive, fuel-free, energy harvesting systems that absolutely depend upon favorable weather. By extension, this same dependency applies to their storage systems as well.

This is why the gas-backed renewables grid planned for California might be better described as a gas-powered grid augmented by wind and solar, which isn't an appealing idea at all. [6] California, for example, burns gas even when the sun is shining and the wind is blowing, because it is cheaper to keep backup gas turbines idling than starting them up and turning them off again several times a day. This "idling fuel" can amount to nearly 25% of the state's overall gas consumption.

Fueled plants are independent systems, with their maintenance downtime and backup from other fueled plants and hydro scheduled well in advance. Managing this kind of prearranged backup energy is a routine matter on any large grid. Fuel-free renewables, on the other hand, are *inter*-dependent systems, with a natural intermittency that requires frequent and unpredictable support from storage or backup. This greatly increases the need to shuffle large blocks of energy around to keep the grid in balance. Even if utility-scale RE farms do have their own short-term backup and storage, they would still have to be interconnected on a regional or national basis to form a fuel-free, power-sharing renewable energy grid.

The only practical way to transmit that much backup energy on a daily or hourly basis—on short notice and without disrupting the regular traffic of the grid—is with a separate "overlay" network of HVDC (high-voltage direct-current) long-distance transmission corridors. Serving as an energy superhighway, an HVDC network could transfer large blocks of energy, from coast to coast if need be, to back up local or regional RE grids. At least, that's the plan.

A low-EROI energy-harvesting system like wind or solar can serve as an off-grid resource, or be used in other important niche applications, [7] but it cannot serve as a stand-alone utility for a modern, developed society. In the US, for example, power interruptions will generally total less than three hours per year. [8] That's a reliability rate of 99.97%.

Rebuilding our national grid around the idea of passive energy harvesting, with daily downtimes imposed by Mother Nature, would require a massive overbuild of fully-interconnected RE farms and enough storage to (hopefully) ensure 99.97% grid reliability during seasonal changes and bouts of unfavorable weather. As we show in *Roadmap to Nowhere,* this would cost trillions of dollars more than a nuclear-centric grid, and require about eighty thousand square miles more elbow room than a national nuclear fleet (about the size of Minnesota). Not to mention all the material throughput and waste from the short-lived wind and solar equipment (see Chapter 18). And all the new long-distance transmission corridors.

Whatever technologies we ultimately use to power the nation, people instinctively know a simple truth of modern life: Utility power should come with an on-switch, not a weather app.

7.2 IF YOU CAN'T PUT SUNSHINE IN YOUR POCKET, IT'S NOT A FUEL

Fuel is stable, portable stuff that natively contains stored energy. Used in the appropriate device, some of that energy can be extracted on demand and put to work. Fuel is not an ambient natural phenomenon conjured into existence by transient weather, the earth's rotation, or the pull of the moon. Fuel, in short, is a substance, not an occurance. Even so, the term "fuel" is frequently used in RE circles to describe wind, water, and sunshine, and too many people take it literally. In reality, renewables are not free-fuel devices—they are *fuel-free* devices. Big difference.

Because this is so, renewables must be firmed with fuel or batteries to function as actual utilities. Imagine if the fire department was ready to roll only 25% of the time. What if your water only worked 40% of the time? The yearly average productive capacity of US solar is typically less than 25% of installed capacity, with maybe 40% for onshore wind. [9] And despite all the tantalizing headlines about energy-efficiency breakthroughs, improvements have not altered these numbers to any game-changing degree, and probably won't in the foreseeable future. [10]

Without substantial backup from storage or fueled power plants (the water behind a dam is storage, and the fuel in a reactor is storage), RE should best be thought of as a fossil-fuel-saving strategy for a grid anchored by nuclear, hydroelectric (water batteries), or fossil fuel (yuck). As we phase out of fossil fuel, prudence dictates that we should phase into carbon-free fuel rather than fuel-free systems.

As the EROI graph (Fig. 21) in Chapter 3 shows, concentrated solar power is the only buffered renewable that (barely) satisfies the 7:1 requirement for a utility-grade power system in a developed society. With such a narrow margin of utility, a CSP farm is essentially useless if it doesn't deliver as intended.

A recent example is the failure of Crescent Dunes, one of the world's largest CSP farms. [11] After six years of lackluster performance, Nevada's NV Energy finally concluded that they had been dealt a losing hand and decided to fold, canceling their power purchase agreement with the facility. [12]

The truth is, renewables should not be solely relied upon to power society in any practical way. Even concentrated solar will barely squeak by with a buffered EROI of 9:1, despite the ideal conditions in the Mojave Desert, which apparently weren't ideal enough for Crescent Dunes. [13]

The failure of Crescent Dunes illustrates one of the inconvenient truths of clean energy: The mass quantities of on-demand, carbon-free energy the modern world requires can only be reliably and sustainably generated, now and for the foreseeable future, by nuclear power or large hydroelectric facilities.

Crescent Dunes Concentrated Solar Power Plant
Credit: Amble (CC-BY-SA-4.0)

Wind and sunshine aren't fuel, and we're not building any more rivers. This means that nuclear is the only existing energy source that can—all by itself, if need be—achieve a carbon-zero future, let alone a carbon-negative one.

7.3 FILL 'ER UP

There is a growing interest in hydrogen as a carbon-free fuel. But as you sort through all the information and hype, keep in mind that just like "renewable energy," "hydrogen fuel" is a term of art. Strictly speaking, hydrogen is an energy carrier and not an actual fuel, since the only energy we can get from it is a portion of the energy we use to make it. [14]

Fuel natively contains stored energy—that's what makes it fuel. The energy that goes into making it has already been expended by Mother Nature (thanks, Mom!).

Hydrogen fuel is something we humans have to make ourselves. We're the ones who have to expend the energy to separate the hydrogen and oxygen that form a water molecule, before we can use the liberated hydrogen as a fuel or a storage medium. And when we do, we can only get back a small portion of the energy expended in the process. The nice thing about actual fuel (wood, coal, gas, uranium, etc.) is that the first step of putting energy into it has already been done. All we have to do is release the energy (please use responsibly).

Because this is so, and because reactors run on such energy-dense fuel, they would be far more practical than renewables for powering a hydrogen infrastructure. Actually, they'd be far more practical at powering nearly anything. When it comes to genuine carbon-free *fuel*, nuclear is the only game in town.

And about that word "renewable." It's true that wind and sunlight are continuously renewed, while fossil fuels are not. The marketing term "renewable energy" gained a certain cachet during the Oil Crisis of the 1970s, and the name stuck. The focus, however, should be on *sustainability*. As we saw in Chapters 5 and 6, there are enough available reserves of nuclear fuel to power the planet essentially forever.

Nuclear power is as sustainable as clean energy gets. And we have plenty of fuel.

CHAPTER EIGHT

How Clean Should Clean Energy Be?

It should certainly be a lot cleaner than the rare-earth tailings at Lake Baotou in China, a mess described by the BBC as "the worst place on earth." [1] Most of the Rare Earth Elements (REE) used by the wind and solar industries (and more worrisome, the US defense industry [2]) have been sourced in China, where working conditions are well below US mine and refinery standards.

Even when care is taken, the waste stream of solar panel manufacturing is a concern: One tonne of polysilicon, for example, generates three to four tonnes of highly toxic silicon tetrachloride. [3] But solar waste has been discreetly kept from view to protect our "green" sensibilities, [4] not to mention our sense of national security (see the video in endnote #2—pretty sobering stuff).

We don't pollute—our suppliers do it for us, way over on the other side of the world so we don't have to think about it. And for what it's worth, the mining and refining to produce a gigawatt-year of wind energy exposes industry workers to five times more radioactivity than the mining and refining for the same amount of nuclear energy (Fig. 44). Even worse, the mining and refining needed to manufacture a GW-year of solar generates *forty times* as much radiation as the mining and refining for a GW-year of nuclear power—amazing, but true. [5]

Industrial Waste Discharge Pipe
Similar to the tailings waste at rare earth mines in Baotou, China (see link #1).
Credit: belovodchenko at Canva

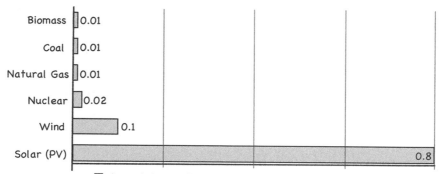

Fig. 44: Radiation in the Solar Industry
*Source: https://www.unscear.org/docs/publications/2016/UNSCEAR
_2016_Report-CORR.pdf (See note 48 and Table 45)*

Currently, carbon-free nuclear power produces nearly 20% of US electricity. And as we mentioned, no member of the American public has ever been injured or killed by nuclear power production. The same can be said for the US Navy's nuclear fleet, after logging more than 5,700 safe reactor-years. [6]

Nuclear is also the only energy production system that accounts for every speck of its waste, securely stored for future use or permanent deep burial (see Chapter 18). By contrast, and quite aside from the rare-earth mess happening overseas, enormous landfills of wind and solar waste are expected in the years to come. That's because any passive energy-harvesting system requires a boatload of energy-gathering equipment. And unless it's a hydroelectric dam, the stuff will only last a few decades before it breaks down or wears down, and it becomes cheaper to get new gear. Making matters worse, many of the early wind farms were simply abandoned, leaving property owners on the hook for cleanup. [7]

Abandoned Wind Turbines
*Credit: Western Area Power
(CC-BY-SA-2.0)*

Until recently, only a few RE manufacturers have included recycling costs in their pricing, but this is starting to change across the industry, and kudos to them. Even so, the industry at large has a legacy of abandoned farms to contend with, even as their volume of waste continues to grow, most of which cannot be effectively recycled.

Consider the heat, or petrochemical solvents, needed to melt the fiberglass resin from a wind turbine blade wider than a transport truck and longer than a football field. Recyclable blade material is being explored, but at present the only practical option is to chop the used blades into smaller, more manageable pieces. Now imagine the US having to recycle, or more likely dispose of, literally *hundreds* of these colossal blades, every single day until forever. This is just one consequence of an all-renewables future seldom mentioned by the RE industry.

Turbine Blades Awaiting Burial in Casper, Wyoming
Source: https://www.bloomberg.com/news/features/2020-02-05/wind-turbine-blades-can-t-be-recycled-so-they-re-piling-up-in-landfills

The big chunks can be used for landfill (fiberglass isn't organic, but it is inert), and the tiny chunks can be used to bulk up concrete and asphalt. Both are good gestures, and more ideas will surely arise. But the sheer volume of wind and solar waste is a guaranteed money pit, reminiscent of our three decades of plastic recycling. Activists are just starting to make noise about RE waste—expect to hear more in the years to come. [8] The RE world might not want to hear it, but plasma-arc furnaces, powered by fast reactors running on used fuel or depleted uranium, could solve their waste problems.

In contrast to renewables, nuclear produces so much consistent power that power plants can make enough predictable money to set aside funds for their own waste management and decommissioning.

In the US, nuclear plants are required to set aside a fraction of a penny per kilowatt-hour on whatever energy they generate. If a reactor runs for its full lifespan, this can add up to more than a half-billion dollars. [9] Prematurely decommissioning reactors, and replacing them with non-nuclear plants, undercuts this funding scenario and distorts the cost of nuclear power.

A case in point: SONGS (San Onofre Nuclear Generating Station) on the California coast north of San Diego was shuttered in 2013, shortly after Fukushima. Decommissioning was projected to be about $700 million for the pair of one-gigawatt reactors, which is about right for a plant that size. However, an additional $3.3 billion was tacked on to compensate San Diego Gas and Electric for lost revenue from prematurely closing the plant. On top of that, the four gas plants that replace San Onofre will cost SDG&E another $4.4 billion; their ratepayers are footing the bill for this as well. [10]

These are the sorts of needless expenses that come from replacing two perfectly good reactors for no discernible reason other than nuclear fear. And California's losing argument with reality continues apace:

In late August 2022, Sacramento announced that by 2035 the state will ban the sale of new gasoline-powered vehicles. Just one week later, a brutal heat wave came along, and they asked EV owners if they could please refrain from charging their vehicles. As you can imagine, this rankled the locals, and just a few days later California's famously anti-nuclear state assembly voted an eye-popping 67–3 to keep Diablo Canyon open until 2030. [11]

In a similar change of heart, Greta Thunberg has voiced support to keep Germany's last three reactors open. [12] Unfortunately, they didn't take her advice, but she is sure to inspire others.

8.1 A FATE WORSE THAN COAL

Nuclear power's contribution to our nation's energy should be much higher than today's 20%. But as the 1970s rolled around, the US nuclear buildout was grinding to a halt. A flurry of new regulations inspired by the thoroughly unscientific "ALARA" principle (Chapter 15) was methodically ratcheting prices out of the market.

With Three Mile Island in 1979, nuclear fell from wide public favor and the buildout of coal was seen as a safe alternative. In retrospect, it was a tragic mistake: From the 1979 TMI accident until now (2023), carbon emissions from those "safe" coal plants have hastened the deaths of over 100,000 Americans (we break down the numbers here: [13]). This entirely preventable tragedy sprang from a misinformed public freaking out over a partial meltdown at TMI that harmed no one.

In simple terms, US nuclear power fell out of favor over a minor steam release that amounted to an unscheduled chest X-ray for some downwind residents. Yes, Three Mile Island was scary, but it wasn't dangerous.

More recently, the same unwarranted concerns have been playing out overseas. In the wake of Fukushima, both Japan and Germany have drastically cut back on nuclear power. Japan we can understand, but Germany? Perhaps they were concerned about tsunamis in the Baltic Sea.

The consequence of this latest global outbreak of nuclear fear has been increased CO_2 emissions in Japan [14] and lackluster CO_2 reductions in Germany. [15] In the first six years after the accident, this bad idea was responsible for an estimated 28,000 additional premature deaths from carbon emissions—the same thing that happened in the US after Three Mile Island.

"If Japan and Germany had reduced coal instead of nuclear after Fukushima, they could have prevented about 28,000 air pollution-induced premature deaths, and 2.6 billion tons of CO_2 emissions between 2011 and 2017. Thus, these countries' post-Fukushima energy choices *have resulted in major levels of avoidable impacts of the accident.*" [emphasis added] [16]

It stands to reason that an additional 26,000 premature deaths could be estimated for the six years since the study. In our view, the phrase we emphasized in the above quote translates to:

". . . have resulted in thousands of wrongful deaths."

This is how deadly nuclear fear can be. Like any concentrated source of energy, nuclear can indeed be scary, but that doesn't mean it's dangerous. The fact is, nuclear fear is far more deadly than nuclear power, and always has been. Fear is the mind-killer.

8.2 DEATH PRINT

This sobering term refers to fatalities per terawatt-hour of electricity generated by the world's energy production systems. [17] As shown in the graph below, nuclear power has the lowest death print of any large-scale energy technology.

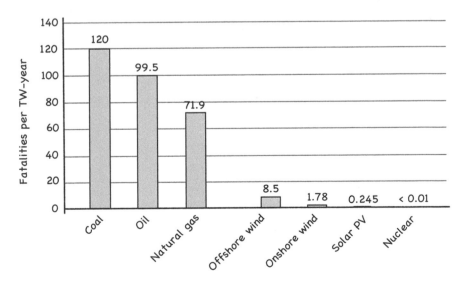

Fig. 45: Nuclear Has the Lowest Fatalities
*Source: https://www.world-nuclear.org/gallery/the-harmony-programme
/energy-accident-fatalities-for-oecd-countries.aspx*

This even holds true with existing Generation-II and -III reactors. When Gen-IV reactors come into use (probably by the late 2020s), they will be walk-away safe, which means exactly what it says. As we saw in Chapter 6, passive safety is one of the defining criteria for Generation-IV.

Here's a close-up of the wind, solar, and nuclear in the previous chart. Nuclear has just 0.01 fatalities per terawatt-year:

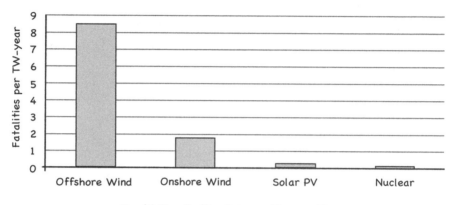

Fig. 46: Detail of Fatalities per Terawatt-Year

The safety record of nuclear power is startling, indeed. Per unit of energy produced, existing Gen-III nuclear technology, imperfect though it may well be, is already:

- 24.5 times safer than PV solar

- 178 times safer than onshore wind

- 850 times safer than offshore wind

- 7,190 times safer than natural gas

- 9,950 times safer than oil

- 12,000 times safer than coal

Wind-turbine casualties are mostly from installation and maintenance. Thus far, solar casualties have mostly been from residential DIY rooftop installations. Solar is now typically installed by professionals, but this doesn't change the fact that any rooftop is a hazardous workplace. And wind will always be a high-wire act. [18]

Wind Engineers
Credit: Northeast Marketing (CC-BY-SA-4.0)

Whatever hazards that nuclear power may entail, they are far outweighed by the consequences of not using it to electrify the nation and power the planet. It bears repeating:

Nuclear has the lowest death print
of any mass-energy production
technology in history.

When nuclear power is objectively examined, its advantages are readily apparent. Even so, some people suspect that science, industry, and government are suppressing the "real truth" about nuclear. Others believe that UNSCEAR (United Nations Scientific Committee on the Effects of Atomic Radiation) is in the pocket of Big Nuke. An impartial observer might wonder why the UN's IPCC reports (Intergovernmental Panel on Climate Change) are widely accepted, while the UN reports from UNSCEAR are rejected out of hand, usually by the same people.

In a similar vein, some believe the WHO (World Health Organization) is under the influence of the IAEA (International Atomic Energy Agency), which is supposedly dissuading the WHO from issuing reports critical of nuclear power. This weird belief persists in spite of a formal written agreement between the two organizations that explicitly declares their publishing autonomy, with each agency free from review by the other. [19]

The Information Age of the 20th century has brought us the Disinformation Age of the 21st century—through the magic of the interwebs, the crazies have found each other. It's now become all too easy to get lost in the weeds; this is another reason why we cite sources for so much of what we say. Don't take our word for anything—we *want* you to be skeptical and check our sources.

But all enthusiasm aside (we freely admit we're nuclear advocates), the simple fact remains that after sixty-plus years of nearly trouble-free operation, very few cancers or deaths from radiation have been credibly linked to American nuclear power. This includes mining, construction, fuel processing, energy production, and our stockpiles of used fuel and depleted uranium. It's also true that nuclear power plants are not only some of the safest industrial worksites in the world, but are radiologically safe as well. In fact, nuclear employee health rates are better than the general population. [20]

Simply put: Nuclear is the safest form of mass-energy production available, with the lowest death print of any energy technology, and one of the lowest life-cycle CO_2 emissions of any form of energy generation. And as we saw in Figs. 45 and 46, the nuclear industry also enjoys some of the lowest radiation emissions in the energy sector, even with renewables taken into account.

Given everything we've covered thus far, why on earth do some people get so riled up about nuclear power?

CHAPTER NINE

How It All Began

WHILE RADIATION CAN INDEED BE DANGEROUS in high doses, the issue has been blown entirely out of proportion by a fatally-flawed model of radiation risk assessment called Linear No-Threshold, or LNT.

More than anything else, the LNT model of radiation safety has distorted public perception about nuclear energy and stymied the buildout of nuclear power. Simply put: LNT is the bottleneck restricting the expansion of nuclear power.

First presented to the world in a 1946 Nobel Prize lecture by Hermann Muller, when nuclear technology was in its infancy and nuclear medicine had just begun, Muller declared in no uncertain terms:

> *"There is no safe dose of radiation,*
> *and all doses are cumulative."*

Science later discovered that both assertions are false. But long before this good news came to light, this pair of erroneous ideas had taken root and spread like crabgrass, weaving themselves into our regulatory regime.

Muller had been a pioneer in the field of radiation-induced genetic mutations, and was horrified by the introduction of nuclear weapons. He received the 1946 Nobel Prize for his work on LNT, even though the evidence he cited in his Nobel lecture was insufficient to support his claims. As the Cold War heated up through the 1950s, LNT's twin-pack of scary misinformation (no safe dose / all doses are cumulative) became fixed in the public mind as immutable law, even though no one had yet developed an explanation for these claims, or an experiment that could be replicated to confirm their existence. [1]

With the gravitas of a Nobel to back him up, Muller's linear no-threshold model was, and still is, regarded by many as a proven theory, even though it is no more than a hypothesis, and a flawed one at that. A proper theory says, "A causes B, and here is how and why." Any genuine theory makes predictions that can be verified and reproduced by independent experiments, or by careful observation of natural phenomena. Muller's lab results had only suggested a hypothesis, and not a theory. A hypothesis says, "It seems possible that A causes B, so now we have to find out if that is true. And if it *is* true, we will then try to explain how it happens." Scientists are cautious when they use the word "theory," while popular media typically isn't. Unfortunately, popular media is where most people get their science.

Muller's declarations to the Nobel audience (and thus the world press) were made just one year after the end of World War II, before a long-term study on survivors of the atomic blasts over Hiroshima and Nagasaki could even begin. Nevertheless, linear no-threshold soon became a principal organizing concept of the anti-nuclear movement.

Muller's test subjects were fruit flies, whose short lifetimes allow for a rapid series of experimental observations. The problem was, his work did not show what he claimed. Regardless, he expressed complete certainty in his Nobel speech: *"There is no escape from the conclusion that there is no threshold."*

Actually, it's quite easy to escape from Muller's conclusion due to the limited scope of his lab work, because the evidence he needed simply was not there (we'll get to that shortly). Furthermore, just weeks before his 1946 Nobel speech, Muller was made aware of a more recent study, conducted by Dr. Ernst Caspari and supervised by Muller's colleague Dr. Curt Stern. The Caspari paper contradicted LNT by showing clear evidence of a radiation safety threshold.

Muller's written response to Stern was that, since Caspari's work had always been impeccable, the linear-no-threshold concept warranted further experimentation:

"However, I see that it [the Caspari paper] is very important and shall do all I can to go through it in a reasonable time, surely before I leave again early in December [to accept the Nobel Prize]. I hope that Caspari can wait that long if necessary. In the meantime, I wonder whether you [Stern] are having any steps taken to have the [threshold] question tested again, with variations in technique.

"It is of such paramount importance, and the results seem so diametrically opposed to those which you and the others have obtained, that I should think funds would be forthcoming for a test of the matter." [emphasis added] [2]

Muller then sailed to Sweden and accepted the 1946 Nobel Prize for Physiology or Medicine, declaring in no uncertain terms that a radiation safety threshold does not exist, and that all doses, no matter how minute, were cumulative. His categorical statements caused some backstage grumbling among his circle of colleagues, who were privy to Caspari's work, but Muller's stunning pronouncement stuck. [3]

Ten years later, Hermann Muller was a distinguished member of the BEAR-I committee at the National Academy of Sciences. (BEAR stands for the "Biological Effects of Atomic Radiation.") And so it was that in the summer of 1956, his linear no-threshold model of radiation safety was essentially declared to be settled science by the Academy's landmark BEAR-I report to the public. This position was taken by the NAS in spite of significant disagreements between members of the committee, which went unresolved. In part, BEAR-I stated:

"There is no minimum amount of radiation which must be exceeded before mutations occur. Any amount, however small, that reaches the reproductive cells can cause a correspondingly small number of mutations. The more radiation, the more mutations." [4]

This unequivocal declaration by the Committee was far from unanimous. Belying the certainty expressed in the official report, people on the committee were writing notes to each other behind the back of their famous board member. Although their views were never published in a minority report, some of the notes have been found. One is from Milislav Demerec to Theodosius Dobzhansky, trailblazers in the field of evolutionary genetics:

"I, myself, have a hard time keeping a straight face when there is talk about genetic deaths and the tremendous dangers of irradiation." [5]

So do we, until we reflect on the fear and paranoia that LNT has stirred up in the decades since, and consider the opportunities the world has lost to clean up our collective energy act. Then it's not so funny.

If Three Mile Island had been understood for what it actually was—a partial meltdown, with the minor steam release of a small amount of non-lethal radioactive material—we would have dealt with the accident like an airliner crash. Although, a highly unusual crash with no casualties, just some chest X-rays.

We would have cleaned up the mess, conducted an investigation, and followed up with industry-wide improvements. The country would have kept on building a fleet of more and better nuclear plants, instead of more and more coal plants, and by the millennium we could have kissed fossil fuels goodbye.

9.1 "I WAS WORKING IN THE LAB, LATE ONE NIGHT . . ." – BOBBY PICKETT

From the time he was in graduate school, Muller had been in a race with other geneticists to demonstrate gene mutation; whoever published first would be a shoo-in for a Nobel Prize. His particular approach was to zap fruit flies with higher and higher doses of ionizing radiation until he induced an observable mutation in the offspring. [6] His attention wasn't even on low-dose radiation. This was just as well, because there were no instruments at the time that could accurately measure low doses, much less detect biological effects (if any) from said doses.

That's because the year was 1927, not 1946. [7] Like many Nobel recipients, Muller published his findings long before he received his prize. His landmark experiments were conducted nearly a century ago, in the early days of radiology, before electron microscopes or high-precision mass spectrometers even existed. [8]

Using the equipment available at the time, Muller zapped fruit flies with more than 2,700 mSv in 3.5 minutes and called it a low dose. That's low compared to ground zero at Hiroshima or Nagasaki, or the dose-rates at Chernobyl (none of which had happened yet), but it's hardly a low dose—2,700 mSv in 3.5 minutes is more than 100 *million* times the average background radiation on planet Earth. [9]

As we mentioned, a human chest X-ray delivers an instantaneous dose of about 0.1 mSv. This means he zapped those poor fruit flies with nearly 27,000 human chest X-rays in less than four minutes. [10] If this seems excessive, that's because it is. Another thing to note: Muller had been using this same approach ever since graduate school, blasting flies with higher and higher doses in an effort to produce an observeable effect. You would think that having to apply more than 100 million times the average background dose found on planet Earth, to induce a genetic mutation in a living organism native to this planet, would have told him he was on the wrong track. But no.

Back then, gene theory and genetic experiments were confined to directly observing offspring for variations and defects. This explains the focus on reproductive cells, since the effects, if any, could be directly observed in the offspring. No one back in the 1920s had a clue about what actually constituted a gene—the A, C, G, T sequences we learned about in high school biology weren't discovered until the 1960s. In Muller's time, genetic research was limited to observing macroscale phenomena (eye color changes, wing size, etc.) and trying to understand how and why they happened.

This was Gregor Mendel's original pea plant methodology from the 1860s, which we also learned about in biology class. Mendel had made his mark on hereditary science some sixty years before Muller's work, and by the 1920s the direct observation of offspring (flowers or flies) was still the only game in town. Muller irradiated fruit flies and examined their offspring, raising the dose bit by bit until mutations were finally observed. He wasn't the only scientist zapping specimens with zoomies, but he was the first to publish, and became a superstar in his field.

The problem was that instead of confining his conclusions to what he could test and observe with the equipment that existed at the time, Muller sweepingly concluded that there was a straight-line (linear, or proportional) relationship between dose and effect, all the way down to zero:

- Twice the dose causes twice the effect.

- Half the dose causes half the effect.

- This relationship holds true down to zero dose/zero effect.

The last point was pure conjecture on his part, but if it were true (it's not) the implications for humanity were profound. Even so, he couldn't experiment on humans to verify his claims. When LNT was declared to be established fact in 1956, biological science still wasn't at the point where human tissue (skin, muscle, etc.) could be grown in a lab and used for experiments. That didn't happen until 1998. [11] Zapping fruit flies and mice, and speculating what the results might mean for human health, would have to do.

9.2 THE DOSE MAKES THE POISON

As the physician Paracelsus explained back in the 1500s: *"All things are poisons, for there is nothing without poisonous qualities. It is only the dose which makes a thing poison."* [12] More than four hundred years of pharmacology, radiology, and common sense have proven him right. But Muller apparently thought that radiation was somehow unique in this regard, an exception to the rule possessing some special kind of awful, to where even a minuscule quantity would inevitably cause harm—subatomic Ebola, if you will. Here's a rough analogy of how he worked through the problem:

It is observably true that drinking two gallons of water in one sitting will almost certainly kill a healthy adult. [13] Thus, we could reasonably speculate that drinking one gallon in one sitting might kill half of those who try. Drawing a line on a graph through those two data points, we could postulate (whether we should or not) a Linear No-Threshold model for the lethality of water, in a straight-line relationship, all the way down to zero dose/zero deaths (Fig. 47). Here's how that would play out:

If drinking two gallons in one sitting kills all of the people all of the time, and drinking one gallon kills half of the people half of the time, then:

- one quart would kill one out of eight people

- one pint would kill one out of sixteen people

- one drop of water would kill one in a million people.

Nonsense like this becomes even more absurd when it's combined with a concept called "Collective Dose," and a group of people is informed that according to statistics, one of them will probably die because the group collectively received one lethal dose.

That's clearly absurd. A 100-count bottle of 325 mg aspirin, if ingested all at once, will kill a healthy adult. [14] But if the pills were equally distributed to a group of 100 people, it would be laughable to predict that one of them will die from taking one aspirin. And yet, this is the conceptual basis of Collective Dose (no, we're not kidding).

Fig. 47: LNT Applied to Human Hydration
Credit: By the authors

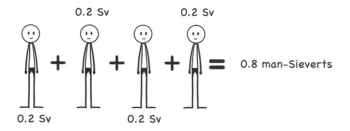

Fig. 48: Collective Dose
Credit: By the authors

The danger of using this approach, calculated in "man-Sieverts" and guided by the wisdom of LNT, is that LNT assumes harm from even the smallest dose. So let's say that in a cohort (group) of 500 people near a radiation release, it was estimated that each person received an average of 0.02 sieverts (20 mSv). This adds up to a Collective Dose of 10 man-Sieverts for the entire cohort (500 × 0.02 = 10).

Using an LNT-inspired estimate of, say, 0.1 death per man-Sievert for people closer to the accident, Collective Dose predicts that one person in this cohort of five hundred will die from radiation (0.1 death /man-Sv × 10 man-Sv = one death). A calamitous outcome, from an average individual dose of just 20 mSv—the equivalent of two CT scans.

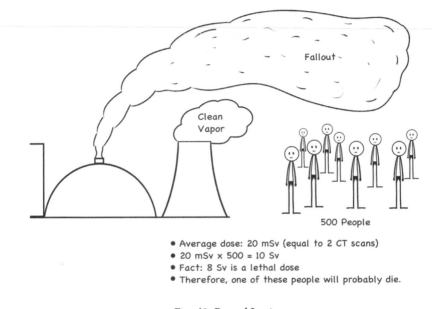

- Average dose: 20 mSv (equal to 2 CT scans)
- 20 mSv x 500 = 10 Sv
- Fact: 8 Sv is a lethal dose
- Therefore, one of these people will probably die.

Fig. 49: Pretzel Logic
Credit: By the authors

9.3 NUCLEAR FEAR SELLS

Falsely predicting death and injury for thousands of people is an especially pernicious use of LNT, and that is exactly what anti-nuclear activists have done since Chernobyl. After thirty-five years of spreading this frankly reprehensible disinformation, the fiction of 4,000 future deaths has become firmly lodged in the public mind. [15]

This is a tragedy, because no one can reach a correct conclusion about anything, including nuclear power, with incorrect information—including all the expectant mothers who aborted after Chernobyl. Thankfully, there have finally been some glimmers of sanity amid all the fear and loathing. In 2010, a year before Fukushima, the US-based Health Physics Society released a position statement:

"The Health Physics Society advises against estimating health risks to people from exposures to ionizing radiation that are near or less than natural background levels because statistical uncertainties at these low levels are great." [16]

In December 2012, less than two years after Fukushima, the United Nations presented their findings on nuclear safety to the Fukushima Ministerial Conference by essentially saying the same thing:

"UNSCEAR [United Nations Scientific Committee on Effects of Atomic Radiation] does not recommend multiplying very low doses by large numbers of individuals to estimate numbers of radiation-induced health effects within a population exposed to incremental doses at levels equivalent to or lower than natural background levels." [17]

This time around, the advice might have (partially) sunk in. Because unlike Chernobyl, no official Collective Dose/LNT mortality predictions have been announced for either the Fukushima cleanup crew or the Fukushima downwinders. Even so, the Japanese government is still keeping the reoccupation limit at a needlessly restrictive 28 mSv per year. [18]

One result of all this misplaced anxiety about Fukushima is that after a decade, some 30,000 Japanese citizens are still displaced from their homes, farms, and businesses, not to mention the 1,600 evacuation casualties or the 600 subsequent suicides. Or the hundreds of billions of dollars wasted on unnecessary cleanup and soil removal, along with relocation expenses for the long-term evacuees. [19]

Given our long life expectancies in this modern world, about one in three adults who reach middle age will eventually die of cancer, with or without excess exposure to radioactive material. [20] Using LNT and Collective Dose to predict the fate of the Chernobyl downwinders is essentially telling them that any cancer mortalities in their area can and should be blamed on the meltdown. That's like blaming the rooster for the sunrise, and makes the accident seem far more tragic than it already was—a death toll of nearly sixty people, most of whom were first responders from the nearby city of Pripyat.

Using Collective Dose in combination with LNT is a patently absurd way to predict long-term health effects from a widespread accident like Chernobyl. Since every smidgen of Chernobyl fallout was considered harmful (no safe dose), the estimate was pegged at 4,000 mortalities. Note that this was an estimate of "statistical deaths," meaning future mortality estimates based on an unproven hypothesis. That's quite different than estimates based on confirmed, real-world deaths from verified, real-world causes. Insurance actuarial tables, for example, are compiled using actual (hence the term) historical data, not inconclusive lab experiments.

Employing the principle of no safe dose to predict 4,000 future casualties has conjured a crippling fatalism in the Chernobyl down-winders, still evident to this day. After the accident, many became convinced they were doomed, even though the fallout dose they received was less than the yearly average background dose of several populated regions on earth. Evidence of their fate was plain to see: All around them, about one out of every three adults were dying of cancer.

CHAPTER TEN

There Is No Safe Dose of BS*

IN THE SUMMER OF 1956, the US National Academy of Sciences put their stamp of approval on Muller's linear no-threshold model, making it the centerpiece of their BEAR-I report to the public (pronounced "Bear One").

Announced with much fanfare in the press, BEAR-I was the first in a series of major NAS reports on the Biological Effects of Atomic Radiation, and it was based on bad science right from the start. Recent investigations have also raised credible concerns of bias and influence behind the landmark report. [1]

One bad sign among many: The BEAR-I committee dismissed direct observational evidence that, ten years after the nuclear detonations in Japan, the offspring of blast survivors had shown no evidence of inherited genetic damage. Upon Muller's strong advice, the committee rejected such "field work" as being unreliable, and focused on lab experiments that zapped fruit flies instead:

> *"We should beware of reliance on illusory conclusions from human data, such as the Hiroshima-Nagasaki data, especially when they seem to be negative [i.e., contrary to LNT]."* [2]

---

* bad science

Aside from declaring that there was no safe dose, the BEAR-I report also took pains to explain that all doses—no matter how small—were cumulative:

> "The total dose of radiation is what counts, this statement being based on the fact that the genetic damage done by radiation is *cumulative*." [3] [emphasis in the original]

This twin-pack of scary misinformation (no safe dose / all doses are cumulative) was widely publicized in the popular press in the summer of 1956, and it created a sensation. [4] It was the early days of the Cold War, a time when the atmospheric testing of hydrogen bombs was a growing concern. It was also a time when nuclear power was just starting its rise: Atoms for Peace in 1953, the *USS Nautilus* in 1954, and the first commercial nuclear power plant at Shippingport, Pennsylvania in 1958.

"No safe dose!" became the mantra of a growing anti-nuclear weapons movement. While we applaud their efforts to stop global fallout with an atmospheric test-ban treaty, there was no evidence to substantiate their claim that there is no safe dose from even the slightest trace of fallout. Indeed, the idea was directly at odds with more than four centuries of what pharmacology and internal medicine had already learned about threshold doses.

In the high-dose range, Muller's Proportionality Rule holds true—no argument there. He correctly found a direct proportional (hence "linear") relationship between ionizing radiation and inherited genetic damage at high doses (more radiation = more damage). Muller's mistake is that he incorrectly extrapolated what he saw at high doses straight into the low-dose range, all the way down to zero dose / zero effect (hence "no-threshold"), and called it his Proportionality Rule.

In Muller's line of work, something like this is more properly called a SWAG:

The LNT Model

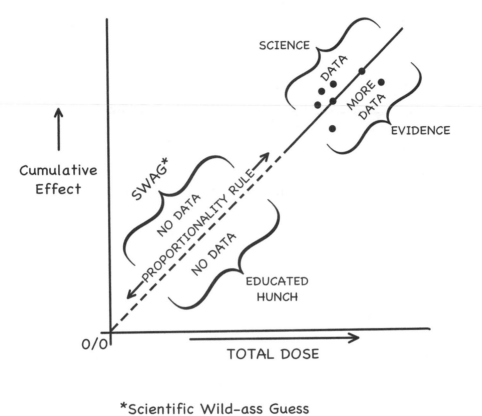

*Scientific Wild-ass Guess
Fig. 50: SWAG
Credit: By the authors*

Sorry, but it's true. Despite his utter lack of evidence of any low-dose effects, Muller's idea that there is no safe dose of radiation became carved in stone as one of those "everybody knows" common-sense realities, even though it's provably wrong and makes no sense.

If LNT is false (which it is), then every health and safety standard, and every nuclear plant construction and licensing standard, that is based on the linear no-threshold model should be reviewed. This includes revisiting the bogus estimate of 4,000 future deaths from Chernobyl.

While updating the regulations on radiation safety may prove embarrassing for those in entrenched positions, and to those who profit from nuclear fear, it will be incredibly painful for others to learn the truth: Remember that in the wake of Chernobyl, bad medical advice based on LNT prompted over 100,000 abortions of *wanted* pregnancies.

The fact is, radiation is such a weak carcinogen that biological consequences from small doses are impossible to detect—if they even exist. To which no-safe-dosers respond: "Well, if it hasn't been proven, then it hasn't been disproven, either." This unscientific dodge is inspired by a sly bit of sophistry called the Precautionary Principle, which we discuss later.

With a misinformed populace, in the midst of an escalating Cold War bristling with nuclear weapons, the anti-nuke button was all too easy to push. Hollywood had a field day scaring us to death with zoomies, mad scientists, and radiation-spawned monsters. At the same time, anti-war activists contrived a false link between nuclear weapons and nuclear power, and made it stick. The consequence of this dust-up (if you will) was a fear of all things nuclear, even for the radiation we routinely encounter in modern society and the natural world. By the late 1950s, the Atomic Age wasn't aging well.

Medical science now understands that all living cells, including reproductive cells, routinely repair their own DNA damage thousands of times a day. The only exceptions are when somatic (body) cells undergo mitosis (cell division), or when a pair of gametes (sex cells) undergoes fusion (the merging of male and female reproductive cells). As you can imagine, these are intense, all-consuming events in the life of a cell, so any repairs conducted during or soon after the process can sometimes be less than perfect. [5]

None of this was known back in 1956. But the public did have plenty of chances to read about BEAR-I in newspapers and magazines, and copies of the report were widely available, including public libraries. It was the first broadly publicized report on the subject—it was even subtitled "a report to the public."

Credit: Wallace Smith and Sara Bancroft

The US, and hence the world, was told in plain language that even low-dose radiological effects on human reproductive cells will lead to reproductive harm, and that this harm will surely manifest in the immediate offspring or in subsequent generations.

10.1 SUPERSIZING THE BS

An important detail that often gets lost in the conversation on LNT is that Muller's work was focused on *reproductive* cells, not on *somatic* cells, meaning all the cells in the body that aren't reproductive cells (from the Greek root *soma*, meaning body). Damage to reproductive cells was scary enough, and a good thing to know if you were planning a family. And then one year later, the news about radiation got a whole lot worse.

A May 1957 paper by Edward B. Lewis, a geneticist at CalTech, rocked the world all over again by declaring that low doses of ionizing radiation can also cause cancer. [6] And not as a malady you could pass on to your children, but a malady that you yourself could develop in your own body.

That got everyone's attention. In a matter of weeks, Ed Lewis was a celebrity, testifying before Congress. But as he spoke, it eventually became clear that he really didn't have any actual *evidence* of low-dose effects, merely an *assumption*. When asked about this, Lewis explained his reasoning as follows:

"The point here, however, is that in the absence of any other information, it seems to me—this is my personal opinion—that the only prudent course is to assume that a straight-line relationship holds here [in the low-dose region] as well as elsewhere in the higher-dose region." [7]

Appearing soon after on *Meet the Press*, he argued with Admiral Strauss, the head of the Atomic Energy Commission (AEC), whose thankless task was to minimize all the noise in the press about atomic fallout. Lewis won the debate, and soon after that he was featured in *Life* magazine:

> "WARNING OF DANGER was sounded by Dr. E. B. Lewis of CalTech. In an article in *Science* he proved that there is a direct relationship between radiation and leukemia. He predicts a five to 10 percent increase in leukemia if strontium-90 level in humans [from bomb test fallout] reaches a figure which the AEC still considers harmless." [8]

A *Washington Post* headline at the time summed up the Lewis Effect: "All Radiation Held Perilous: Nation's Top Geneticists Unanimous In Opinion, Fallout Produced Now Will Shorten Lives in Future, Congress Is Told" [9]

Our nuclear fears were no longer confined to DNA damage that could affect our offspring. The no-safe-dose scare of the summer before was now a direct threat to our own lives as well, with fallout from nuclear bomb tests causing leukemia and who knows what else. This hit home hard in the summer of '57, redoubling the anti-war movement's effort to seek a total ban on atmospheric weapons testing—a noble cause, but unfortunately based on bad science.

In the next chapter, we'll explain how Ed Lewis went off the rails, and took the whole world with him. Unfortunately, his error wasn't fully revealed until 2015, and by that late date the bad science of LNT had enjoyed a head start of nearly sixty years, institutionalized by BEAR-I as far back as 1956, supersized by Lewis in 1957 and reaffirmed in subsequent NAS reports.

Some twenty years after this Fifties sci-fi double feature starring Muller and Lewis, the newly-formed Environmental Protection Agency needed a cancer risk yardstick, and the National Academy of Sciences recommended LNT:

"Given the prestige of the National Academy of Sciences, the[ir] recommendation to use the LNT model was adopted quickly in the US and elsewhere, and [was] *generalized from the narrow area of genome risk, to those involving somatic cells, with application to cancer risk assessment.*

"When the US Environmental Protection Agency (EPA) cancer risk policy was first developed in 1976, the EPA turned to the NAS for a suitable model for risk assessment and subsequently *adopted the LNT model as its centerpiece for its cancer-risk policy,* providing the key foundation for cancer-risk assessment guidelines starting in 1977, and continuing to the present day." [emphasis added] [10]

And so it was that the EPA enshrined Muller's no-safe-dose LNT model as official regulatory doctrine, expanded by Lewis to encompass every cell of the body.

Two years later, in March 1979, a minor steam release at the Three Mile Island nuclear power plant was exploited by anti-nuke activists, reviving the twin memes of "no safe dose" and "all doses are cumulative." With the EPA legitimizing the narrative, nuclear fear gripped the country all over again, with aftershocks that can still be felt today. And it's all been based on bad science.

Unfortunately, TMI happened before science had definitively established that, contrary to LNT, living cells do in fact routinely repair their own DNA damage, whether from radiation, oxidative stress, or chemical damage. The fact is, cells have been doing this ever since cellular life became fashionable, around three to four billion years ago. [11] Massively high dose rates are one thing, but low-dose radiation is not an existential threat to any lifeform, including us.

10.2 PLENTY OF BLAME TO GO AROUND

The lack of effective pushback against the misinformation on nuclear energy is partially due to the unfortunate position the

US nuclear industry has been in, ever since the advent of commercial nuclear power. That's because "US nuclear industry" is a misnomer. Each player is actually the nuclear division of a much larger conglomerate.

General Electric, for example, also makes wind turbines, gas turbines, and natural gas drilling equipment. And many nuclear power plant operators like Southern Company, Entergy, and NextEra also own and operate fossil, wind, and solar plants. So their nuclear divisions are in no position to advertise how much more efficient and effective a reactor would be compared to gas plants or renewables. It would be like Ford's Electric Vehicle division throwing shade on carbon-powered trucks and SUVs.

This has to be incredibly frustrating for the folks who work in the nuclear industry (for lack of a better term). One of their top priorities should be to debunk LNT, once and for all. It's the linchpin of every overwrought regulation and every uptick in cost, with no solid science to back it up. Instead, reactor builders have been silent, or have tried to engineer away people's fears with ever more redundant safety features. Meanwhile, distortions and lies about nuclear kept building in the public mind, complicating the conversation on clean energy. The facts on radiation should have been common knowledge to the general populace decades ago, as familiar as any other industrial or household hazard. But for too many people, it's still voodoo.

A word about Hollywood and the press: On the one hand, they have to make a buck. On the other hand, most of the people working in media are genuinely concerned about the environment. These two ambitions do not have to be mutually exclusive. Nuclear fear sells, but the math is the math, and the reality is that renewable energy simply does not stack up to nuclear power, especially when it comes to all-weather, around-the-clock, mass-energy production. Local wind and solar are clearly useful, but the inherent limitations of these energy-harvesting technologies are indisputable. Narratives based on bad science should stay where they belong—in fantasy, comedy, and camp science-fiction. The public deserves to see a lot more pro-nuclear cameos than *Iron Man* and *The Martian*.

It's time to normalize nuclear power. With such a remarkable track record, and a century of science to back it up, there is a wealth of stories to tell. Radiology, for example, came into common use as early as the 1920s; it was "the wonder of the age!" The British X-Ray and Radium Protection Committee was formed shortly thereafter, which became the International Committee on Radiological Protection (ICRP). [12]

In 1902, the workplace standard for radiation was a hazardous 30,000 mSv/year. [13] By the late 1920s, negative health effects were being observed. In one infamous example that most people know, some watch-dial painters were sickened from the glow-in-the-dark radium in their paint. It turned out they had a habit of licking their brushes to keep a fine point for their handiwork, and since radium chemically mimics calcium the radium gradually accumulated in their bones.

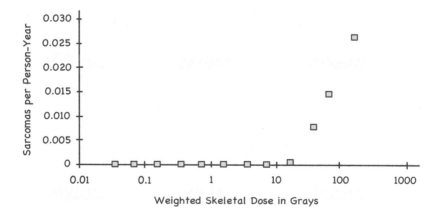

Fig. 51: Sarcoma Incidence In Radium Watch-Dial Painters
Source: https://inis.iaea.org/collection/NCLCollectionStore/_Public/28/000/28000733.pdf

The part of the story that most people *haven't* heard is that no cancers were observed in the painters who ingested less than a cumulative dose of 10 Grays, or 10 Sieverts (10,000 mSv). That is a substantial amount of accumulated radiation; four to five Sieverts received at once time is rated LD50—a lethal dose for 50% of recipients. [14] Ten thousand mSv, or ten Sieverts, is the equivalent of two LD50 doses, ingested and retained over a number of years.

Here's another view of the data. Note that many of the women who received even the highest, long-term cumulative doses (upper left portion of graph) did not develop malignancies:

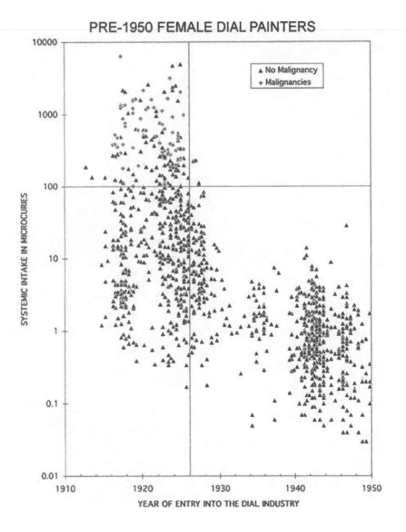

Fig. 52: Sarcoma Incidence in Watch-Dial Painters (Scatter Graph)
Source: http://www.rerowland.com/dial_painters.htm

145

LNT adherents still haven't explained this compelling evidence of a threshold. Neither have they explained the equally compelling evidence against cumulative dose. Taken together, these two points clearly refute their model. Unfortunately, the nuclear industry hasn't been able to make a peep about any of this to the public. And Hollywood has hardly bothered.

By the 1930s, the ICRP had tightened their workplace safety standard to 500 mSv /yr, and casualties declined to nearly zero. In 1990, the International Committee tightened the standard to a stringent 20 mSv per year. [15] This is in spite of the fact that British radiologists have always been, and still are, a healthy lot:

> "Mortality rates in British radiologists who were registered after 1954 were *significantly lower* relative to those in other medical professional groups; this was true *for both cancer mortality and all death causes combined.*" [emphasis added] [16]

The British study was repeated in the US, with a cohort of more than 43,000 radiologists and a control group of nearly 65,000 psychiatrists. The results were the same: Working with and around low-dose radiation was not only safe, it might even be good for you.

Cell repair was confirmed more than forty years ago, and has since become a well-understood reality in biology, genetics, and nuclear medicine. Yet to this day, neither the National Academies of Science or the Environmental Protection Agency have rejected 1950s-vintage LNT. The only advantage of using LNT as a regulatory guideline is that it's a simple concept anyone can use, regardless of whether it makes any sense. Surprisingly (or perhaps not), this is actually one of the stated justifications for its continued use. [17]

Simplicity is not a good reason to use a bad idea. Besides which, anyone who has ever taken a prescription drug, or an over-the-counter medicine, can wrap their head around the idea of a threshold dose. It's not a highfalutin concept, by any means. Even so, such an extreme "precautionary" attitude about nuclear power has taken hold that people routinely assume any zoomie could be a deadly serious threat, and proceed from there.

The zeitgeist shift that began in the summer of '56 kicked into high gear in the summer of '57, wilting the salad days of the optimistic Atomic Age. By the end of the decade, we were living on the New Frontier and digging backyard bomb shelters.

CHAPTER ELEVEN

How Ed Lewis Went Off the Rails

(and took the whole world with him)

IN 2015, JERRY CUTTLER AND JAMES WELSH published a paper [1] in the *Journal of Leukemia* that took a fresh look at Ed Lewis's 1957 bombshell, "Leukemia and Ionizing Radiation." [2] Lewis, you recall, had rocked the world with his analysis of the first ten years of leukemia data from the Life-Span Study (LSS) in postwar Japan. The only problem was, his analysis was grossly in error. First, a bit of background:

Conducted from 1947 until 2012, the Life-Span Study [3] was a massive project that monitored the health of almost 97,000 atomic blast survivors and their offspring, including some 33,000 non-irradiated survivors from the Hiroshima and Nagasaki areas. With countless interviews and on-site observations, the LSS staff carefully determined the disposition of the *hibakusha* (the survivors) at the moment of the blast.

This included their age and health, exactly where they were and what they were doing, what they were wearing, what if anything they were carrying, how far they were from a window or wall, whether there was any other shielding in the radiation path, and so on. By establishing the circumstances in such granular detail, they could estimate the dose received by each survivor.

In addition to more than 200,000 immediate injuries and deaths (mostly from the shock waves and subsequent fires), the intense radiation from the blasts had consequences for several hundred thousand survivors of both sexes, spanning a wide spectrum of age and health. It was a unique, though ghastly, opportunity to study the effects of ionizing radiation, both immediate and long-term, on a large cohort of living human beings and their offspring.

By the late 1950s, the ever-expanding LSS data set had amassed a wealth of information on leukemia rates among the survivors, along with their specific distances from the hypocenter at the time of the blasts (when a bomb detonates in midair, the ground-zero point below the blast is called the hypocenter). This was the most comprehensive data available on leukemia and ionizing radiation, and became the principle source for Lewis's paper.

Hermann Muller rocked the world the summer before with the BEAR-I report. Now Ed Lewis's paper, featured in the May 1957 issue of the prestigious journal *Science*, claimed there was no safe dose of radiation for leukemia, a cancer of the blood. These of course are somatic cells, and not the reproductive cells Muller had studied. If Lewis was right (he wasn't), it was a short step from his paper to conclude that low doses of ionizing radiation could cause other cancers as well.

And that's how Muller's no-safe-dose radiation scare about reproductive cells metastasized into a looming threat to every cell of the human body. The error Cuttler and Welsh discovered was that Lewis combined two important cohorts of blast survivors into one, and called them his control group. This takes a bit of explanation, but it's the key to understanding how the world came to worry about low-dose radiation, a gnawing nuclear fear that persists to this day. First, we need to understand hormesis and the J-curve.

11.1 PARADOXICAL EFFECTS

Multiple investigations suggest that low-dose radiation may have a "hormetic" effect. This is where a low dose of a potentially dangerous substance or stress is beneficial, by prompting the body to rally its defense mechanisms. Low-dose radiation apparently stimulates an immune response, much like vaccinations or other immunotherapies. [4]

Notice the beneficial effect of the hormetic model's J-curve in figure 53, shown by a dip below the "control line" or safety threshold. On this graph, lower is better, so the dip shows that low doses can have an opposite or "paradoxical" effect by stimulating the body in beneficial rather than harmful ways. Potentially harmful effects only begin to manifest in the higher-dose range, as the J-curve rises above the threshold.

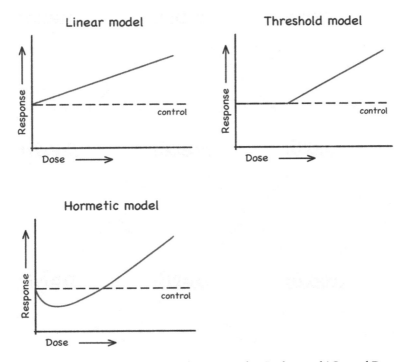

Fig. 53: Radiation Response Models Compared to Background / Control Data
Source: https://www.the-scientist.com/vision/challenging-dose-response-dogma-49062

Hormesis is a well-accepted concept in wellness and nutrition. The experts in these fields stress (no pun intended) that within certain limits, physical challenges and even toxins can be beneficial to our health by prompting an adaptive response in the body.

A simple example is exercise. While a sedentary lifestyle is debilitating, a daily bit of physical challenge promotes health and wellness, even with something as simple as a daily walk. Beyond the modest measures that promote general wellness, increasing degrees of stress also build strength and resilience. While hard work or vigorous exercise does slightly damage the organism, it recovers by building back stronger. Hence the mantra, "No pain, no gain." [5]

But like anything else, too much of a good thing can cause more damage than our bodies can repair. Beyond this point of diminishing returns, more stress just causes more damage. Increasing levels of harm and increasing degrees of benefit can be plotted on mirror-image graphs, but the concept is the same.

Biochemist TD Luckey suggests that a yearly hormetic dose for adults may be in the range of about 60 mSv per year. [6] Indeed, low-dose radiation "hot spots" have long been known for their therapeutic effects, such as the radon baths in Claremore, Oklahoma. [7]

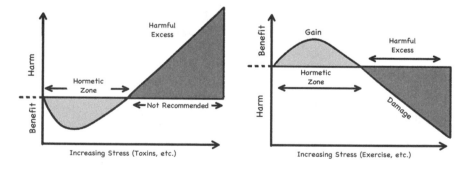

Fig. 54: Hormesis and Health – Harm vs. Benefit From Two Perspectives
Credit: By the authors

11.2 CUTTLER AND WELSH

In their 2015 paper, Cuttler and Welsh show that when the full data set on leukemia rates in the *hibakusha* is properly viewed, there is a clear threshold and a strong J-curve hormetic response. This is starkly different than the straight-line, proportional response Lewis claimed to see. The question is: How did Lewis develop such a distorted view of the data?

Cuttler and Welsh's graph of the full data set on leukemia in blast survivors is shown in figure 55. Heads up: This is a logarithmic graph, squeezed down from the top and in from the right. It's a trick that keeps a large data range shrunk down to a single page, while the low-dose range is still easy to see.

The dashed curve on the right and its data candlestick are from Cuttler and Welsh (statistical candlesticks show the high, low, and average, all in one). For clarity, we penciled in the Average Cases line and the various zones used in the Life-Span Study. The zones denote distance from the hypocenter and thus the dose range received by the survivors in each zone. (The hypocenter was at the far right side of Zone A.)

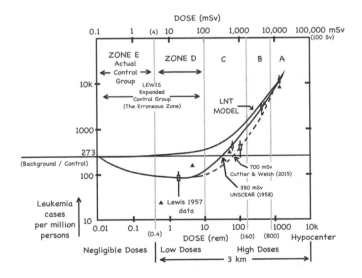

Fig. 55: Cuttler and Welsh 2015 Graph (with Zones and Average Cases Line added)
*Source: https://www.researchgate.net/publication/288992343
_Leukemia_and_Ionizing_Radiation_Revisited*

In 1958, just months after Lewis published, UNSCEAR released a major paper on the first ten years of the Life-Span Study, using the same data Lewis worked with. [8] As you can see in the graph above, UNSCEAR's curve crosses the threshold at about 350 mSv (scaled along the top of the graph).

When Cuttler and Welsh reviewed the LSS numbers from their 2015 perspective, they found the threshold for leukemia was 700 mSv, much higher than what the French Academy of Sciences reported in 2005. As we quoted in Chapter One: "Epidemiologic studies have been unable to detect in humans a significant increase of cancer incidence for doses below about 100 mSv."

BTW, this doesn't mean there is a discrepancy between UNSCEAR 1958 and Cuttler and Welsh 2015. Rather, it shows the evolution of knowledge over time—a graph will change as science advances and data accumulates. Given the fact that UNSCEAR only had ten years of LSS data to work with, while Cuttler and Welsh had the full sixty-five years at their disposal, and the benefit of five additional decades of accumulated knowledge on radiation and health effects, their curves are in good agreement. Both are well away from the no-safe-dose LNT line, and both show a significantly lower-than-average incidence of leukemia in Zone D—paradoxically *lower* than the non-irradiated control group in Zone E.

This is the crucial point that Lewis missed in the 1958 data. He should have noticed that the survivors in Zone D, who received an average of 20 mSv, had subsequent rates of leukemia that were one-third the rate of leukemia in the Zone E, where people received virtually no radiation at all. Not one-third less, but *one-third*. This alone effectively refutes LNT, and strongly suggests a beneficial, or hormetic, effect in the low-dose range.

Note that all three lines start on the *y*-axis at 273 cases per million, the average rate of leukemia in postwar Japan. This is represented by the 33,000 non-irradiated people in Zone E. Also note that leukemia cases only begin to rise above 273 per million in zones C, B, and A, the three zones closest to the hypocenter.

Being so close to ground zero, the survivors in zones C, B, and A had subsequent rates of leukemia directly proportional to the doses received. This is Muller's Proportionality Rule, which is entirely valid in the high-dose range. But he improperly extrapolated this same relationship into the low-dose range, and Lewis followed his bad example.

11.3 ED CALABRESE

Cuttler and Welsh weren't the first researchers to see leukemia's J-curve in blast survivors, but they were the ones who finally mainstreamed this important issue, along with Ed Calabrese, a professor of toxicology at the University of Massachusetts at Amherst.

Over the past several years, Dr. Calabrese has assembled the most complete history of the long, strange tale of LNT. This chapter is a condensed version of the Ed Lewis thread in the greater LNT tapestry that Calabrese has stitched together. We try to do justice to his work in *The LNT Report*, a short companion book to this volume.

Intrigued by the Cuttler and Welsh paper, Calabrese took a fresh dive into the historical record and discovered that several studies have discussed the same paradoxical J-curve pattern in the low-dose range of the Life-Span Study data, but the papers had been lost in the sea of LNT orthodoxy. Thanks to Cuttler, Welsh, and Calabrese, this important issue is finally on the table for review.

A metachart compiled by Calabrese (Fig. 56) shows the results of nine different studies using the LSS leukemia data available at the time. Taken together, they show a clear refutation of LNT that's been buried in the literature since the 1950s. Note that the second study on the list is the NAS 1956 BEAR-I Pathology Panel. This was the medical panel on the BEAR-I committee, outvoted and ignored by the BEAR-I Genetics Panel that was chaired by Muller himself.

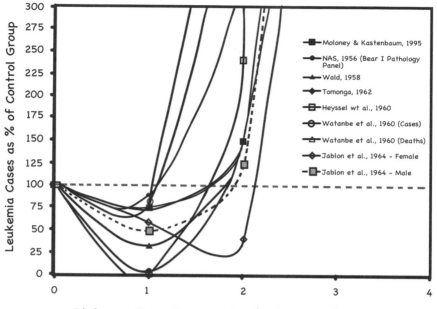

Fig. 56: Studies Showing J-Curve Response In The Low-Dose Range
Source: https://www.sciencedirect.com/science/article/abs/pii/S0013935121003194

The key data table in Lewis's 1957 paper was his Table 2. In 2015, Cuttler and Welsh corrected the table using the same Life-Span Study numbers that were available to Lewis (Fig. 57). For reasons we'll explain later, Lewis improperly dismissed the fact that Zone E was a distinct and separate cohort that served as the control group for a very good reason: They were about 33,000 non-irradiated Japanese from the Hiroshima and Nagasaki suburbs who were still alive in 1950–1957. Altering this control group would alter the results of the study, which is exactly what Lewis had done.

Let's look at figure 57. We're going to concentrate on the right column, listing the number of leukemia cases per million in each zone, which manifested due to the doses recorded in the third column, which we will translate to mSv as we go.

Zone	Distance from hypocentre (m)	Dose (rem or cSv)	Persons exposed	Number of cases of leukemia	Total cases per million
A	0 - 999	1300	1,241	15	12,087
B	1000 - 1499	500	8,810	33	3,746
C	1500 - 1999	50	20,113	8	398
D	2000 - 2999	2	33,692	3	92
E	over 3000	0	32,963	9	273

Table 2: Leukemia incidence for 1950-57 after exposure at Hiroshima (adapted from UNSCEAR-1958, Annex G, Table VII) [8].

Fig. 57: Cuttler and Welsh's 2015 Revision of Lewis's 1957 Table 2
Source: https://www.researchgate.net/publication/288992343
_Leukemia_and_Ionizing_Radiation_Revisited

NERD NOTE: The effective doses in the third column are expressed in rems, the conventional radiation measurement unit used in the 1958 UNSCEAR report. One rem equals 0.01 Sievert, or 1 "centiSievert," an antiquated way of saying ten milliSieverts.

In SI units, Zone D received a 20 mSv average dose, Zone C 500 mSv, Zone B 5,000 mSv, and Zone A 13,000 mSv.

As shown in the bottom row, the non-irradiated people in Zone E had a leukemia rate of 273 cases per million. According to LNT, the rate in Zone D, which was closer to the blast, should be proportionally higher, with the rates proportionally higher still for zones C, B, and A. But that's not what happened.

The Hiroshima and Nagasaki detonations were in August 1945, and twelve years later, in 1957, the non-irradiated people from Zone E had a leukemia rate of 273 cases per million people. But the people from Zone D, who received an average of 20 mSv, had just 92 cases per million, while the people from Zone C, who received an average of 500 mSv, had 298 cases per million.

That dip—from 273 down to 92, and back up to 298—is a classic J-curve hormetic response to a relatively mild dose of radiation: less than what those who were closer to the blasts received, and more than those who were farther away.

The J-curve dip changes the entire picture of radiation health effects.

11.4 DON'T BE LEWIS!

Lewis improperly blended zones D and E together, and called the combined cohort his control group, meaning the group to which other groups are compared in an experiment. Combining the original D and E cohorts obscured a critical distinction:

Being closer to the blast, Zone D received an average of 20 mSv more radiation than Zone E. And yet, by 1957 the Zone D alumni had one-third the leukemia rate compared to those in Zone E—not one-third less, but one-third.

This "reduced effect" fades and then disappears as we get closer to the blast in Zone C, where people received an average of 50 rem, or 500 mSv. From the Life-Span Study data available to Lewis in 1957, and reported by UNSCEAR in 1958, two things had already become apparent:

- Any dose below 350 mSv (where the UNSCEAR curve crosses the threshold) is apparently safe for leukemia.

- Hormetic benefits may lie somewhere between 350 mSv and the average background dose.

As we saw in Fig. 55, Cuttler and Welsh's 2015 review of the complete 65-year LSS data set (1947–2012) found an even higher threshold, at 700 mSv (0.7 Sv). This was twice what UNSCEAR was reporting in 1958, just ten years into the LSS project.

Since a reduced effect at higher doses is impossible to explain with LNT, Lewis had apparently dismissed the contrary evidence instead. His error was so remarkable that Cuttler and Welsh quote his rationale from his 1957 paper:

"Since the majority of the population in Zone D (from 2000 meters on) was beyond 2500 meters, the average dose is under 5 rem [50 mSv] and is thus so low that *Zone D can be treated as if it were a 'control' zone.*" [emphasis added]

No, it can't. And this is where Ed Lewis went off the rails.

The point in all of this is that the actual safety threshold is much higher than our regulatory agencies acknowledge. Even so, getting them to raise the threshold to even 100 mSv is going to be a formidable task; raising it to 350 mSv may be expecting too much. Pushing for a regulatory threshold of 700 mSv may be entirely unrealistic, but it is reassuring to know that even 350 mSv would still ensure a substantial margin of safety.

Our "Don't Be Lewis!" graph (Fig. 58) puts all the pieces together. First off, notice how Lewis viewed the data. He isn't even looking at one of his own data points, the triangle below his right foot. This is the reduced-effect data from Zone D, showing lower rates of leukemia than the Average Cases Line he's standing on, at 273 cases per million.

Instead, Lewis lumped Zone D and Zone E together to form his control group. With this Erroneous Zone behind him, and ignoring the data right below his feet, the data looming before him looked alarming, indeed: His three points (triangles, actually) in zones C, B, and A conformed quite well to LNT. Following Muller's lead, Lewis extrapolated this proportional relationship into the low-dose range.

From his erroneous perspective, he concluded that Muller's no-safe-dose warning for reproductive cells and genetic mutation must also apply to somatic cells and cancer. Leukemia had manifested in direct proportion to the doses received in zones C, B, and A, and since LNT (wrongly) dictated that this linear relationship extends into the low-dose range, Lewis concluded that even low doses of ionizing radiation can cause leukemia, and probably other cancers as well.

But when we assume a proper vantage point and consider all the data (our "Not Lewis" stick figure in Fig. 58), we can see clear evidence of four things:

- A high-dose proportional response.

- A threshold of at least 350 mSv, if not 700 mSv or higher.

- A J-curve response below that dose.

- LNT is BS (bad science).

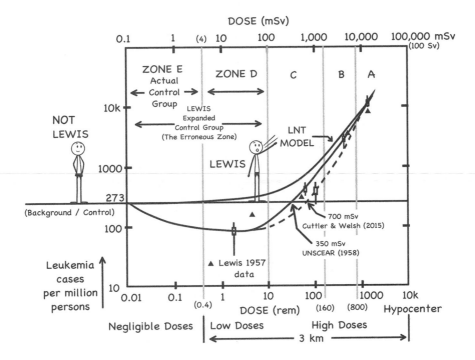

Fig. 58: Don't Be Lewis!
Credit: By the authors, with apologies to Cuttler, Welsh, and Calabese

CHAPTER TWELVE

Who Framed Nuclear Power?

CALABRESE, CUTTLER AND WELSH confirmed a long-standing inconvenient truth for no-safe-dosers: Not only has LNT never been proven, but the hypothesis has been flatly contradicted by peer-reviewed science. [1] It would be nice if the nuclear industry could at least mention this to the public—no wonder nuclear fear persists.

In 2006, the National Academy of Sciences ever-so-slightly backed away from their position on LNT in the official brief of their BEIR-VII report (Biological Effects of Ionizing Radiation), stating that the smallest dose has "the *potential* to cause a small increase in *risk* to humans." [2]

In a nutshell: *There is no dose without risk.*

Compare that to what they said in 1956: "There is no minimum amount of radiation dose which must be exceeded before any harmful mutations occur." [3]

In a nutshell: *There is no safe dose.*

Subsequent reports do refer to risk rather than harm, but it was the alarming and well-publicized BEAR-I report, written for public consumption and widely promoted in the media, that stuck in the collective consciousness. Even people who didn't follow the news got the basic idea: *Radiation = harmful mutations.*

This assurance of inevitable harm apparently stuck in the minds of NAS as well, despite their later, more moderate language. But just like BEAR-I, their later BEIR reports also ignored the mounting pile of evidence being gathered by the Life-Span Study in Japan.

NERD NOTE: Don't let the name change confuse you. The first two BEAR reports were focused on the biological effects of *atomic* radiation. But after publishing BEAR-II in 1960, the NAS concluded that ionizing radiation was a more accurate description of the committee's concern. This is why subsequent publications were named BEIR reports, focusing on the biological effects of *ionizing* radiation. (Unlike BEAR, BEIR is pronounced "beer.")

The 1980 BEIR-III committee did initially consider the LSS evidence coming in from Japan, which by that point had been accumulating for more than thirty years and had yet to show any signs of inherited genetic effects in the *hibakusha* offspring. But the committee ultimately rejected the evidence, relying instead on a massive US study being simultaneously conducted on a large population of mice at Oak Ridge National Lab, which is a story in itself (see *The LNT Report*). The study was later refuted, [4] but by then it was too late—official doctrine and regulatory policy were tightly organized around the LNT model as a sort of self-affirming reality loop.

The 1990 BEIR-V committee took another bold step into left field by accepting the notion of Collective Dose. [5] Here's an extract from BEIR-V's introduction. The first paragraph reasserts linear no-threshold, while the second paragraph shows how they improperly paired LNT with Collective Dose:

"The cancers that result from radiation have no special features by which they can be distinguished from those produced by other causes. Thus the probability that cancer will result from a small dose *can be estimated only by extrapolation* from the increased rates of cancer that have been observed after larger doses, *based on assumptions about the dose-incidence relationship at low doses.*

"In this report it is estimated that if 100,000 persons of all ages received a whole body dose of 0.1 Gy (10 rad) [100 mSv equiv.] of gamma radiation in a single brief exposure, about 800 *extra cancer deaths would be expected to occur* during their remaining lifetimes, in addition to the nearly 20,000 cancer deaths that would occur in the absence of the radiation. Because the extra cancer deaths would be *indistinguishable from those that occurred naturally,* even to obtain a measure of how many extra deaths occurred is a difficult statistical estimation problem." [emphasis added]

Let's break that down: First, they *assume* that cancer from low doses can be estimated by extrapolating the proportional effects observed at high doses into the low-dose range. (As they say in the world of construction: "Assume makes an a-s-s out of u and me.")

Then, they predict an extra 800 cancer mortalities in 100,000 people if each person receives a whole-body (absorbed) dose of 0.1 Gray, or 100 mSv. As we have seen, this is the threshold below which cancer risks have not been detected. [6]

Further, these additional deaths would be lost in the statistical noise of the "nearly 20,000 cancer deaths that would occur in the absence of radiation." But rest assured, there will indeed be 800 radiation casualties among the deceased, even though the radiation victims could not be identified as such.

This was Muller's SWAG, supersized by Lewis into a cancer scare and reaffirmed as official doctrine. And note that the BEIR-V report came out nearly a decade after it was proven that cells routinely self-repair DNA damage from low-dose radiation.

12.1 SECOND NATURE

After nearly four billion years of evolution on this nuclear planet, you would think that self-repair would be the starting assumption about any living cell, rather than assuming there was no safe dose of radiation. But apparently not.

Framing radiation in terms of risk rather than harm may seem like little more than an anodyne way of rewording the boilerplate, and this may well be true—"risk" is a less triggering way of saying "harm" or "risk of harm." But a simple edit, intended or not, can illuminate a crucial distinction. In this case, it highlights the fact that the public's attitude between risk and harm is immense. To show what we mean, here are the same two statements applied to commercial aviation:

There is no flight without risk.
There is no safe flight.

If "*No Safe Flight!*" was chanted by sign-carrying activists, blocking access to airports and calling for all of them to be shut down posthaste, and if their rallying cry was internalized by enough of the public, it could cripple the commercial airline industry. Prices would skyrocket from excessive safety protocols, carriers would start going out of business, and far-less-practical alternatives for travel would seem much more appealing.

That's pretty much what happened to American nuclear power. For over half a century now, the common-knowledge misinformation on all things nuclear has continued to revolve around the false assumption of no safe dose, vintage 1956. Science has moved on; we should all do the same. This would include abandoning LNT's evil twin—the supposed cumulative effect of low-dose radiation. Like the linear no-threshold model, "cumulative effect" is an unproven assumption that all doses, no matter how small, are micro-aggressions that accumulate in the body. And if you build up enough of them, you'll end up with cancer.

These assumptions are false. Nevertheless, it explains (for some people) why about 30% of adults die of cancer. It's "obviously" from a lifetime accumulation of zoomies. This is also why some people opt for a full-body search at the airport. Others have sworn off eating sushi ever since Fukushima. (They're probably the same people.)

And yet, many others are lining up for Botox treatments, and Brotox treatments for the discerning gentleman. This bizarre "beauty" treatment consists of microinjections of botulinum toxin, one of the most acutely lethal substances known to science. Each year, thousands of people get the procedure, even though there have been reports of as many as sixteen deaths since it came into use in the 1990s. [7]

> That's sixteen more fatalities than Three Mile
> Island and Fukushima *combined*.

As with most things in life, Botox patients weigh the risks and benefits, take the best precautions they can, and keep a stiff upper lip. But when it comes to radiation, too many people are convinced that even the lowest dose can pose a threat, so why take chances?

12.2 "GET THE BALANCE RIGHT." – DEPECHE MODE

Despite some evidence to the contrary, most people are actually rather sensible creatures. They can assess risk versus benefit if they have the facts, and if they can separate fact from fiction. Granted, those are two big ifs, but that's how life on Earth survives and prospers—by assessing and responding to the risks and benefits of everything we encounter. This includes the evolutionary challenge of adapting to, and learning to utilize, a caustic substance like oxygen.

Since those adventurous days of yore, we lifeforms have become keenly attuned to detect and defend against threats to our survival. Whether we're any good at it or not depends on many factors, but the orientation is so primal that framing something in terms of harm, rather than risk vs. benefit, can dramatically alter our perception and acceptance of said something. That's how we're wired, and that's how we got this far, so the distinction between risk and harm is crucial.

Mortalities from aspirin and other NSAIDs average about fifteen deaths per 100,000 users. [8] And yet the drug is still judged, by both science and society alike, in terms of risk vs. benefit rather than degrees of harm. Contrast that with the fact that no member of the US public has ever been injured or killed by either new or used nuclear fuel. When the subject is viewed dispassionately, the overwrought safety standards for commercial nuclear power seem to be inspired by little more than misinformation and fear.

Thalidomide, released in 1957, was originally marketed to alleviate anxiety, insomnia, and morning sickness. But after four years of widespread public use, the drug was responsible for about 4,000 late-term fetal deaths, and severe abnormalities in roughly 6,000 surviving infants. [9] This single pharmaceutical has had more ill effect on global human health, in the form of genetic damage and crippled lives, than Chernobyl ever had or ever will—and that's even if we include the false prediction of 4,000 eventual "statistical" deaths from the meltdown.

Back in the 1950s, it wasn't known that some drugs are chemical "enantiomers" (sorry, chemists use lots of weird words). This just means that some drug molecules have mirror-image states, known as left- and right-handed drugs, which perform in much different ways. A simple example of a left- and right-handedness is black licorice and spearmint—they're mirror images of the same molecule, with two wildly different flavors. It was the thalidomide disaster that led to this discovery, and the development of safe "enantiopure" medications. [10]

With the mystery resolved, thalidomide is now being used to successfully treat certain cancers, HIV-related issues, and even Hansen's disease (leprosy). That's because a pharmaceutical meltdown like the thalidomide disaster of the early 1960s did not spark a sustained call to eliminate any and all pharmaceuticals.

It can be entirely appropriate for a particular drug, or class of drugs, to be restricted, halted, or phased out. In the same way, a bad reactor design like Chernobyl should be, and has been, cut out of the herd. It truly was the Pinto of reactor designs—even though the actual Pinto cost more lives.

This is how medicine and industry improves, and the public knows this. So despite the risks of using pharmaceuticals, even as prescribed, there has never been a sustained public sentiment for the total phase-out of pharmaceuticals, coupled with a rapid transition to 100% alternative medicine. Along with surgery and most other medical procedures, pharmaceuticals are at least grudgingly accepted by all but the most doctrinaire of health nuts.

Except when it comes to vaccines. For a lot of reasons that don't add up, anti-vaxxers are trying to frame any and all vaccines as too risky to consider—*"No Safe Vaccines!"* Robert Kennedy Jr., an ardent anti-vaxxer, is also a big fan of gas-backed solar farms, which he candidly describes as gas plants. [11] It's interesting to note that in spite of having gas backup, the concentrated solar plant he touted in Ivanpah, California, was a flop. So was its neighbor, the Crescent Dunes CSP plant in Nevada. To add insult to injury, Ivanpah needed so much gas backup they had to pay an undisclosed sum to Pacific Gas and Electric (PG&E) for transmitting too much gas-generated power to California, and not enough clean electric sunshine as contractually agreed. [12]

Chernobyl has long been the ultimate bogeyman in the effort to dissuade us from nuclear power, which, again, is like using the Pinto to scare us about every car that has ever been built, or ever will be. *Technology improves.* And even if no advances in nuclear technology had been made since the 1980s, no one will *ever* build a reactor like Chernobyl again, for the same reason that no car company will ever build another Pinto. The public knows this without having to be told, and yet we're all supposed to worry that some reactor company might build another Chernobyl. [13] Behold the power of nuclear fear.

For all their drawbacks, the public knows that pharmaceuticals, like vaccines, when properly prescribed and taken, confer far more benefit than whatever harm they may cause—which, we shouldn't have to mention, is far more harm than has ever come from nuclear power.

But we do have to mention it, and other things like it, until nuclear is finally reframed. Framing shapes perception and positioning, which strongly influences our acceptance or rejection of anything. [14] Cultures, belief systems, and Madison Avenue have employed this technique since forever. Framing can even precede information—scary movie music tells us who the bad guys are before we know a thing about them. Mention the word nuclear and it can be an uphill battle from there.

Framing something in terms of harm, rather than risk, is usually reserved for things that are inherently bad, with little or no upside. American nuclear power has been saddled with this false frame since the unscheduled chest X-rays of Three Mile Island.

12.3 FEAR AND PARANOIA ARE THE TWO MOST COMMON FORMS OF RADIATION SICKNESS

While the general public may not know that LNT has never been proven, the opinion leaders of the anti-nuclear movement ought to. If they don't, they're misinformed about the very thing they're trying to ban. On the other hand, if they accepted the peer-reviewed science on radiation safety, their central argument would collapse.

The fear, uncertainty, and doubt (FUD) stirred up by anti-nuclear groups has impeded the effort to tackle air pollution and climate change—not to mention ocean acidification, rising ocean temperatures, and acid rain. And now the world's future may hang in the balance. [15] To which some people respond: *But nuclear is high carbon!"* The rationale (if you could call it that) behind this laughable falsehood goes something like this: "Nuclear power leads to nuclear weapons, which leads to nuclear war. And all the firestorms the bombs set off would release billions of tonnes of carbon into the atmosphere. Therefore, nuclear is high carbon."

Set aside the fact that nuclear power does not, in fact, lead to nuclear weapons (see Chapter 19). Hiroshima and Nagasaki suffered the firestorms they did because they were built almost entirely of wood and paper, while our modern population centers are not. In fact, the US shamefully developed napalm during the war (at Harvard, of all places) to firebomb Dresden, Tokyo, and other civilian population centers. [16]

There are plenty of reasons why napalm and nuclear weapons should never be used, but claiming that nuclear energy leads to nuclear war makes about as much sense as claiming that petroleum leads to napalm, so therefore every gas station is a potential napalm factory. Nonetheless, "high-carbon nuclear power" is an actual talking point of Mark Jacobson, whose all-renewables Roadmap we unpack (and dismantle) in *Roadmap to Nowhere*.

Even when skeptics do accept that nuclear is low carbon, their fallback position is that an abundance of carbon-free power, available on demand, regardless of weather, season, or time of day, is a really bad idea "because zoomies." They further contend that because there is no safe dose, radiation casualties will inevitably occur, and inexorably increase, with the continued use of nuclear power. They also believe that used fuel will be a toxic hazard for hundreds of thousands of years, with no possibility of safe storage, recycling, or composting.

All these convictions are flatly incorrect. But with nuclear power thus sidelined from clean-energy discussions, the foregone conclusion is that a mix of wind, water, and sunlight (WWS) is the only possible, and ethical, option we have. This is the "no nukes" box in which we are all supposed to think about clean energy, even though LNT—the very essence of the box—has been roundly refuted by science. The problem is, the general public has never heard the full story on either radiation or nuclear power.

This will change over the next few years because as irony would have it, the initial anti-nuclear posture of some who support the Green New Deal has inspired a strong and sustained pushback from the pro-nuclear environmentalist community. [17] Nuclear is finally back on the table for discussion, where it's always belonged.

CHAPTER THIRTEEN

LNT vs. the 100 mSv Threshold

In 2013, two years after Fukushima, UNSCEAR (UN Scientific Committee on the Effects of Atomic Radiation) called out the error of using LNT for assessing a large population. The UNSCEAR chairman's speech to the Fukushima Ministerial Conference on Nuclear Safety, which we quoted on page 132, made their position clear.

Apparently, it was an attempt by UNSCEAR to put LNT to rest, or to at least place it in proper perspective. But like some of the strange ideas that wander through the public as a kind of zombie politics, LNT is a conventional "wisdom" that is still alive—even though the popular ideas of no safe dose and cumulative dose are dead wrong. [1] A wealth of evidence has accumulated since the linear no-threshold model was first conceived, demonstrating that a threshold does in fact exist, and can be (very) conservatively set at 100 mSv per year.

The linear no-threshold model should have been put to rest ages ago, but we air-conditioned cavemen are programmed to respond to perceived immediate threats. One drawback of this rudimentary survival trait is that some threats are impossible to detect with our senses. By stoking our imagination and fear, these undetectable dangers can all too easily grow into Godzilla-sized bogeymen, whether they're an actual threat or not. That's why when it comes to radiation, there is no safe dose of BS.

Tribal wisdom, both ancient and modern, tells us "this is dangerous" or "this is safe," and sometimes these pronouncements stick more
firmly than they should. We busy people accept much of it on faith,
as part of a cohesive narrative that shapes our tribal identity. Most
of us are like that; it's part and parcel of being a social animal. And
this is why science must be the final arbiter, because science doesn't
care what anyone thinks—even a tribe of scientists.

The advertising, media spin, and social impact of the 2019 HBO
series *Chernobyl* is a recent example. Although it told a gripping story,
the series played fast and loose with some of the important facts. In
real life, for example, the people on the bridge in the last scene did
not die from radiation, as the finale implies.

The show's creator insisted that his series was not anti-nuclear,
even though several stereotypical anti-nuclear fear buttons and movie
tropes were used. His intended point was that a toxic bureaucracy
can be deadly, regardless of the kind of power it wields. No argument there, and the Chernobyl accident is a visually dramatic vehicle
to bring that story to life. But for many who watched the series, it
also confirmed their conviction that nuclear power was the actual
bad guy that just so happened to be in the service of a toxic Soviet
bureaucracy. [2] Thinking otherwise would be akin to tribal heresy,
like a surfer watching *Jaws* and rooting for the shark.

Back in 1946, in his lecture at the Nobel ceremony, Muller
emphatically stated that there would always be some harmful effect,
however low the dose may be:

*"Both earlier and later work by collaborators showed definitely
that the frequency of the gene mutations is directly and simply
proportional to the dose of irradiation applied."* [3]

Imagine what his audience was thinking. They were probably
well-enough informed to appreciate the implications of no safe
dose, and the news that all doses are cumulative. Muller had their

undivided attention, and in light of his findings the road ahead was clear: Minimize or eliminate all radiation, because you never know. You yourself might survive a high dose, and your children may turn out fine, but their children could wind up with genetic anomalies.

13.1 PERCEPTION VS. REALITY

Simply put, Muller was wrong. And his correspondence suggests that he probably knew he was wrong, or at least had his doubts (see *The LNT Report*). The most charitable explanation is that he fudged the facts to help rid the world of nuclear weapons. The problem is, he and his LNT acolytes tried—and are still trying—to throw the baby out with the bath water by ridding the world of nuclear power as well. Dr. Helen Caldicott is but one colorful example. [4]

Muller was instrumental in the push for an atmospheric test-ban treaty for nuclear weapons, and for that he deserves both credit and thanks. Unfortunately, he promoted LNT as the basis for his reasoning. Many people accepted the Nobel laureate's statements on his authority alone, and have been leery of radiation ever since (*Death by a zillion zoomies!*). As a result of Muller's LNT scare, followed a year later by Lewis's cancer scare, we are two decades into the 21st century and nowhere close to meeting our clean energy goals.

When you stop and think about it, the public belief in "no safe dose" and "cumulative effect" is a mighty strange leap of faith. Because quite aside from all the peer-reviewed studies, and all the white papers released by credible sources, direct observation has shown that cell repair exists, and has functioned quite well right from the start. After all, here we are, living on a nuclear planet awash in radiation, evolved from cellular life that got zapped by way more zoomies than whatever we're encountering now, some two billion years hence.

Since natural radiation is nearly impossible to mitigate, the man-made radiation in science, medicine, military, and industry—especially the nuclear power industry—became the target of our overly-stringent standards. Even though some of those standards, in the context of real-world conditions, are entirely unrealistic.

For example, the Nuclear Regulatory Commission has set the maximum radiation level for decommissioned nuclear power facilities at less than *one-tenth* of the 3 mSv average yearly background radiation levels found across the United States. [5] This is just one example of why decommissioning an American reactor has become so time-consuming and expensive.

Note that this is even more stringent than Japan's soil standard, which is 1 mSv above background. Further note that both standards contradict everything that science, including the National Academy of Sciences, has known since at least 1983—that all living cells repair most of the zoomie damage they encounter, even the damage incurred during cell division. And, that life has been doing this for billions of years. [6]

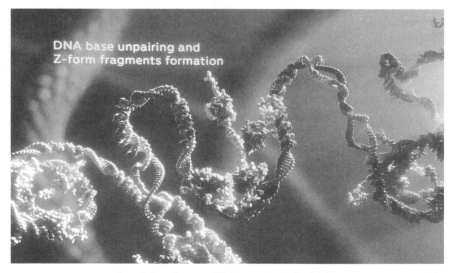

DNA base unpairing and
Z-form fragments formation

Artist's Rendering of Enzymes Repairing DNA
Credit: Shutterstock

Curiously, the tourism industry has managed to avoid such overbearing rules and regulations. More than a few famous vacation spots are touted for having levels of background radiation comparable to portions of the Chernobyl and Fukushima exclusion zones. Guarapari Beach in Brazil is but one example. People flock there for the health benefits of its radioactive monazite sand, which is rich in thorium. [7]

Radiation count on Guarapari Beach, Brazil
33.48 μSv per hour is about 300 mSv per year.
Source: https://www.youtube.com/watch?v=RvgAx1yIKjg

People also frequent the radon baths in Germany, and have been for 800 years; today the visits are covered by anti-nuclear Germany's national health insurance, so go figure. [8] Though they didn't know anything about radiation back then, they knew there was something beneficial in the water, aside from the heat and minerals available at other baths.

Simple anecdotal evidence suggests that LNT doesn't work in the real world; MIT recently showed that it doesn't even work in the lab.

13.2 ZOOMIES AND CHEESE

In a 2012 experiment at MIT, laboratory mice were irradiated for thirty-five days with a daily dose equal to 400 times background radiation. That's more than an entire year's background dose, delivered each day for more than a month. The result: No lasting DNA damage was observed.

> "DNA damage occurs spontaneously even at background radiation levels, conservatively at a rate of *about 10,000 changes per cell per day*. Most of that damage is fixed by DNA repair systems within each cell. The researchers estimate that the amount of radiation used in this study produces *an additional dozen lesions per cell per day*, all of which appear to have been repaired." [emphasis added] [9]

Let's break that down:

- A living cell repairs its own DNA about 10,000 times a day.

- After thirty-five days of continuous, high-dose radiation, the number of daily repairs in each cell increased by twelve repairs per day.

- Not 12%, but twelve—from 10,000 repairs per day to 10,012.

This amounts to an increased risk of 0.12%—a bit more than one-tenth of one percent—after five straight weeks, day and night, of being zapped with 400 times the average background dose.

Two conclusions are clear:

- Even substantial radiation above background dose causes no lasting damage.

- The radiation safety threshold is apparently much higher than many believe.

Some scientists contend that the 100 mSv threshold could be set much higher, in the range of 700 or even 1,000 mSv per year or more. [10] However, a conservative threshold dose of 100 mSv per year is frequently cited in recent literature. Nuclear medicine has been successfully applying the pharmacological principles of thresholds and dose-rates since 1946, when the field was born—ironically, the same year Muller got the Nobel for LNT. And yet, after all these years, the regulations for nuclear power still regard every last zoomie as a threat. [11] The cognitive dissonance is remarkable.

CHAPTER FOURTEEN

LNT vs. Common Sense

W<small>HEN</small> A<small>MHERST TOXICOLOGY PROFESSOR</small> E<small>D</small> C<small>ALABRESE</small> read the Cuttler and Welsh paper in 2015, he initially thought the authors were "seeing something that wasn't there." [1] But the paper had been peer-reviewed so he took another look. Calabrese had already been investigating the LNT controversy, and a fresh search through the literature found several early papers discussing the phenomenon as well. This is his J-curve graph compilation from Chapter 11:

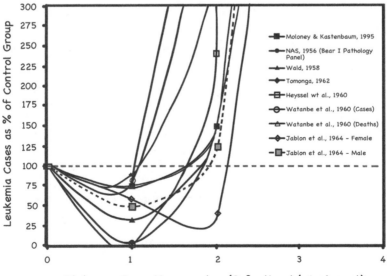

Digging deeper into Lewis's early work, Calabrese discovered something in an unpublished draft of his '57 paper, which had been reviewed by the BEAR-I committee. (Conveniently, the committee chair was also Lewis's department chair at CalTech.) The Lewis draft cited a 1954 study by Furth and Upton of radiation experiments using thousands of mice and pigs in the A-bomb tests in the Nevada desert, [2] in which they found clear evidence of a safety threshold in the irradiated mammals. Any reference to this conflicting data was excluded from Ed Lewis's published paper.

The data trail amassed by Calabrese on the history of LNT includes thousands of letters acquired from the Muller estate, between Hermann Muller and his lifelong friend and pen pal, fellow geneticist Edgar Altenburg. Calabrese also acquired the papers and personal correspondence of Curt Stern, Ernst Caspari, and other scientists in Muller's circle. Studying these letters alongside the work of the respective parties, as well as the works they were citing (and at times misrepresenting) in their own papers, Calabrese uncovered what he contends is a clear and convincing case of scientific misconduct. By 2013, he had already published the details in the *Archives of Toxicology*, with a timeline of important events. [3] Then along came Cuttler and Welsh.

In March 2015, Jerry Cuttler brought the flaws in Lewis's 1957 paper to the attention of Marsha McNutt, then-editor of the journal *Science*. Cuttler had attached a copy of the Calabrese timeline to his email, and asked that the matter be reviewed. If McNutt found that he and Calabrese were correct in their research and conclusions, Cuttler asked her to consider an official retraction of the BEAR-I genetics panel report, published by *Science* on June 29, 1956. [4] This is the technical paper that underlies the key pronouncements of the BEAR-I public report, and forms the foundation of Lewis's paper as well.

Dr. McNutt rejected Cuttler's request. Undaunted, Cuttler forwarded their exchange to Ed Calabrese, whose work on LNT was becoming known in academic circles. In August 2015, Calabrese wrote to McNutt, responding to several points she made in her reply to Cuttler, and asked her to reconsider the issue. They had a series of exchanges that went nowhere, her last email to Calabrese concluding with: "Please respect that the matter is closed."

Nevertheless, Calabrese persisted, continuing his compilation and research, culminating in his October 30, 2017 paper "Societal Threats from Ideologically Driven Science," which included his exchange with McNutt. [5]

In 2016, McNutt was appointed president of NAS, now known as NASEM—the National Academies of Sciences, Engineering, and Medicine. Six years later, in June 2022, NASEM issued a 300-page proposal [6] calling for a multi-decade study to explore the harmful effects of low-dose radiation, with a recommended budget of a hefty $100 million a year.

While the proposal does acknowledge a strong and growing push-back against LNT, it also points out that there is increasing evidence of possible harmful non-cancerous effects from low doses, such as cardiovascular and brain function issues. While nothing definitive has been found, there are hints. Fair enough, and further research is entirely appropriate. But the precautionary "principle" of assuming that harm exists until proven otherwise is not appropriate at all.

To be perfectly blunt, the Precautionary Principle is fear-based public policy; it is not science. Worse, it's sophistry, because it slyly assumes the ability to prove a negative—in this particular instance the ability to prove the absence of harm. Which, of course, cannot be done. And once you buy into *that* squirrel cage, science ends and nuclear fear begins. Overwrought regulations are sure to follow—and boy, howdy, have they ever.

In the real world, absolute safety can never be "reasonably" achieved. And yet, this has become the default approach to nuclear power, ballooning costs and extending construction times to outlandish degrees. All of which somehow proves that nuclear power is unworkable. And it all comes back to LNT, with no science to back it up.

Even so, the 2022 NASEM proposal makes clear that until the issue is thoroughly re-examined (over the next few decades, and if they can find the money), the principles of LNT and ALARA (see next chapter) shall remain in effect as official policy.

Sorry, NASEM, EPA, DOE, NRC, *et alphabeta*, but we don't have a few decades. Please respect that the matter is far from closed.

Indeed, a pile of evidence has been building over the last several decades that clearly refutes LNT; the situation has warranted attention since the 1950s and there has been scant improvement since then. As late as 2006, the most recent BEIR report from the NAS once again affirmed LNT as the model for radiation risk assessment:

"A comprehensive review of the biology data led the committee to conclude that the *risk would continue in a linear fashion at lower doses without a threshold*, and that the smallest dose has the potential to cause a small increase in risk to humans. This assumption is termed the "linear no-threshold model." [7]
– BEIR-VII Report, 2006 [emphasis added]

Others in the field have a much different take on things:

"Based upon these findings and those cited by others, it becomes apparent that the LNT model cannot be scientifically valid." [8]

What is scientifically valid is that any exposure to radiation can carry a *potential* risk of harm. What is equally valid is that risk must always be kept in perspective, and that no one can prove a negative. Even so, the no-safe-dose / no-threshold doctrine of LNT continues to be the underlying basis of nuclear safety rules and regulations: Any non-zero risk is supposed to be a really big damn deal, even though it almost never is.

More troubling, the tenets of LNT have been successfully used to scare people away from nuclear power. And yet, the truth about radiation can be explained to anyone by this simple example:

Getting a little sun is fine, but being out in the sun day after day, all day long, until you develop melanoma, is not a good idea. Neither is staying out of the sun entirely to avoid any possible risk.

14.1 THE NOSE UNDER THE CAMEL'S TENT

The BEIR-VII report did not take pains to explain to the reader that, from the LNT point of view, even an utterly insignificant risk must still, technically, be regarded as an indication of potential harm. This is probably because, unlike BEAR-I, the later BEAR and BEIR reports were written for professionals in the field. This being the case, they didn't explain to the reader that if the chances of harm are eleventy-nine gazillion to one, there is still—technically speaking—a non-zero degree of risk.

Your lifetime risk of dying in a plane crash is one in eleven million. [9] This means there is some risk involved; it does not mean that flying is harmful or dangerous. Similarly, the biological sciences deal with probability, statistical risk, and percentages. Absolute answers are typically sought in the world of pure mathematics. In the real world, there is a non-zero risk in nearly everything we do, even the simple act of breathing. [10]

Everything in nature is on a probability spectrum, and can only approach, but never reach, 100% certainty. The inability to realize absolute certainty, in the realm of real-world science and technology, has long been used to generate misperception, fear, and doubt. Because no matter how safe something is, it can never be absolutely 100% safe, no matter what happens, forever and ever times infinity. That's not how reality works.

So why are we going on about this? Because the inherent difference between biology and math has long been the camel's nose under the nuclear tent. Media sensationalists make a big deal out of the fact that the acute risk of cancer doubles with, say, a tritium leak from a nuclear plant affecting the groundwater. They don't bother to mention that doubling that risk would cut the margin of safety from, say, a comfortable ten billion to one, down to a nail-biting five billion to one.

In a nutshell: *Risk does not equal harm.*

Especially infinitesimal risk. Do you worry about the fact that you could be struck by a meteor? We didn't think so. Such a minuscule risk is never weighed in terms of potential harm, much less in terms of probable harm. Conflating the two—the risk of being hit, and the potential harm from being hit—wouldn't make sense.

In the same way, using LNT as a harm prediction tool is entirely unreasonable, and just plain old bad science, especially for dose-rates below 100 mSv per year. It's long past the time for no-safe-dosers to either move on, or present clear and unambiguous evidence of adverse biological effects at low doses. Nuclear power has been in widespread use for more than sixty years now, with three infamous incidents and a handful of work-related accidents. Surely some convincing evidence would have cropped up by now.

Focusing on risk, even when that risk is infinitesimal, amounts to fearmongering, which is all too easily monetized into fear lobbying. This phenomenon is not unknown in Washington, nor absent from state legislatures. In the case of nuclear power, the result has been a raft of unhelpful policies that advance the buildout of a weather-dependent national grid, even as we face a future of less predictable weather.

While facts are stubborn things, so too are the beliefs we hold in things that are unproven, or even disproven. Because even more entrenched in the psyche, and stronger than any particular belief, is the all-too-human desire to be right—especially when the evidence is lacking, or suggests otherwise.

14.2 WORDS ARE POWERFUL THINGS

Public agencies tend to err on the side of caution, which generally bestows a public good. It's been said that speed limits are set for the safety of a school bus on a dark and stormy day. But just like anything else, caution can be overdone.

Regrettably, the National Academies of Sciences, Engineering, and Medicine offers no guidance on where to place a safety threshold, or that one may even exist. They simply contend that any dose carries some degree of risk, and leave it for regulators and lobbyists to parse. Government and industry can work with that assumption, if they can agree on a threshold below which the risks are negligible to non-existent, and if they also take care to never conflate risk and harm.

Unfortunately, fearmongering and misinformation usually derive their potency from an underlying kernel of truth. And on some deep, visceral level, people intuitively sense that the power to save the world can also be used to destroy the world. This is an enduring theme in religion, literature, movies, TV, and video games. Spencer Weart explores the phenomenon in his excellent book *The Rise of Nuclear Fear.* [11]

As Weart explains, the subject of nuclear energy can conjure a host of archetypes deep within the human psyche: the alchemist tinkering with the building blocks of nature, the high priest seeking infinite power, the mad scientist wielding death rays. All of them are variations on the underlying theme of Man Playing God, and the stories seldom have a happy ending.

The way this can manifest in everyday life is that you can explain the facts until you're blue in the face, but sometimes the fear won't dissipate:

> *"Yeah, well, who knows? Maybe UNSCEAR is right, and maybe they're not. Maybe Fukushima will turn out to be as bad as Chernobyl."*
>
> *"Bad like, where nothing happens to the downwinders?"*
>
> *"No! Sometimes cancer takes years to develop. And with all that fallout, maybe 4,000 unlucky people really <u>did</u> swallow the proverbial drop of water. You just never know."*
>
> (Sigh.)

14.3 TAKE TWO ASPIRIN AND READ THIS IN THE MORNING

When people are gripped by nuclear fear, evidence-based risk assessments do little to reassure them. A zero-tolerance policy becomes the only acceptable standard. Because you never know . . .

Actually, you could know if you examined the data, and learned from people who study the subject. (To reiterate: We're the writers; they're the scientists. That's why all the endnotes.) Even something as simple as waving a Geiger counter over a banana will show that we do indeed live on a nuclear planet.

An absolutist approach, such as zero tolerance or no safe dose, is usually reserved for things that are judged to be irredeemably evil. In too many minds, nuclear energy falls in that category. Because radiation is so potentially awful, the rationale goes, it is better to assume that harmful effects will occur from extremely low doses, and cautiously proceed from there—even though such effects have never been found after a century of observation.

Radioactive material itself, on the other hand, is quite easy to find, and that's actually part of the problem: A single, solitary, harmless atom can be detected if it's having a becquerel moment. Plus, the atom's signature will be unique—by their emissions shall ye know them.

This is why minute traces of Fukushima cesium could be detected in Southern California seawater, which flat out rattled some of the locals. But detecting its presence doesn't mean it's dangerous. As with any substance, it all depends on the dose.

No one is arguing for less safety, but it's clear that instilling fear and overdoing safety precautions serves no one and costs a needless fortune. Here's a simple example: A foot of armor plating will stop a bullet. Adding a second foot of armor won't make you any safer, and eliminating that extra foot won't put you in any danger. And yet, we instinctively have misgivings when someone suggests that we dial back on what is clearly an over-cautious standard. It's human nature, just like the fear of flying.

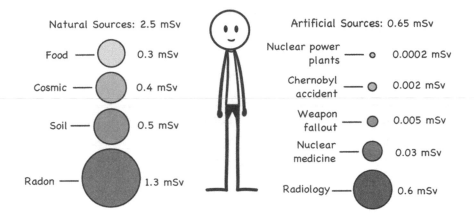

Fig. 60: Comparison of Yearly Average Doses
*Source: https://www.sageadvisory.com/wp-content/uploads/2019/09/ESG
-Perspecitves_October-2019_Nuclear-Power.pdf*

Yes, Three Mile Island had safety and performance issues. And just like investigations in the aviation industry, the TMI investigation resulted in several important changes that have now become standard operating procedure in the nuclear industry. [12] Which is great, but what the industry should have *also* been doing since TMI is loudly and pro-actively debunking LNT at every opportunity. Without the freedom to weigh in on this vital issue, no safety measure will ever satisfy misplaced concerns.

The public's fear of flying pales in comparison to their nuclear fear, despite the far greater risk of air travel. And yet, for all the safety improvements and tighter regulations in the nuclear industry, [13] efforts to reassure a misinformed public have mostly backfired. "If it wasn't so dangerous," the suspicion goes, "why do they keep focusing on safety?"

Fair point, up to a point, because the same could be asked of any industry—safety truly is Job #1. And the best way to resolve safety issues is to establish tight but rational standards and stick with the science. Then keep educating the public, and debunking the bad information over and over again, as the healthcare industry has done, until the facts about their product became common knowledge, displacing the distortions and falsehoods. [14]

One example of how such distortions play out is with the soil removal at Fukushima. Most of it has been removed to calm public fears, and nothing more. And a decade after the accident, they're *still* removing bags of topsoil from cultivated farmland [15] to achieve a ridiculously low reading of one mSv/year above pre-accident levels. [16] The average background dose in Japan is about 2 mSv per year. [17] If the displaced people just up and moved to Denver, their yearly background dose would skyrocket to a reckless, death-defying 7 mSv/yr. [18]

As a Japanese government scientist explained: "We know we don't need to reduce radiation levels for public health. We're doing it because the people want us to." [19] Cynical translation: "We're doing it because the people have been propagandized into a state of nuclear fear, by the same government that is housing them in shelters and spending $25 billion to remove 5 cm (2 inches) of topsoil." [20]

Fukushima, Japan—Bags of Removed Topsoil
Credit: Keow Wee Loong

The problem now, of course, is what to do with all the soil (putting it back where it came from comes to mind). Fukushima cleanup and relocation costs are estimated to exceed $250 billion, because instead of educating their citizens, Japan is catering to their fears and wasting their money. Like the bumper sticker says, if you think education is expensive, try ignorance.

Relocation expenses for the displaced population are expected to cost $60 billion. Add $25 billion for soil removal, and that's one-third of the cost right there. [21] And then there's all the tritiated water they've been storing in giant tanks. You may have heard about the 1.3 million tonnes (thus far) of tritiated water, and the debate about what to do with it.

The water was used to keep the reactors cool after the meltdowns, so now it has traces of tritium, a faintly radioactive isotope of hydrogen. Years of storing it, and eventually dispersing it into the ocean, is expected to cost nearly $3 billion, instead of just piping it offshore to begin with. [22] And people wonder why nuclear power is "expensive." It's not, but nuclear fear costs a bloody fortune.

Just how grave a threat is this water to the world? None whatsoever. And as time goes by, with the radioactivity of the tritium gradually approaching a big fat zero, it's become even more of a nothingburger without the bun. Radiation experts correctly advised in 2020 that the water should be released into the ocean. [23] But misinformed people all over the world pushed back at the idea, incorrectly assuming that the water was, and still is, dangerous stuff.

As of August 2023, 1.3 million cubic meters of fresh water has been flushed through the damaged reactors. The water was stored, and treated to reduce radionuclides far below the regulatory standard for drinking water. The only "harmful" substance that remains is tritium, with a radiative intensity of about 1.06 million Becquerels per liter. (For the math on this, see Part Three of our supplement in endnote #25 below.)

Before this water is released to the sea, a bit at a time over the next few decades, it will first be diluted with seawater to reduce its tritium activity to just 190 Bq per liter. This is about fifty-three times less than the WHO's tritium standard for drinking water. [24]) Except it'll be seawater.

Fukushima, Japan – Tritiated Water Storage
Credit: IAEA Imagebank (CC-BY-SA-2.0)

To put these numbers in perspective: Your body's store of dietary potassium gives you a daily radiation dose of 0.453 μSv (microSieverts). Drinking 25 ml (milliliters) of Fukushima tank water on a daily basis—straight from the tank, and prior to seawater dilution—would gradually match the same daily internal radiation you're already getting from your stored potassium. And how much is 25 ml? A jigger is 44 ml, so 25 ml is a splash more than a half-shot of tank water, neat or on the rocks, taken as a daily "maintenance dose." (See our supplement Dietary Potassium vs. Fuku Water [25])

Now, before you go flying off to Japan to give this a try, consider that a round-trip flight from LA to Tokyo would give you *160 times* more radiation than downing a half-shot of Fukushima tank water (see the Generation Atomic Calculator in endnote #24.)

Here's another way of looking at it: With the tank water's estimated long-term total radioactivity of one quadrillion Bq, the total ocean discharge of all Fukushima water will equal about six days of natural tritium formation in our planet's upper atmosphere, resulting from cosmic rays zooming down to earth. [26] That's what all the fuss is about.

The solution to pollution is dilution, and the Pacific Ocean holds about 700 *trillion* cubic meters of water. However well-intentioned, much of the money and effort would be better spent on returning Japan's displaced citizens to their lives and livelihoods.

Overwrought concerns are nothing new when it comes to nuclear power. By the time of the Three Mile Island accident in 1979, the anti-nuclear weapons movement had expanded its horizons to include commercial power reactors. TMI was a watershed moment for their cause, even though it was a total non-event from the perspective of public health. Fear, not facts, turned a non-lethal industrial accident into an apocalyptic scare story.

The entire region was glued to their TVs watching wildly inaccurate news alerts, such as an AP reporter's unfortunate phrasing that equated a hydrogen gas explosion at TMI with a hydrogen bomb. A media echo chamber quickly ensued, with breathless reports amplifying the rumors and disinformation. It was your typical disaster news cycle, with a twist:

Just two weeks prior, *The China Syndrome* had opened in theaters nationwide. The free publicity of a real-life nuclear accident, even though it was a mere shadow of the disaster in the movie, turned the film into a blockbuster and made nuclear power the national fear du jour.

Jane Fonda, the female lead, went before a thoroughly rattled public and shaped the moment into a cause celébre. [27] The hit movie / real-life mashup was followed by a series of No Nukes concerts around the country, attended by more than 300,000 newly minted no-safe-dosers. One of the California concerts even featured a speech by then-Governor Jerry Brown. Public opposition to nuclear power was fine with him—his family was (and still is) heavily invested in oil and natural gas. [28]

Some of the concert stars still donate their time and prestige to the anti-nuclear movement. [29] The anti-nuke / celebrity nexus now includes Leonardo DiCaprio and Mark Ruffalo, the latter of whom is also a board member of Mark Jacobson's Solutions Project foundation, promoting Jacobson's 100% renewables roadmap.

1978: Jerry Brown Campaigning at UCLA.
Jerry Brown had the ear of young people, in California and across the nation. One year later as the Governor of California, he spoke at the No Nukes rally near the Diablo Canyon nuclear power plant in the wake of the Three Mile Island incident.
Credit: Joe Kennedy, Los Angeles Times (CC-BY-SA-4.0)

14.4 THE BURDEN OF PROVING A NEGATIVE

After a half-century of conflating nuclear power with nuclear weapons, anti-nukers have added so much counterfeit moral weight to their no-safe-dose mantra that it can sometimes be difficult to even discuss the issue. If something is so irredeemably bad, what's there to discuss?

A zero tolerance for radiation puts any scientist or nuclear advocate in the impossible position of having to prove a negative. It's like searching the entire back country of North America to prove the absence of Bigfoot. Then another blurry picture is published in the tabloids, and you have to do it all over again.

Logic and fair play dictate that the burden of proof should rest with those who claim there is harm, rather than those who claim there is not. The determining standard is objective reality, not what anyone believes or fears. And yet, despite ample evidence that doses below 100 mSv /year are safe, the Precautionary Principle is routinely invoked in nuclear debates. [30]

This sentiment of "guilty until proven innocent" fits LNT like a glove; it also fosters a climate of perpetual probation for nuclear power: In any mishap, the default position is "dangerous until proven safe." [31] Applying the Precautionary Principle would make sense with, say, a new molecule, or an untested technology with potentially dire consequences, like bioengineering or artificial intelligence. That's not the case with nuclear power.

Nuclear has been in commercial use now for almost seventy years, with about 450 reactors in operation worldwide. Nearly 100 are in the US, some of them in continuous service for over four decades. The world's navies are using hundreds more, shipboard versions of what are now called small modular reactors (SMRs). And as we mentioned, the US Navy has logged more than 5,700 reactor-years without a single radiation incident. This is by no means a new or untested technology.

During those seventy years, the potential hazards of radiation have been explored with epidemiological studies, which track health anomalies in large populations and correlate them to suspected sources. Tracking has also been done by monitoring the health of nuclear workers in science, medicine, and industry. So far, so good: The report is titled "The Healthy Worker Effect and Nuclear Industry Workers." [32]

From 1980 to 1988, this comprehensive health study was conducted on US shipyard workers, from which three large cohorts totaling 70,000 workers were screened and selected. The final report, titled "The Nuclear Shipyard Worker Study," was completed three years later. [33] This is from the abstract:

"Although the NSWS was designed to search for adverse effects of occupational low dose-rate gamma radiation, few risks were found. The high-dose workers demonstrated *significantly lower* circulatory, respiratory, and all-cause mortality than did unexposed workers. *Mortality from all cancers combined was also lower in the exposed cohort.* The NSWS results are compared to a study of British radiologists." [emphasis added]

In 2008, another massive study of 250,000 general nuclear workers [34] came to the same conclusion:

"The relationship of cancer mortality with radiation dose received by the exposed group is of prime importance. For example, the results from 250,000 nuclear workers were summarized by T.D. Luckey: '*The average mortality of nuclear workers . . . was substantially lower than in control groups.*'" [emphasis added]

Precautions are always appropriate with any hazardous substance. But the Precautionary Principle, justified in anti-nuclear circles by the erroneous ideas of no safe dose, cumulative dose, and Collective Dose, is a whole other can of worms.

14.5 YOU MAY HAVE NOTICED

In this entire discussion on radiation safety, we have refrained from saying that there is zero harm from low doses of radiation. This is an important distinction to keep in mind, for three important reasons:

- Scientific consensus is based on a preponderance of evidence, not unlike a civil suit.

- Mathematical calculation, on the other hand, is absolute. If the input information is correct, the calculated result has no shadow of a doubt, not unlike a criminal case.

- We are discussing biology, not math.

Hardcore skeptics of nuclear power insist on 100% certainty, much like those who doubt human-induced climate change or vaccine safety. This amounts to insisting that cellular biology (or climate science, for that matter) perform like pure mathematics, isolated from the uncertainties of the real world. Most science, including biological science, doesn't work that way. Don't take our word for it:

> *"If you thought that science was certain, well, that is just an error on your part."*
> – RICHARD FEYNMAN, PHYSICIST

Modern science has observed that small exposures to cellular threats will stimulate the immune system, whether those threats are biological or radiological. Indeed, the vast majority of immune system stimulation comes from these external threats: viruses, bacteria, injury, food, drink, air pollution, vaccines, radiation, accidents, the list goes on.

We can now measure the smallest doses of radiation, their exact type, and the precise rate of delivery in real time. However, measuring the *biological effect* that this radiation may—or may not—have on living cells (the thing that actually concerns us) is another challenge entirely, because whatever radiological effects that do occur would not be happening in isolation.

The real world in which we live and breathe is replete with cancerous hazards. On the microscopic level, this generates an ongoing and continuous "background noise" of cell damage and repair—to the tune of 10,000 repairs per cell, per day. Also keep in mind that the human body is about 60% water, and water slows down zoomies, reducing their effect. This means the mass of the body itself is a form of protection, along with our skin. So whatever radiative effects we seek to measure would have to be relatively large, to get past our natural defenses and not be undone by the continual busyness of cell repair.

Cells are awfully crowded places, and all that stuff in there is constantly wiggling around. Further complicating our search for the elusive low-dose zoomie event is the fact that any effects—if they occur, and if we can find them—will almost certainly not be in direct proportion to the applied dose.

With enough study, patterns may emerge in the mosaic of information, the same as in any other field of sustained and careful obser-

Inside A Cell (artist's conception)
Credit: Shutterstock

vation. For example, the weight of evidence now supports the AGW hypothesis (anthropogenic global warming) to a greater than 99% certainty. [35] In much the same way, the weight of evidence clearly supports a 100 mSv /year threshold. Even so, they both have their share of principled detractors, as well they should. Science wouldn't be science without disagreement and debate.

With either issue, it is the consensus of evidence we are referring to, and not the consensus of opinion. One is science, the other is confirmation bias.

You would think the knowledge gained from a century of radiology and nuclear medicine would have closed the case by now on the issue of low-dose radiation in nuclear power. Since zoomies are zoomies, our confidence in nuclear power, after all this time, should be similar to our confidence in nuclear medicine—or in commercial flight, for that matter. It can be scary if you start dwelling on all the what-ifs about flying, but science, statistics, and objective reality assure us that air travel isn't dangerous. Slightly risky, yes, but it's not dangerous. And yet, with just a bit of a mind flip, you can give yourself the heebie-jeebies about flying.

History has shown that with our active imaginations, we humans are uniquely adept at freaking ourselves out. Because this is so, the utter inability of biological science and medicine to prove a negative has given the antis a certain demagogic advantage—they can always play the no-safe-dose trump card. One tactical success they've achieved is that by invoking LNT, Cumulative Dose, and the Precautionary Principle, nuclear power is currently being regulated through a safety framework called ALARA—As Low as Reasonably Achievable. [36]

Sounds reasonable, but . . .

CHAPTER FIFTEEN

That Depends on What You Mean by "Reasonable"

THE CDC (CENTERS FOR DISEASE CONTROL AND PREVENTION) tells us that "as low as reasonably achievable" is the official US guiding principle on radiation safety. This is in spite of the fact that the entire rationale behind ALARA comes down to a sci-fi double feature from the 1950s—a horror show that convinced the world there is no safe dose for any cell of the body, and that all doses are cumulative.

> "This principle [ALARA] means that even if it is a small dose, if receiving that dose has no direct benefit, you should try to avoid it. To do this, you can use three basic protective measures in radiation safety: time, distance, and shielding." [1]

The beneficial doses the CDC is referring to are those administered in nuclear medicine, and in that context the guideline seems reasonable enough. At the same time, it's not reasonable to establish elaborate and expensive procedures to ensure that nuclear workers stay below a jobsite threshold of 50 mSv per year, a dose that is well below any threshold of concern. [2] De-ratcheting that standard to 100 mSv /year would save billions of dollars, all by itself.

It's also not reasonable for the EPA to recommend that tritium levels in potable water stay below a 740 Bq per liter limit. [3] Especially when Brookhaven National Lab has determined that the No Observable Effect Level (NOEL) for tritium in potable water is *37 million* Bqs per liter. [4]

It's curious that the EPA standard is exactly 50,000 times more stringent than the radiation limit set by one of our premier federal science labs (37,000,000 ÷ 740 = 50,000). Especially curious, since federal agencies like the EPA that set the rules and regs for public safety are obliged to base their decisions on the science coming from, or verified by, our federal labs, and not on popular misconceptions.

Neither is it reasonable to insist that public exposure to man-made radiation stays below 1 mSv per year, like the long-term soil requirement in Fukushima. [5] Particularly when so many other inhabited places around the world have much higher background doses and appear to have normal rates of cancer. As we mentioned, the US Department of Energy's regulations are even more overwrought, with a cleanup criterion for decommissioned nuclear power plants of 250 microSieverts per year, or 0.25 milliSieverts. [6]

That is one-quarter of one milliSievert per year, which is less than 10% of the average US background dose of 3 mSv /year—a preposterous standard. Indeed, background doses in the US can range from the mid-2s to the low-7s, depending where you live. It makes no sense to set a national standard for decommissioning reactors at one quarter of one milliSievert (0.25 mSv) above background radiation, when the background dose in the NRC's sphere of jurisdiction varies by nearly *twenty times* that amount. The only possible "benefits" from such an unduly stringent regulation is that it scares people, slows the buildout of nuclear power, and raises the cost of compliance.

And yet, at the same time, the NRC has the good sense to assure us that if we drank water for an entire year from a well contaminated by what they term a "significant" tritium leak from a nuclear plant, it would still give us far less internal radiation than the potassium stored in our own bodies. [7] Don't these guys compare notes?

As a side note, tritium is a total lightweight in the realm of radioactive substances. With a short half-life of just 12.3 years, this faintly radioactive isotope of hydrogen has a rapid decay rate. This of course makes it easy to cite sky-high becquerel numbers (decays per second) and scare the pants off of uninformed people. You'd think the people who wrote the decommissioning standards would have consulted with the people who wrote the tritium standards, and that both teams would have checked with our national labs. But apparently not.

Our point in all of this is that the EPA and the NRC are regulatory bodies, not science labs. This being the case, the regulations they devise are supposed to be based on solid science from our national labs. Unfortunately, their regulatory standards don't seem to spring from any unified body of up-to-date knowledge.

Things aren't much better in Europe. The European ALARA Network published their "Optimization of Radiation Protection" guidebook, prompting major changes in a French smelting plant that works with low-level radioactive waste. After all the expense and bother, the exposure level of their workers dropped from an already super-safe 7 mSv /year to an even more perfect 5 mSv /year. [8]

It's not reasonable to alarm the public over something as benign as a faint trace of Fukushima iodine-129 in British Columbia's groundwater. That particular whisper of meltdown radiation is literally *millions* of times less radioactive than our internal dietary supply of potassium. (See our supplement on Iodine-129 from Fukushima: [9]) Alarms over such minuscule doses are especially unreasonable, given the fact that 2,000 residents in the Ramsar region of Iran live with a background dose of about 150 mSv /year, while enjoying a slightly *lower* incidence of cancer than the general Iranian population. [10]

The 500-kilometer stretch of beach we mentioned in the Guarapari region of Brazil has a radiative output of about 33.5 μSv (microSieverts) per hour, which totals 300 mSv/yr for year-round residents. Tourists come to bury themselves in the sand for the therapeutic effects. They haven't been dropping like flies, and neither have the locals. [11]

Some areas of the heavily-populated Indian state of Kerala have background levels up to 70 mSv per year. [12] A dose-rate like this would trigger a compulsory evacuation notice in Japan. And yet, for the last 5,000 years, Keralans have grown most of their own food in the region's radioactive soil—including bananas and avocados. As a matter of fact, a Keralan's diet has ten times the radiation of food deemed fit to eat in the UK. [13] And yet, a Keralan's life expectancy at birth is 74.9 years, [14] the highest in India, and their cancer rates are normal. [15] (Here's a wild idea: Maybe Japan can send them some free topsoil! ☺)

It is not reasonable to permanently close an entire nuclear plant, with a pair of perfectly good one-gigawatt reactors, over a set of flawed steam generators that improperly resonated at 100% power. The San Onofre nuclear generating station (SONGS) in Southern California could have been safely run at 70% power until its less-than-perfect generators were replaced, as the vendor had offered to do. [16]

But in the aftermath of Fukushima, a fresh case of nuclear fear was ginned up along the US west coast, and San Onofre was shuttered, even though Fukushima, Onagawa, and every other reactor in Japan showed that a 9.0 earthquake will not damage a reactor.

For all the squabbling about steam leaks and tsunamis, a regulatory tangle doomed the plant as much as anything else. If the state had its priorities straight, it could have taken the plant into receivership as a vital part of the infrastructure, and made sure the steam generators were replaced. In the meantime, the 2-gigawatt plant could have operated at 70% power and produced a steady 1.4 GWs, which is nothing to sneeze at. But up and down the coast, sun-bronzed noses were raised at the very idea. So Sacramento stood by as San Diego Gas and Electric shut down the nuclear equivalent of Hoover Dam operating at full capacity, in favor of renewables backed by natural gas. Which are actually gas plants augmented with renewables.

Speaking of Hoover Dam, a long-standing concern of Southern California has been the potential for chronic water shortages brought on by climate change. In response, a methane-powered desalination plant has been built in Carlsbad, a few miles down the coast from San Onofre. Most Californians don't know that the SONGS reactors could have powered *forty* of the same-sized desalination plants—with none of the CO_2 emissions and natural gas leaks. [17]

If California and Germany had kept their reactors operating in the wake of Fukushima, and built more reactors instead of more renewables, they could have been well on their way to 100% clean power by 2018. Instead, after nearly a decade, Germany's *Energiewende* program of transitioning to an all-renewables grid has been a bust. With $600 billion projected to be spent between 2011 and 2025, their carbon footprint as of this writing (2023) has gone down by a measly 4.6%. [18]

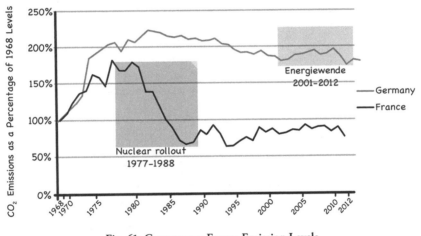

Fig. 61: Germany vs France Emission Levels
Source: https://me.me/i/250-germany-vs-france-co2
-emissions-as-a-percentage-of-4149482

Germany could have fully decarbonized their entire national grid with existing nuclear technology for about $340 billion (we break down the numbers here: [19]). That's a little more than half of what they've already spent on wind, solar, and "biomass"—the polite term for cutting down forests and burning wood pellets. [20] The graph below shows that solar and biomass are about equal in Germany's renewable energy portfolio.

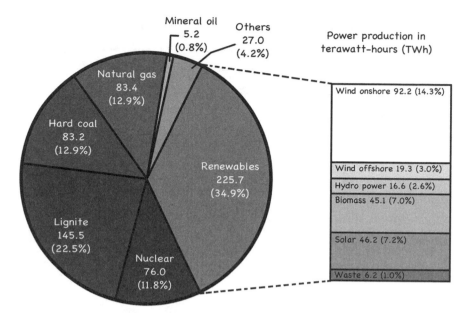

Fig. 62: Energy sources in Germany Power Production
*Source: https://www.nextbigfuture.com/2019/11/france-spent-less-on-nuclear
-to-get-about-double-what-germany-gets-from-renewables.html*

It is also unreasonable to have called for the permanent shut-down of the 620-MW Vermont Yankee nuclear plant because of a minor tritiated-water leak from a cracked discharge pipe. [21] This provided a moment of unintended comedy during the tritium pro-tests at Vermont Yankee. The antis refreshed themselves by eat-ing bananas (free-range and vegan, no doubt), blissfully unaware that every banana contains a radioactive micro-aggression of potassium-40. [22]

In 2010, a cracked discharge pipe at the plant had developed a water leak, with a radiative count peaking at 2,500,000 picoCuries per liter. [23] This seemingly large number alarmed the locals, trig-gering an outbreak of nuclear fear. While the Vermont Yankee plant wasn't closed because of the leak (which was eventually fixed), the tritium issue went a long way toward hardening public opposition to the plant. Premature closure was starting to sound like a reasonable idea, and four years later the activists got their way. But what these concerned citizens didn't bother to tell the locals was that the Vermont Yankee water was completely harmless. An adult would have had to drink 9.7 fluid ounces (about 0.27 liters) of this water every day of their lives to match the internal radiation they already get from their dietary potassium.

We've already seen that Fukushima tank water is harmless, even prior to dilution; the Vermont Yankee water was 11.5x less radioactive than that. Here's the math:

Fuku tank water has 28,789,649 picoCuries per liter (see the drop-down menu in the top left window of the Atomic Calculator [24]) and Vermont Yankee's tritiated water was 2,500,000 picoCuries per liter. And 28,789,649 ÷ 2,500,000 = 11.5. (So there.)

Assuming there is risk, even from small doses, is prudent. Assuming that all radiation is harmful is entirely unreasonable. Doing the latter has inspired a long history of misguided efforts, like shutting down a perfectly good reactor because it wasn't quite perfect enough.

CHAPTER SIXTEEN

The Forbidden Zone

Lᴏᴏᴋ ᴛʜᴇ ᴏᴠᴇʀᴡʀᴏᴜɢʜᴛ ʀᴇᴀᴄᴛɪᴏɴ to Vermont Yankee's tritium leak, it has been equally unreasonable to keep people out of the Fukushima and Chernobyl exclusion zones.

Why? Because on a two-hour flight from, say, Switzerland to Kyiv, you would earn more frequent-flyer zoomies than taking a two-day tour of the accident site. [1]

Russian troops occupied Chernobyl when they invaded Ukraine, and though they churned up a lot of dust in the process, local radio-activity levels remain the same. When the war is over and the facility inspected, tours will probably resume.

The average background dose at Chernobyl is down to about 26 mSv per year, [2] and any hot spots in the Exclusion Zone have long been identified and cordoned off for public safety. This is easy to do when even the faintest bit of radiation can be detected.

In April 1986, Chernobyl experienced the worst disaster that any reactor could have. After a major meltdown, a steam explosion blew the reactor vessel apart, and a hydrogen explosion blew the roof off the building. Then the graphite blocks in the core caught on fire and burned for ten days—flammable graphite, rather than pressurized water, was used to moderate the reactor's neutron flux. To top it off, there was no containment dome housing the reactor, a design decision as fantastically stupid as building a car without brakes. These are just a few of the reasons why no one will ever build a reactor like Chernobyl again—an unmitigated disaster that schooled the entire industry.

With the reactor split open and the roof blown away, tonnes of burning graphite sent radioactive particles downwind for thousands of miles. In the immediate aftermath, twenty-eight first responders died of radiation exposure from fighting the fire. Twenty survivors eventually died over the next eighteen years, though it should be noted that none of these deaths have been specifically linked to radiation. In addition, nearly two out of three of the first responders who received doses above 800 mSv were still alive and healthy in 2006. [3]

Fifteen downwind children also succumbed to thyroid cancer. [4] They would have been protected if Soviet authorities had promptly admitted that a catastrophe occurred on their watch, and dispensed iodine pills to satisfy the body's need for the element, which is greater in a growing child. This would have prevented the dietary uptake of radioactive iodine from the environment, which can affect the thyroid in concentrated amounts. With a flammable, uncontained reactor, it was the least they could have done for their citizens.

The Chernobyl accident was a significant blow to the Soviet regime's legitimacy, and the world's confidence in nuclear power. Even so, the Chernobyl area has since become a thriving wildlife habitat—the largest wild animal refuge in Europe. [5] Human abandonment of the region has enabled a thirty-five-year ecological study of low-dose radiation on the local flora and fauna, along with some off-grid humans as well: Soon after the disaster, several thousand locals ignored Soviet authorities and moved back into the zone.

Since then, they have lived full, healthy lives, growing most of their food in the local soil. About one hundred of the original *babushkas* (elderly women) are still kicking, and seem to have normal life expectancies. [6] Those who remain are doing well, and so are the plants and animals. While some deformities have been detected in the local flora and fauna, they are well within the range of expected natural mutations.

The local mushrooms do have trace amounts of Chernobyl isotopes, but there is no evidence of harmful effects from eating them. To put this in perspective: The average European consumes about 5 kilograms of mushrooms per year, which in Yankee math is about one pound of mushrooms per month. If Europeans consumed only Chernobyl-area mushrooms, it would add about 0.12 extra mSv per year to their yearly average background dose—an increase of 4%. [7]

In 2006, the World Health Organization (WHO) wrongly estimated that there would be about 4,000 future statistical deaths from Chernobyl. This LNT-inspired number has stuck in the public's mind ever since the accident. Here's an excerpt from their infamous document. (Note that as of 2006, only nine children had died.):

"The total number of deaths already attributable to Chernobyl or expected in the future over the lifetime of emergency workers and local residents in the most contaminated areas is estimated to be about 4,000. This includes some 50 emergency workers who died of acute radiation syndrome and nine children who died of thyroid cancer, and an *estimated total of 3,940 deaths* from radiation-induced cancer and leukemia among the 200,000 emergency workers from 1986-1987, 116,000 evacuees and 270,000 residents of the most contaminated areas. *These three major cohorts were subjected to higher doses of radiation amongst all the people exposed to Chernobyl radiation.*" [8] [emphasis added]

Let's dig into this statement: The estimate of 4,000 deaths, minus those who have already died, is 3,940. About one-third of those (1,300) are supposed to occur among the 200,000 emergency workers ("liquidators") who cleaned up the mess. They accounted for one-third of the people who comprise the "three major cohorts

[that] were subjected to higher doses of radiation amongst all people exposed to Chernobyl radiation."

So it stands to reason that if the estimate of 4,000 subsequent deaths was even vaguely correct, some substantial number of radiation-induced maladies or mortalities would surely manifest in the liquidator cohort, in the neighborhood of about 1,300 mortalities. They were the first responders, not the residents or evacuees. So if any of the three cohorts was going to suffer from radiation effects, it would likely be the liquidators.

But that's not what happened—in fact, *quite the opposite.* In 2010, four years after the WHO released the document quoted above, the journal *Dose-Response* published a sweeping analysis of the literature, [9] undertaken by respected radiation researcher Zbigniew Jaworowski. Contrary to the WHO's 2006 statement, he found that the liquidators were significantly healthier than the other two cohorts:

"In comparison to the Russian general population, a 15% to 30% *lower mortality* from solid tumors [was found] among the Russian Chernobyl emergency workers [the liquidators], and a 5% *lower average* of solid tumor incidence [was found] among the population of the Bryansk district, the most contaminated in Russia.

"In the most exposed group of these people (with an estimated average radiation dose of 40 mSv) a 17% *decrease* in the incidence of solid tumors of all kinds was found.

"In the Bryansk district the leukemia incidence is *not higher* than in the Russian general population. According to a report titled "UNSCEAR 2000b," [10] *no increase in birth defects, congenital malformations, stillbirth or premature births could be linked to radiation exposures caused by the Chernobyl fallout.*

"The final conclusion of UNSCEAR 2000b is that the population of the three main contaminated areas, with a cesium-137 deposition density greater than 37 kBq per m^2, *need not live in fear* of serious health consequences, and forecasts that *generally positive prospects* for the future health of most individuals should prevail." [emphasis added]

In plain language, actual medical evidence, rather than statistical projections based on bad science like LNT and Collective Dose, shows that the people everyone thought were the most likely to get cancer from Chernobyl—the liquidators who cleaned up the destroyed reactor—have actually been *healthier* than their comrades ever since the accident. This is the polar opposite of the LNT-based prediction of 1,300 deaths in the liquidator cohort.

But wait—there's less! Just like the *hibakusha* (bomb survivors) in Japan, it turns out that the Chernobyl liquidators have also failed to bequeath any radiation-damaged DNA to their children. According to Stephen Chanock at the U.S. National Cancer Institute:

"People who had very high-dose radiation [i.e., the Chernobyl liquidators] didn't have more mutations in the next generation. That's telling us that if there's any effect it's very, very subtle and very rare." [11]

This is yet another good-news story that the nuclear industry should be shouting from the rooftops, with fancy commercials like those splashy ads the fossil fuel industry routinely puts on TV. For the last thirty-five years and counting, the general public has been grossly misinformed about both Chernobyl *and* fossil fuel. And if you didn't know, well now you know.

16.1 GROUND CONTROL TO MAJOR TOM

Scientists have never administered low doses of radiation in experiments on primates. NASA considered experimenting on monkeys in 2010, but dropped the idea after harsh criticism. [12] Aside from the ethical considerations of animal testing, the public outcry was also a clear indication of widespread "radiophobia"—the fear of radiation. We say we want to go to Mars, but we won't test ourselves, or any other primate, with the radiation dose-rates that would be encountered on the voyage, or on Mars itself. This awkward situation stems from the conviction that all radiation is categorically harmful.

Nuclear fear being what it is, NASA will have to rely on "ecological studies" until public opinion changes. This means they will restrict themselves to passively observe the ambient radiological effects to human health here on Earth, along with the health of our current and former astronauts, and extrapolate as best they can.

The problem with this approach is that, as always, any possible effects from low doses are lost in the statistical background noise of other ongoing risks for cancer. These include things like smoking, drinking alcohol, eating red meat, fossil-fuel pollution, petrochemical and industrial toxins, repeated sunburn, and if you're in California, drinking coffee.

WARNING: Drinking coffee can expose you to acrylamide, which is known to the State of California to cause cancer or reproductive toxicity.

The California Coffee Experience
Source: https://oehha.ca.gov/proposition-65

Despite all evidence to the contrary, no safe dose continues to be a foundational belief in anti-nuclear circles, a fear card they don't hesitate to play. The activism, politics, and legislation driven by this discredited idea have had a tremendously corrosive effect on the public acceptance of nuclear power, and the efforts to green the grid by mid-century have suffered as a result—in spite of the fact that LNT has been contradicted, yet again, in 2018:

> "Epidemiological data provide essentially *no evidence for detrimental health effects below 100 mSv*, and several studies suggest beneficial (hormetic) effects. Equally significant, many studies with in vitro and in animal models demonstrate that *several mechanisms initiated by low-dose radiation have beneficial effects*." [13] [emphasis added]

What's striking is that while NAS (2006), UNSCEAR (2012), and NIH (2018) all acknowledge possible *risk* from low doses, none of them assumes *harm*. Taken together, their reports reveal a half-century of (slowly) evolving mainstream scientific thought and positioning on radiation health and safety, refuting the central thesis of the anti-nuclear power movement. [14] That's all well and good, but it's not good enough. Congress, the Department of Energy, and the Nuclear Regulatory Commission need to resolve the regulatory bottleneck inspired by Muller's bad science.

16.2 "TIME IS FLEETING." — FRANK-N-FURTER

Science, government, and industry have more than enough evidence to establish a sensible safety threshold for ionizing radiation based on modern science, and they don't need another twenty years to figure it out, as NASEM is now proposing. The problem is, our brain stems grow restless whenever a long-established safety threshold is relaxed. It can even happen with such mundane changes as lowering the backyard fence to get a better view. Even though you're still living in the same gated community, there's that thing in the back of your mind about how a burglar can get in easier now.

Supplied with the correct information, and fortified with a thorough debunking of bad science and misinformation, the public's view of nuclear power can evolve. We the authors, for example, were dismissive of the entire subject until we finally started reading up on it; many other nuclear advocates have told us the same thing.

In fact, according to a survey conducted just four years after Fukushima, a strong majority of Americans are actually in favor of nuclear power—see the graphs below. But after a decades-long barrage of scare stories and propaganda, that same majority also thinks that a majority of their fellow Americans are against it. As it turns out, fully two-thirds of Americans are "in favor of" or "strongly in favor of" nuclear. [15] And yet, they think that less than one-third of their neighbors feel the same way.

This is a classic example of closeted behavior, after a lifetime of normalized social opprobrium about all things nuclear.

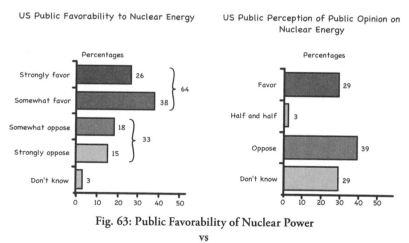

Fig. 63: Public Favorability of Nuclear Power
vs
The Public's Perception of Nuclear's Favorability (2015)
*Source: https://thebulletin.org/2016/04/public-opinion
-on-nuclear-energy-what-influences-it/*

Surveys also show that the more a person knows about nuclear technology, the more they support it. Even in a country like Japan, whose people have experienced the worst that nuclear has to offer. [16]

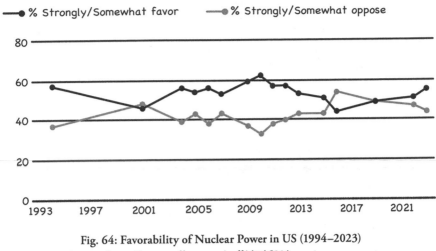

Fig. 64: Favorability of Nuclear Power in US (1994–2023)
Source: https://news.gallup.com/poll/474650/americans-support
-nuclear-energy-highest-decade.aspx

CHAPTER SEVENTEEN

What Are the Odds?

Your LIFETIME RISK OF DYING IN A PLANE CRASH is one in eleven million. [1] That means there is risk involved; it does not mean that flying is unsafe. This same distinction is fundamental to our understanding of nuclear power. With any type of energy system, as with anything else in life, risk comes with the territory. The ability to keep risk in perspective is what got us out of the cave and into our cars.

LNT and ALARA, along with a misapplication of the Precautionary Principle and the resulting squirrel cage of having to prove a negative, have combined to push the nuclear industry's already stringent standards to nearly impossible levels. Evidence shows that this was, and still remains, intentional. [2]

If there was no safe dose of radiation, nuclear medicine and radiology would not be possible, and we naked apes would routinely die of sunburn (the high-frequency UV portions of sunshine are ionizing). Granite would be categorized as a hazardous substance, and we would have to evacuate Denver and the halls of Congress (see next section). Bananas, avocados, and Brazil nuts would be banned, along with smoke alarms, X-rays, and CT scans—especially CT scans. They can be up to 30 mSv apiece!

Science and objective reality, along with a library of statistics, tell us that radiation concerns should be framed in terms of acceptable risk rather than absolute safety. This is yet another shortcoming of LNT, even when it's used to assess risk rather than harm. To be clear:

Any assessment of risk is meaningless
unless it is placed in perspective.

Below 100mSv/year, the health risk from ionizing radiation is so remote that the chance of harm verges on the "imposserous," as the Cowardly Lion put it. More Americans die *each year* from falling out of bed than have ever died from sixty-five years of US nuclear power. [3]

In fact, nuclear actually saves lives. In a paper co-authored by James Hansen, the world-renowned scientist at the forefront of global warming awareness and activism since the 1980s, Pushker Kharecha at Columbia University Earth Institute has estimated that between 1971 and 2009 nuclear power has saved 1.8 million lives that would have otherwise been lost to fossil-fuel pollution. Peter Lang of Australian National University has estimated that the rollback on nuclear power since its heyday in the 1960s and '70s has been responsible for about 18% of the world's excess carbon and about 9.5 million lost lives. [4] Ironically, we can thank the 1979 TMI accident for prompting policy revisions and inspection protocols that have increased reactor safety, while also raising the output of our nuclear plants from about 65% average capacity to more than 90%.

In any fair assessment of nuclear's benefits and hazards, its benefits—both the lives it saves and the clean power it generates—should be given due credit. Like any other energy-generating technology, nuclear power should be judged by risk/benefit calculations, and not prejudged by the absolutism of No Safe Dose or the Precautionary Principle.

Each year, more than 5,000 American men die from aspirin, [5] over 300 people die from drowning in the bath, [6] and some die playing high school football. [7] And it's worth repeating that about once a year, an American will actually be shot by their own dog. [8]

Maybe we should ban high school football, and maybe we should only use analgesics with no known lethal dose. But should we also ban bathtubs and dogs, and make people sleep on the floor? Even if it's a frosty three-dog night?

Junior
Credit: Mike Conley

17.1 "COME FLY WITH ME . . ." — FRANK SINATRA

Our first stop is Denver. Nestled against the granite slopes of the Rocky Mountains, the mile-high city of Denver, Colorado, has a background radiation level of about 7 mSv /yr, roughly double the global average and about triple that of Japan. Even though all that granite is radioactive, and even though residents have a mile less atmosphere above them to shield the cosmic rays, local cancer rates are lower than national norms. [9]

Working in the granite monoliths of Washington, DC, or New York City's Grand Central Station will give you a larger dose of radiation than working at a nuclear power plant. [10] Savannah, Georgia has so much thorium in its monazite beach sand that five loads of a pickup truck contain enough thorium to power the city for more than a day. (See our supplement on Savannah thorium: [11])

As we mentioned, populated regions of India, Iran, and Brazil register higher radiation levels than most portions of the Chernobyl and Fukushima exclusion zones. [12] This last factoid should come as a relief to the retired personnel of the other three Chernobyl reactors, which were untouched by the 1986 meltdown and fire in the reactor next door. A few weeks after the accident, the three remaining reactors were brought back online with a full contingent of personnel. They ran for several more years, the last one shut down in December 2000. [13]

The workers at these three nuclear plants could only put in two weeks at a time before they were required to take some days off, but not because of radiation. They needed to decompress from the psychological effects of working near the ghost town of Pripyat—overgrown and crumbling, and littered with everything its 50,000 citizens hastily left behind.

Abandoned Amusement Park in Pripyat
Credit: Shutterstock

17.2 "PARANOIA STRIKES DEEP." — STEPHEN STILLS

The adage is particularly true when facts are manipulated for propaganda's sake, and even more so when outright falsehoods are concocted for public deception. A case in point: This is a NOAA map (National Oceanic and Atmospheric Administration) of the maximum wave heights generated by the Fukushima earthquake:

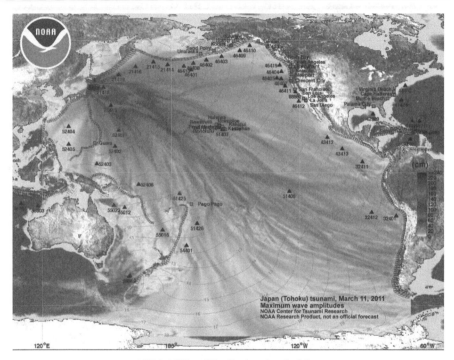

NOAA Wave Height Amplitude Map
Source: https://nctr.pmel.noaa.gov/honshu20110311/Energy_plot20110311_ok.jpg

Anti-nuclear groups used this image as much as five years after the accident to falsely depict the spread of ocean-borne radiation from Fukushima. In a bit of disaster voyeurism, the popular blog ZeroHedge used this headline: "Fukushima radiation has contaminated the entire Pacific Ocean, and it's going to get worse." [14] (*Zoomies! We're doomed!*) The scare was so effective that some health-conscious people on the

US west coast are still concerned about traces of Fukushima cesium in the local seawater, even though the beaches of Fukushima have (finally) been reopened to the public. [15]

Another whopper appeared in the form of a forged map falsely attributed to the Australian Radiation Services Company. [16] If the fallout depicted in the bogus ARS map (see below) had actually drifted to the US, it would have killed most of the lifeforms on the west coast. Despite a thorough debunking by the ARS, anti-nuclear activists (such as Helen Caldicott) were using the bogus map as late as September 2014, three-and-a-half years after Fukushima. [17]

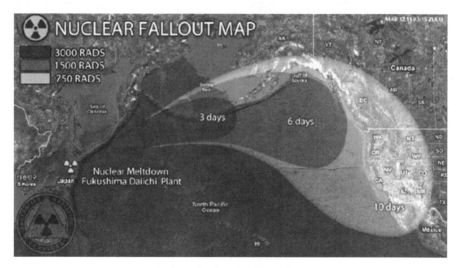

Fake Fukushima Fallout Map
Source: https://www.snopes.com/fact-check/nuclear-fallout-map/

If you listen to some of the fearmongers long enough, you'd think that any day now the surf at San Onofre and Diablo Canyon will start glowing in the dark, even though it's common knowledge (or should be) that just like anyplace else on Earth, the oceans themselves are already radioactive—mostly due to potassium-40, the same stuff we store in our bodies.

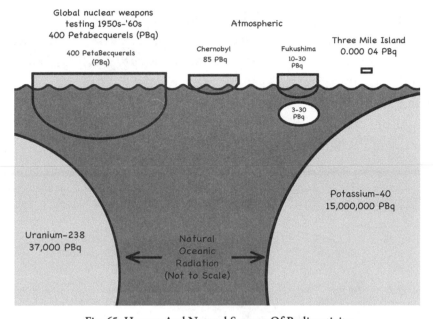

Fig. 65: Human And Natural Sources Of Radioactivity
Source: Source of Radioactivity in the Ocean – Woods Hole Oceanographic Institution
A portion of some radioactive releases have landed in the ocean. This sketch depicts the approximate proportions.

Yes, Fukushima radiation did leak into the ocean, and yes, some of it is still detectable in trace amounts—remember, we can detect activity down to a single becquerel. But oceanic levels of Fukushima radiation are far below any rational safety concerns, like about a half-million times below the safety threshold. To be perfectly precise:

Fukushima cesium has raised the becquerel count in one cubic meter of Southern California seawater from an average of 12,000 becquerels to a hair-raising 12,004 becquerels—a nearly undetectable 0.03% (we break down the numbers here: [18]). As we saw earlier, adults have a constant internal radioactivity of 4,000 to 5,000 Bqs from their dietary potassium. [19] And yet, we're supposed to be concerned about an extra four Bqs of Fukushima cesium in a cubic meter of seawater?

It is true that biotoxins can concentrate at the top of the food chain; the mercury in tuna (a forever toxin) is one unfortunate example. Even

Northern California coastal fog contains tiny amounts of mercury, traces of which have been detected in the local cattle. [20] Southern California mountain lions have elevated mercury levels as well. [21] But despite any scare stories to the contrary, the Fukushima-sourced cesium found in Pacific tuna is simply not a concern. The mercury might be, if you ate tuna every day, but not the cesium. In fact, a serving of Pacific tuna has less radiation than a banana.

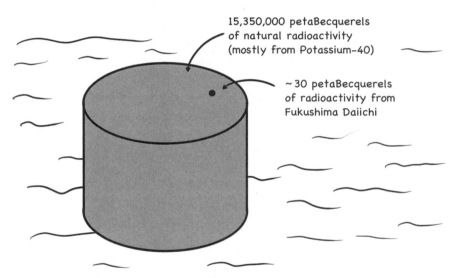

15,350,000 petaBecquerels of natural radioactivity (mostly from Potassium-40)

~30 petaBecquerels of radioactivity from Fukushima Daiichi

Fig. 66: Radioactivity in the World's Oceans (total volume)
Source: https://www.whoi.edu/multimedia/source-of-radioactivity-in-the-ocean/

To put the risk of "Fuku fish" in perspective: Suppose you ate coastal-caught Southern California fish every evening for an entire year. Further suppose that each day's catch had somehow magically ingested, and retained, every cesium atom in fifty cubic meters of seawater (the volume of a large bedroom.) Eating this tainted fish as a daily part of your diet would give you an internal radiation dose of just 0.016 mSv. (See our supplement on Fuku Fish: [22])

That's about five one-thousandths of the yearly global average background dose of 3 milliSieverts. Put another way, the potassium in our Western diets gives us a constant internal dose of about 0.43 mSv/yr. That's about thirty times more radiation than you'd get from a daily diet of magical SoCal fish (we break down the numbers here: [23]).

Bon appétit! And be sure to post a pic of your meal.

Cesium fallout from Chernobyl wound up in Germany's wild mushrooms, and wild boars like mushrooms. So now the boars in Germany are radioactive. Except you'd have to eat about 7 kilos (15 lbs) [24] of their meat in one sitting to equal the radiation you'd get on a flight from Southern California to Oktoberfest. [25]

17.3 THE BIRTH OF NUCLEAR FEAR

In August 1945, the weapons dropped on Hiroshima and Nagasaki killed more than 130,000 people. Some died from the blast itself, some from the intense heat, and others from radiation effects, but most victims died from the ensuing fires.

The Atomic Age had begun. The power that once existed only in science fiction had suddenly become all too real. It was a world-changing event, an existential threat incomparably greater than 9/11. In just three days, August 6 to August 9, 1945, the world became a starkly different place—the genie was out of the bottle.

But coexisting with nuclear fear was nuclear hope. One pound of low-enriched uranium fuel (LEU) has the equivalent energy of more than 100,000 pounds of coal or a million cubic feet of natural gas. (See our twin supplements "Energy Density of LEU" and "Energy Density of Natural Gas": [26]) The ability to harness such stupendous amounts of energy fired the imagination. The prospects of abundant clean energy propelled an expansive confidence that we could rebuild a devastated world into a modern utopia, powered by the peaceful atom. [27]

Meanwhile, something was happening on the far side of our post-war world. In 1949, the Soviet Union tested their first nuclear weapon. It was a troubling development, but it was all happening so far away, even farther away than Hiroshima and Nagasaki. And our own tests were being conducted way out in the Pacific Ocean, or in the desert north of Las Vegas. [28] The locals didn't panic, so why should we?

U.S. postage stamp, 1950s
*Source: https://voicesofdemocracy.umd.edu/teaching
-eisenhowers-atoms-for-peace-speech/*

As the Atomic Age progressed through the 1950s, its world-of-the-future vision shared the cultural stage with radiation-spawned monsters—*Tomorrowland vs. Godzilla*. And though radioactive fallout was known to be nasty stuff, it was all "over there." So long as it stayed over there, it didn't seem like much of a concern.

At the time, the average person's understanding of atomic weapons was formed by what happened in Japan. And as horrific as those detonations were, they were localized events—about five square miles of total destruction in each city, with fallout that drifted downwind for perhaps 100 miles.

By the mid-1950s, however, the sense of safety that came with distance began to disappear. The new H-bombs (hydrogen fusion bombs) being tested by the US and the Soviets were very different weapons. Far more powerful, their mushroom clouds lifted tonnes of radioactive dust and debris high into the stratosphere.

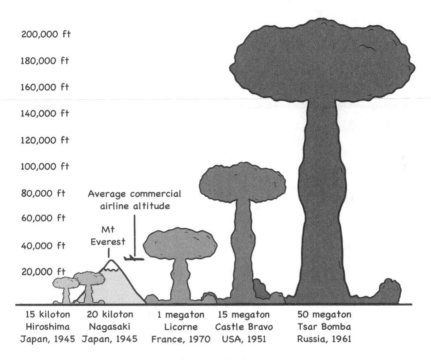

200,000 ft

180,000 ft

160,000 ft

140,000 ft

120,000 ft

100,000 ft

80,000 ft — Average commercial airline altitude

60,000 ft

40,000 ft — Mt Everest

20,000 ft

15 kiloton	20 kiloton	1 megaton	15 megaton	50 megaton
Hiroshima	Nagasaki	Licorne	Castle Bravo	Tsar Bomba
Japan, 1945	Japan, 1945	France, 1970	USA, 1951	Russia, 1961

Fig. 67: Mushroom Cloud Comparisons

With each new test, more fallout was drifting to every corner of the globe. The decade ended with the theatrical release of *On the Beach*, a post-World War III drama about an approaching cloud of lethal fallout. The movie inspired more than a few people to dedicate their lives to anti-nuclear activism, which to their way of thinking included the shutdown of commercial nuclear power.

By the end of the decade, our Atomic Age optimism was overshadowed by the harrowing reality of The Bomb. It loomed over everything and everyone as a nagging, inescapable threat, generating a free-floating anxiety in the public psyche that remains to this day. Because the fallout wasn't over there anymore. It was everywhere.

CHAPTER EIGHTEEN

Contain Yourself

WITH THE DISTURBING COLD WAR REALITY of widespread nuclear fallout, the world finally agreed to a ban on atmospheric weapons testing; the 1963 Nuclear Test Ban Treaty was a hopeful advance in international relations. But when the Three Mile Island power plant suffered a partial meltdown in 1979, anti-nuclear activists pushed the same nuclear fear button to bolster their case, and it worked.

Media stars and celebrities alike told the world how awful TMI was, even though the average dose received by anyone within ten miles of the plant was about as serious as a chest X-ray. [1] Only years later did the nuclear community confirm what they suspected all along, that a planned part of anti-nuclear strategy was to deliberately scare people with lies. [2]

Due to the supposed potential for catastrophic harm (*"No safe dose!"* / *"All doses are cumulative!"*), America was exhorted by media, politicians, and celebrities alike to shut down every reactor we had, and abandon the technology altogether. Because the next accident could endanger the entire world, rendering vast swaths of land uninhabitable for centuries on end . . .

Contrast this bit of fearmongering with a real, ongoing danger that everyone is familiar with: automobile safety. Like the thalidomide meltdown of the 1960s, the Pinto meltdown of the 1970s didn't scare people away from an entire industry. The US automobile mortality rate at the time was about 22,000 a year. Of those mortalities, exploding Pinto gas tanks had killed perhaps 100 people. [3]

That's 100 more casualties than TMI and Fukushima *combined*, and about twice the confirmed casualties from Chernobyl.

No doubt about it, Fukushima was an expensive, disruptive mess. But there is also no doubt that it was made far more awful, and far more expensive, by the Japanese government's overreaction—which, as we mentioned in Chapter One, was inspired by our own Gregory Jaczko, the anti-nuclear former head of the US Nuclear Regulatory Commission.

18.1 THINKING IT THROUGH

When we consider the public's reaction to Three Mile Island, Chernobyl, and Fukushima, it becomes clear that the actual concern in many people's minds is not the toxicity of radioactive material *per se*, so much as the material's potential to spread.

As long as radioactive material stays inside a reactor, a medical device, or a smoke alarm, and does its job without getting out and threatening anyone, either now or in the future, it's a non-issue. The same holds true for any hazardous substance, radioactive or not. So long as used nuclear fuel is given a proper burial, or stored for eventual burnup in a fleet of fast reactors, it's no big deal, either.

Given the stringent storage, transport, and recycling standards of the nuclear industry (see next section), the small, tightly managed waste stream of nuclear power is not a reasonable disincentive, much less the dealbreaker it's become for some people. With rigorous but sensible standards set for nuclear power and radiation safety, we as a nation and we as a planet would actually have a good chance of making our carbon reduction targets. The place to start is by publicly and thoroughly debunking LNT.

18.2 BUT WHAT ABOUT THE WASTE?

We're glad you asked. First off, it's not waste unless we waste it. Secondly, it's not the green goo like you see in *The Simpsons*. Spent nuclear fuel (SNF) consists of hundreds of solid uranium oxide pellets stacked inside a long zirconium tube called a fuel rod (see image below). Dozens of these rods are packed together in a framework called a fuel assembly. (They use zirconium because it's "invisible" to radiation.)

When a used assembly is removed from a reactor, it looks much like it did when it first went in, except for a bit of pellet expansion (Fig. 79). This is due to the buildup of gaseous fission products in the fuel pellets, such as xenon and krypton. As we saw, one of the many advantages of a molten-salt reactor is that these neutron-capturing gases simply bubble out of liquid fuel, and are sequestered until they decay. (Not to worry: Xenon and krypton are chemically inert and biologically inactive, so if anyone happens to breathe it in, they'll just exhale it.)

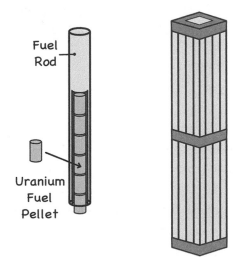

Fuel
Rod

Uranium
Fuel
Pellet

Fuel assembly

Fuel assemblies are typically 14 feet long and contain about 200 fuel rods for PWRs and 80–100 fuel rods for BWRs.

Fig. 68: Fuel Pellet, Rod, And Assembly

Source: https://www.nrc.gov/reading-rm/doc-collections/infographics/fuel-assembly.png

Every twenty or so months, an LWR is shut down and the fuel assemblies are shuffled around, much like logs in a fire, with the oldest ones swapped out for fresh assemblies. After a five-year run in the reactor, used assemblies spend the next five years chilling out in the plant's spent fuel pool.

Spent Fuel Pool
Credit: US Nuclear Regulatory Commission (CA-BY-SA-2.0)

This "spent" fuel is then placed in massive steel-and-concrete dry casks for on-site storage, [4] until it can be moved to its final repose in a DGR (deep geologic repository; see next section), or to a new life at a fuel recycling facility, and from there to full burnup in a fast reactor. Once a fast reactor is through composting the stuff, it will revert to background radiation levels in about 300 years. Our used fuel decay graph is worth a second look:

If this "recycling ver-sus long-term storage" issue isn't resolved by the end of the century, the much-less-radioactive assemblies can be trans-ferred into new casks, giv-ing our intrepid leaders another century to ponder this perplexing problem.

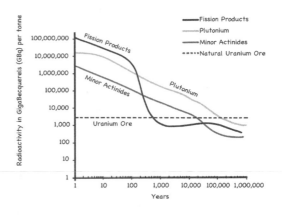

But whether we re-use spent fuel or bury it, its volume is tiny considering the enormous amount of clean energy that comes from it. The picture below shows twenty-eight years of used fuel from the 620-MW Connecticut Yankee nuclear power plant, after generating more than 110 billion kWh of carbon-free energy with a single light-water reactor. That's enough for 350,000 households, or 1,300,000 people, for 28 years.

28 Years Of Spent Fuel, Connecticut Yankee Power Plant
The casks are about 6 meters tall and 2.5 meters in diameter. The pad is 1,480 m²
Source: http://www.connyankee.com/assets/pdfs/Connecticut%20Yankee.pdf

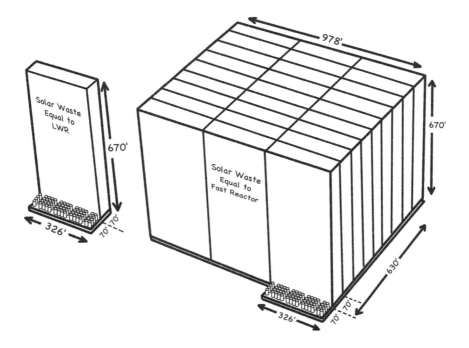

Fig. 69: Solar Panel Waste vs Reactor Waste
Credit: By the authors

For comparison, solar panels that would produce the same 110 billion kWh over the course of their service life would form a 204-meter (670 ft.) high stack of waste. If the used fuel in these casks was fed to a fast reactor for complete burnup, the panels needed to generate the same total energy would form 27 stacks of panel waste—about the volume of twelve Empire State Buildings. If each panel was crushed flat, our stack of used panels would only shrink by 20% to 160 meters, or 530 feet.

It gets worse: The Connecticut Yankee plant only occupied about ten acres (that's the good part). The bad part is that a solar farm producing the same power would need 4,000 acres—about forty times the real estate. (See our supplement on used panels vs. used fuel: [5]) This is what energy density is all about: A pile of solar waste nearly as tall as Hoover Dam, compared to one small storage pad with less than fifty dry casks.

18.3 DEEP GEOLOGIC REPOSITORIES (DGRs)

If used fuel is slated for long-term storage, the assemblies are craned out of their dry casks and placed in transport casks for safe hauling over public roads. Transport casks are incredibly tough (more cool videos: [6]), and the security teams guarding the convoys are on par with US Special Forces.

There are several designs for DGRs, but most are variations on the same concept: a deep shaft dug into stable rock, far below the water table, with several horizontal storage tunnels. This is *deep* rock, where flowing water hasn't been for literally millions of years, and where none is expected for millions of years to come (more on this below).

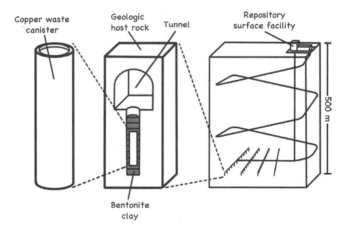

Fig. 70: Used Fuel to Deep Storage
Source: https://s2.docgo.net/uploads/DP2017/03/06/9JjszuG8XT/cf65e049ee9374c2cd
8d8dE1c83b91e4.pdf

When the fuel assemblies arrive at a repository, they're transferred into burial casks and taken deep underground. The casks are placed in tunnels, and in some DGR designs the casks are also packed in tightly with bentonite clay. A plentiful and naturally-occurring substance, bentonite is favored because it expands when wet, sealing the tunnel should moisture somehow intrude in the far-flung future. (The bentonite must first be "calcined," or baked, at 400° C, to drive off any existing water and oxygen. This is cheap and easy to do.)

While some water may *seep* below the water table, it doesn't *flow*. Instead, it undergoes an excruciatingly slow process of migration through solid rock, something geologists call hydraulic "conductivity" (electrical engineers want their term back). During this ultra-slow process, the bentonite in a compromised repository may take thousands of years to become fully saturated. And once it was, it would hermetically seal the placement rooms and access tunnels.

A major Canadian study has produced a detailed 3-D model of the deep rock below their Atikokan Research Area in Ontario. The study concludes that if a DGR were excavated below the local water table to a depth of 730 meters, any migrating water that came in contact with a compromised burial cask, and somehow made it through the bentonite clay after that, would then need more than 500,000 years to reach the surface. Which means that long before it reaches ground level, any radioactive material carried by this water would have already decayed to less than natural background levels. [7]

The solid Reference Case line in the graph below (the jagged line rising from 1.E-11) is the conservative model in this spread of data points. The square data points are from the Whiteshell Research Area (WRA) in Manitoba. The data show that hydraulic conductivity below Whiteshell would be 100 times slower than in the rock below Atikokan. This is indicated by the line at 1.E-13 versus the line at 1.E-11.

And just how slow is slow, exactly? Deeper than 700 meters below ground level, the hydraulic conductivity along the 1.E-11 Reference Case line amounts to one meter of migration every 3,170 years. (See our supplement on Hydraulic Conductivity: [8])

NERD NOTE: The capital E is the math shorthand for "exponent." A negative exponent denotes a number smaller than one. So the shorthand "1.E-11 m/s" means 0.000 000 000 01 meters per second, a tiny fraction of a number with ten zeros and a numeral one, for a total of eleven digits to the right of the decimal point, hence the minus eleven. A positive exponent would represent a number greater than one, with digits to the left of the decimal point.

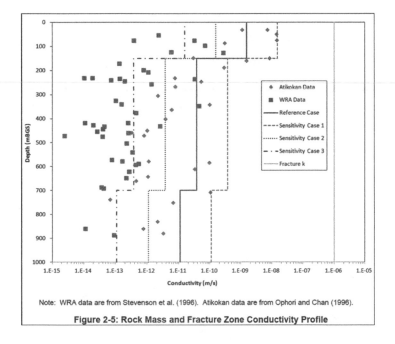

Note: WRA data are from Stevenson et al. (1996). Atikokan data are from Ophori and Chan (1996).

Figure 2-5: Rock Mass and Fracture Zone Conductivity Profile

Fig. 71: Hydraulic Conductivity Estimates for Two DGR Sites in Ontario
Source: https://www.nwmo.ca/-/media/Site/Reports/2018/01/19/15/54/
NWMO-TR_2017_02_Sixth-Case-Study-Report.ashx (frame 81)

The shorthand "1.E-11 m/s" means an infinitesimally small fraction of a meter—namely, one one-hundredth of a nanometer (a billionth of a meter) of hydraulic conductivity (water migration) each second. Another way of looking at it:

At the speed of 1.E-11 meters per second, it takes ten seconds to travel *the width of a hydrogen atom*. Another way of looking at it:

There are 31.55 million seconds in a year: (365.25 days × 24 hr per day × 3600 sec /hr = 31.55 million sec /yr). So 1.E-11 meters per second (0.000 000 000 01 m/s) would be a hydraulic conductivity rate of approximately one meter about every 3,200 years: (1 second /1E-11 meter × 1 yr /31.55E6 sec = 3,170 yr /m).

It gets better: The graph below shows that even if all the burial casks (containers) in this hypothetical repository failed after just 60,000 years, the radioactivity of the exposed fuel would already be less than the average background dose-rates encountered in nature. And, the material would still need another half-million years to migrate up to the biosphere, releasing the occasional zoomie as it continued to lose whatever bits of radioactivity it may have left.

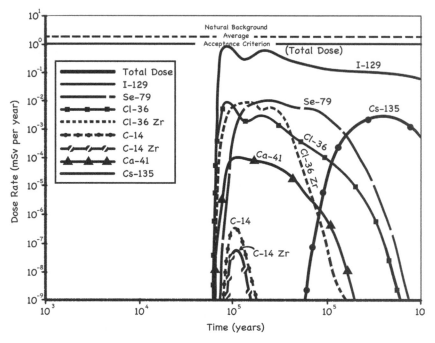

Radionuclides If All Containers Fail at 60,000 Years.

Fig. 72: Cask Failure at 60,000 Years
Source: https://www.nwmo.ca/~/media/Site/Reports/2018/01/19/15/54/
NWMO-TR_2017_02_Sixth-Case-Study-Report.ashx (frame 608)

Once the decision is made that the used fuel stored in a DGR will not be retrieved for further use, the repository shafts can be filled with concrete aggregate, at which point the repository becomes a depository. As you can see, there is such a wide margin of safety in DGR design that even if every cask fails as soon as it's buried, and even if the repository immediately floods with water, the material would *still* need a half-million years to migrate to the surface. So let's all relax about buried radioactive "waste," shall we? When the subject is examined dispassionately, the concern about a properly built repository, dug into the right rock, is seen for what it is—yet another example of nuclear fearmongering.

Engineers are also exploring the possibility of borehole repositories / depositories. The borehole method would use the horizontal drilling techniques developed for oil and gas (see the sketch below). Individual fuel assemblies would be placed in their own slender casks and lowered into the boreholes, far below the water table into ancient, stable rock (video: [9]).

If the casks were fitted with a retrieval mechanism, the depository would be a repository. This will come in handy in the years ahead: Gen-IV reactor companies will be wanting this "waste" to convert or process into more fuel, so a repository / retrieval system would be much preferred over a simple depository. Even though it would be a bit more expensive up front, it would be more beneficial in the long run than throwing away thousands of tonnes of gently-used fuel.

Not surprisingly, people have concerns, the main one being: "But what if lots of water somehow finds its way deep down below the water table, where it hasn't been for millions of years, and where geologists say it won't be for at least another million years, but let's say it gets down there anyway

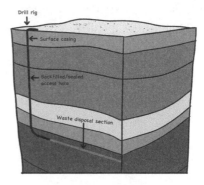

Fig. 73: Borehole Schematic
Source: https://www.world-nuclear-news. org/Articles/Estonias-geology-suitable-for-deep-borehole-reposi

and corrodes the casks, and carries all that nasty stuff back up to the surface? *Then* what?"

That's why all the bentonite clay. Because quite aside from taking a half-million years for hydraulic conductivity to bring anything up to the surface (cracks in the host rock are factored into the migration estimates), the first thing that any actinide would encounter coming out of a failed burial cask is bentonite clay, which serves as more than a packing material—it also chemically "sorbs" radioactive metal oxides. In layman's terms, bentonite is flypaper for nuclear waste.

Besides which, sintered fuel pellets are a ceramic blend of solid oxides that is virtually insoluble in water. [10] And, what little bits that do happen to dissolve wouldn't get very far, due to the super-slow rate of hydraulic conductivity in the surrounding rock. Long before they even reach the rock walls of the tunnel, these heavy metals would stick to the bentonite packed around the cask from which they were leached. In addition, bentonite sorbs metallic particles even better when it's wet, which it would be if the tunnel was compromised. Fission products like cesium oxide, strontium oxide, and other compounds would stick as well.

So let's say that after just 1,000 years, a burial cask in a deep geologic repository somehow failed. How bad would it get? In 2009, Finland's Radiation Safety Authority analyzed what could go wrong with their new deep geologic repository at Onkalo. [11] Here's an infographic on the study from Janne Korhonen, a noted energy analyst in Finland:

What happens if nuclear waste repository leaks?

According to analysis of the deep geologic repository in Finland, assuming that:

- Nuclear waste canisters start leaking after a mere 1,000 years
- A city is built upon the repository site by people who...
- Eat *only* food produced locally and...
- Drink *only* water from local sources and...
- Spend *all* their time (24/7/365) in the most contaminated spot...

0.00018 mSv =

two bananas
(= 0.00009 mSv each)

...it's possible that *one* person living in AD 12,000 *might* be able to receive the highest single dose: **0.00018 mSv per year.**

Fig. 74: When DGRs Go Bad: A Banana Equivalent Dose (BED) of 2
*Source: https://jmkorhonen.net/2017/03/10/what-does
-research-say-about-the-safety-of-nuclear-power/*

18.4 MOTHER NATURE'S OPEN-AIR REACTORS

The sorption ("sorbing") of radioactive isotopes in rock and soil has not only been demonstrated in the laboratory, [12] but Mother Nature has provided a real-life demonstration as well. Remember that way, way back in the day, natural uranium had a much higher concentration of U-235—about two billion years ago, it was over 5%. That's higher than what is now used for most reactor fuel, which is typically enriched to around 4%. This is why some two billion years ago, under the right conditions, natural reactors would form by pure chance and the gathered uranium would spontaneously fission. Amazing, but true.

A dissolved form of uranium-235 was consumed by bacteria, converting it to an insoluble uranium salt, which was then gathered as silt by underground streams. [13] The water slowed down the zooming neutrons that were spontaneously ejected by the unstable U-235, which enabled the material to fission. In fact, this interplay between soluble and insoluble uranium salts inspired an aqueous reactor design in the early 1950s, using a soluble uranium fuel. [14]

While one form of uranium is soluble (see endnote 13), yellow-cake powder, a complex mix of UO_2 and U_3O_8, is only sparingly soluble, and is not soluble at all once it's been sintered into fuel pellets. Think of pottery clay, which is somewhat soluble, versus a kiln-fired coffee cup, which is not. When dry yellowcake powder is sintered with high heat and pressure, it becomes a dense, insoluble ceramic that will remain that way for centuries, if not virtually forever.

So anyway, about two billion years ago, in the region of what is now the country of Gabon in west Africa, underground streams concentrated enough elemental uranium (with a hefty 5% U-235 content) to form a number of natural nuclear reac-

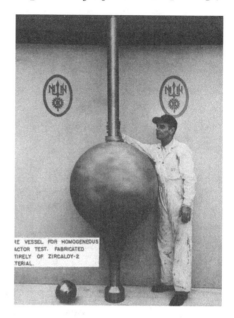

Aqueous Homogenous Reactor
(Reactors don't have to be big.)
Source: https://en.wikipedia.org/wiki /Aqueous_homogeneous_reactor

tors, today known as the Oklo site. These hot spots functioned, on and off, for several hundred thousand years. Science has since determined that the fission products from these open-air, water-moderated reactors have since moved a grand total of *a few inches* (see the videos: [15]).

Natural Reactor in Oklo region of Gabon
Source: https://www.osti.gov/servlets/purl/1118342

"It is estimated that nuclear reactions in the uranium in centi-meter- to meter-sized veins consumed about five tons of U-235 and elevated temperatures to a few hundred degrees Celsius. Most of the non-volatile fission products and actinides *have only moved centimeters* in the veins during the last 2 billion years. Studies have suggested this as a useful natural analogue for nuclear waste disposal." [emphasis added] [16]

Let's break that down: The estimated five tons of U-235 that fissioned at this one Oklo site could have powered a modern city of one million people for five years, at US standards of extravagant electricity consumption, while generating temperatures over 200° C—enough to broil a Steakasaurus.

When these natural aqueous reactors finally ran out of steam, portions of their fission products and transuranic isotopes remained.

Fig. 75: Migration At Oklo Of Uncontained Nuclear Waste
Source: https://thoughtscapism.com/2017/11/04/nuclear-waste-ideas-vs-reality/

All told, the transuranics and fission products the uranium left behind have moved *mere centimeters* from where the fission took place (for all you inch and foot folks, 2.54 centimeters = 1 inch). In much the same way, material from a damaged dry cask, or a burial cask in a repository, wouldn't get much farther than that. And even if it did, it would *still* take another half-million years to reach the surface.

But in defiance of science, anti-nuclear activists have brushed all this aside, making it nearly impossible to transport SNF on public roads. They apparently never saw the Finnish study, or the cool train wreck videos (always a crowd pleaser). They probably didn't read up on Oklo, either. As a result, it's been all too easy to spread scare stories to justify restrictive legislation, and it's been happening for decades (see next section).

California is a good example. You basically can't transport used fuel on the Golden State's highways (soon to be renamed the Golden Brown State, in light of the worsening drought). So even if the Yucca Mountain repository north of Las Vegas ever does open for business (don't hold your breath), you couldn't transport California's SNF there anyway.

Sacramento legislators also won't allow the construction of any new reactors until the state certifies a plan for spent fuel disposal, a moratorium upheld by the Ninth Circuit court. [17] Combined with the state's transport restrictions, and the failure of Yucca Mountain (which was dug into problematic rock, a decision based on politics and not geology), this seemingly sensible safety rule amounts to a squeeze play designed to drive the industry out of business.

Without access to a functioning repository, used nuclear fuel has been accumulating at Diablo Canyon, prompting the antis to disingenuously complain that there is no solution for this hideous waste—other than shutting down Diablo Canyon, of course. Even though, part of decommissioning a reactor includes transporting its used fuel over public highways to permanent storage.

With increased global warming and the onset of long-term drought in the American west, California's headlong rush into gas-backed renewables has clearly become a roadmap to nowhere. Shutting down San Onofre in 2012 was bad enough; closing Diablo would have further ensured a future of flex alerts, and even more rolling blackouts, on a methane-powered grid augmented by renewables and woefully inadequate storage.

With Diablo on the job until at least 2030, California will have a fighting chance of keeping the lights on *and* the EVs charged. And in another five years, who knows? Even the most diehard anti-nukes might learn to appreciate two gigawatts of always-on carbon-free power. Change is good.

18.5 DEFINE "SAFE"

We modern humans use a variety of hazardous materials in our everyday lives. Most of them are kept under the kitchen counter or the bathroom sink, in the garage or maybe the garden shed. Our safety depends on how well they stay bottled up, and how well we control their use.

The chlorine bleach and ammonia under your kitchen sink could harm your entire family (*Do not mix!* Seriously—read the labels). However, you are confident of their containment, so it doesn't keep you awake at night. If these chemicals had a reputation for escaping their containers and forming a cloud of dangerous gas that could drift through the house, you'd store them separately or you wouldn't even have them around.

Do Not Mix!

Bleach + Vinegar

This mixture produces chlorine gas. Inhalation produces coughing and breathing difficulties. Vapors can produce watery eyes and eye burns.

Bleach + Ammonia

This mixture produces chloramine, a toxic gas. Inhalation symptoms include shortness of breath and chest pain.

Bleach + Rubbing Alcohol

This mixture produces chloroform, a highly toxic gas that can induce unconsciousness.

Hydrogen Peroxide + Vinegar

This mixture produces peracetic (peroxyacetic) acid, a strong corrosive.

Fig. 76: Household Hazardous Material
Source: https://discover.univarsolutions.com/fr-ca/blog
/cleaning-products-when-to-avoid-mixing/

As bad as Chernobyl was, the actual consequences of nuclear's three notable accidents (TMI, Chernobyl, and Fukushima) have been far less harmful than conventional wisdom would have us believe. That's because the conventional wisdom about nuclear energy has been deliberately distorted for years. [18] This wasn't always the case. Back in the 1950s, California's venerable Sierra Club supported nuclear power. Among the proponents was one of their most famous members, the renowned nature photographer Ansel Adams.

"Nuclear energy is the only practical alternative that we have to destroying the environment with oil and coal."

– ANSEL ADAMS, RENOWNED PHOTOGRAPHER
AND SIERRA CLUB MEMBER

"Nuclear power is one of the chief long-term hopes for conservation. Cheap energy in unlimited quantities is one of the chief factors in allowing a large, rapidly-growing population to preserve wild lands, open space, and lands of high scenic value. With energy we can afford the luxury of setting aside lands from productive uses."

– WILL SIRI, SIERRA CLUB DIRECTOR (1966)

But as the 1970s unfolded, some of the club's directors feared that an abundance of cheap and clean nuclear power would attract too many people to the west coast. The rugged beauty of the Sierras would be suburbanized, and the magnificent Pacific coast would start looking like Coney Island.

Their solution was to dissuade migration by keeping electricity expensive, and a simple way to do that was to stir up fears about nuclear power—the obvious solution to providing newcomers with plenty of cheap, clean electricity. The club exhorted their members to drum up concerns about radiation safety and nuclear waste, with the aim of regulating reactors out of existence, a strategy that soon became part of Sierra Club orthodoxy. This is hardly a conspiracy theory—their efforts to limit growth by shutting down nuclear are well-documented and ongoing. [19]

The same thing was happening in Gerhard Schröder's Germany in the 1990s. Jurgen Tritten, his minister of environment and a member of the Green Party, worked to make nuclear plants unprofitable by increasing their safety requirements:

"It was clear to us that we couldn't just prevent nuclear power by protesting on the street. As a result, we in the governments in Lower Saxony and later in Hesse tried to make nuclear power plants unprofitable by increasing the safety requirements." [20]

The international movement to ban nuclear power made common cause with the ongoing effort to ban The Bomb, conflating reactors and weapons into a single issue, even though the two are only tangentially related. The anti-nuclear movement also made common cause with the larger anti-Vietnam War / peace movement of the 1960s, and later with the environmental movement of the Seventies. In 1974, a confidential memo by the Sierra Club's executive director proposed a deliberate strategy of fearmongering to make nuclear power expensive. Since then, their stance against nuclear power has become an enduring article of faith.

And talk about a fate worse than global warming: The Sierra Club has also been calling for replacing reactors with natural gas (the polite term for methane). As they declared in a 2019 handout: "No need for nuclear—3 new 900+ MW gas plants are in the works in Ohio." [21] So now we can all breathe a smoggy sigh of relief.

Friends of the Earth, a splinter group of the Sierra Club, was bankrolled by a former oil executive. [22] Unlike the Sierra Club, FOE is also dead-set against any form of fossil power, including natural gas. [23] We at least can agree with them on that. But it would be great if they would consider the fact that since we live on a nuclear planet, we should use this happy circumstance to our benefit.

"Our campaign stressing the hazards of nuclear power will supply a rationale for increasing regulation . . . and add to the cost of the industry."

– MICHAEL McCLOSKEY,
SIERRA CLUB EXECUTIVE DIRECTOR (1974)

18.6 "NO MORE CHERNOBYLS" SOUNDS PRETTY GOOD TO US, BUT . . .

In April 1991, a meeting took place on the fifth anniversary of Chernobyl. *Conference for a Nuclear-Free 1990s—No More Chernobyls* was a two-day confab of 500 anti-nuclear strategists in Washington, DC. This was serious stuff, co-sponsored by Greenpeace with a keynote speech by Ralph Nader. The messaging and tactics from the conference were assembled in a policy paper. [24] Among the talking points:

- "Develop a strategy now to take advantage of the next severe nuclear accident [in order] to kill nuclear power."

- "Reject any attempts to reach compromise on nuclear issues; any conciliations signal weakness of the environmental community in terms of their bottom-line position that nuclear power must be eliminated."

The purpose of the convention was to launch an on-going, multipronged effort to squeeze nuclear out of the market. This would be done through fearmongering on the one hand, and demands for ever higher (and ever more expensive) safety standards on the other, with the regulatory power to make them stick.

It's no wonder that reactors, weapons, and war remain linked in the public mind as an interlocking triangle of doom, even though Israel and North Korea have shown the world that if a country wants to build a bomb, it doesn't need a civilian nuclear power industry to hide behind.

In fact, if a nation-state wants to develop a clandestine weapons program, the worst way to proceed is to try to mask the scheme with a public utility—reactors and fuel enrichment facilities tend to attract a lot of attention. A corollary of this truism is that the world could ban all nuclear power and still not block a path to the bomb. That deserves an instant replay:

The world could ban all nuclear power
and still not block a path to the bomb.

The flip side of this coin is also true: A rollout of nuclear power does not increase the likelihood of nuclear weapons proliferation. [25] South Korea is but one example among many—a major player in the nuclear power industry and they haven't built a single bomb.

Since the 1945 strikes on Japan, information on how to build a bomb has been in textbooks and public libraries, all around the world. [26] In late 1945, the US even published a report titled "Atomic Energy for Military Purposes" that could have been subtitled "How We Did It." While not revealing any classified science *per se*, the paper explains in detail how the Manhattan Project worked out the science and engineering behind the bomb. [27] The report was valued as a guidebook by scientists in other countries, helping them focus their research and development while avoiding dead-ends.

The genie got out of the bottle a long time ago, and now he's online—and that's not counting whatever's on the dark web. This is why stopping commercial nuclear power won't accomplish a thing, other than ensuring that we won't get to Carbon Zero by 2050, or even by 2100.

Nuclear's bad framing came about through a simple twist of fate: It had the wretched luck of being initially used to develop a weapon during a global war. And then, shortly after the war, the first successful nuclear power plant was installed on the *USS Nautilus*, a submarine built to project US power in the new Cold War. So nuclear power was framed as a technology of harm, right from the start.

But that was decades ago. A well-informed public would have sorted out the distinction between weapons and commercial power long before now. Like the difference between gasoline and napalm, the distinction between nuclear power and nuclear weapons is something that anyone can understand.

It's high time for the commercial nuclear industry and the Department of Energy to step up and educate people so it sticks. Lady Bird Johnson, the First Lady in the Johnson administration (1963–1969), changed the nation's perception of litter in just a few short years with her Make America Beautiful campaign. [28] We could surely do the same with nuclear power. Fear dissipates with knowledge.

CHAPTER NINETEEN

You Can't Get There from Here

THE DETAILS FOR A HIROSHIMA-TYPE URANIUM BOMB are available in old textbooks; [1] no one has to work under the cover of a civilian energy program to learn how to build a bomb. A nation-state can certainly try to do them both in parallel, as Iran has apparently been doing all along. An unarmed Hiroshima device (without the uranium "pit") can be built in a well-equipped metalworking shop. However, arming it with highly-enriched uranium would require a nation-state's enrichment infrastructure and supply chain. But other than that little detail, the bomb itself can be built by a skilled craftsman.

A long-haul trucker in Wisconsin with a college course in physics did just that, building an unarmed Hiroshima device in his garage just to prove that he could. He even invited the Department of Energy to check it out. Needless to say, they were aghast. [2] It looked and functioned exactly like the original, right off the shelf and ready to go, vintage 1945. All it lacked was seventy kilos of 85%-enriched uranium-235 (parts H and S in the drawing below), and the requisite four bags of cordite powder (Part W) to propel one half of the pit towards the other, down the length of a military-surplus Howitzer barrel, just like in the original bomb (Part N).

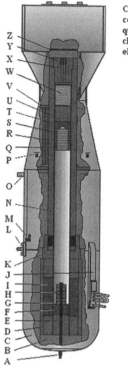

Cross-section drawing of Y-1852 Little Boy showing major mechanical component placement. Drawing is shown to scale. Numbers in ( ) indicate quantity of identical components. Not shown are the APS-13 radar units, clock box with pullout wires, baro switches and tubing, batteries, and electrical wiring. (John Coster-Mullen)

Z) Armor Plate
Y) Mark XV electric gun primers (3)
X) Gun breech with removable inner plug
W) Cordite powder bags (4)
V) Gun tube reinforcing sleeve
U) Projectile steel back
T) Projectile Tungsten-Carbide disk
S) U-235 projectile rings (9)
R) Alignment rod (3)
Q) Armored tube containing primer wiring (3)
P) Baro ports (8)
O) Electrical plugs (3)
N) 6.5" bore gun tube
M) Safing/arming plugs (3)
L) Lift lug
K) Target case gun tube adapter
J) Yagi antenna assembly (4)
I) Four-section 13" diameter Tungsten-Carbide tamper cylinder sleeve
H) U-235 target rings (6)
G) Polonium-Beryllium initiators (4)
F) Tungsten-Carbide tamper plug
E) Impact absorbing anvil
D) K-46 steel target liner sleeve
C) Target case forging
B) 15" diameter steel nose plug forging
A) Front nose locknut attached to 1" diameter main steel rod holding target components

"Atom Bombs: The Top Secret Inside Story of Little Boy and Fat Man," 2003, p 112. John Coster-Mullen drawing used with permission

Fig. 77: 1945 Gun-Type Nuclear Device ("Little Boy")
Source: Credit: Howard Morland (Wikimedia GNU free documentation license version 1.2)

Some people are concerned that a bomb can be made from the plutonium-239 that forms in a power reactor. Happily, this is not the case. The plutonium isotopes in used fuel are hopelessly blended together, rendering it useless for weapons. Visualize a melting bowl of Neapolitan ice cream, and all you want is the chocolate.

With such a wide array of plutonium isotopes in used fuel, ranging from Pu-238 to Pu-243 and beyond, there is no practical way to isolate the Pu-239 from its isotopic brethren. This is because all isotopes of any element are chemically identical and thus cannot be chemically separated. Other methods must be employed.

The concentration of Pu-239 must be greater than 10:1 to make a weapon. Specifically, the purity for a plutonium weapon is 94% (we break down the numbers here: [3]). Compared to the other Pu isotopes present, the ratio of Pu-239 in used fuel is less than 60%, with no practical way of enriching it to a higher concentration. [4]

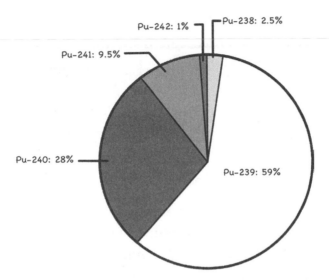

Fig. 78: Plutonium Ratios in Used Fuel
Source: https://beyondnerva.com/radioisotope-power-sources/
radioisotope-selection-for-rhu-fuels/americium-241/

Note that the ratio of Pu-239 to Pu-240 is about 2:1, five times the allowable 10:1 ratio for Pu-240 contamination in weapons-grade plutonium. This alone "denatures" (contaminates) the plutonium-239 in used fuel, making it useless for weapons, though it is still quite useful as reactor fuel.

You might be thinking: How come they can use a centrifuge to separate U-235 from U-238, but they can't separate Pu-239 from Pu-240? The answer is simple: Centrifuge separation exploits the tiny difference in weight (mass) between isotopes. There is a three-neutron difference between U-235 and U-238—enough to exploit in a centrifuge. But there is just one neutron's difference between Pu-239 and Pu-240, making them nearly impossible to tease apart.

In order to isolate the Pu-239 that forms in a reactor's fuel pellets, the reactor must be shut down less than sixty days after start-up, before the Pu-239 absorbs more neutrons and turns into Pu-240, Pu-241, etc. The once fresh and now highly radioactive rods must be removed from their fuel assembly, and the pellets removed from the rods, before chemical separation of any newly minted plutonium can even begin.

Designed to produce energy rather than isotopes, a power reactor cannot be subverted into a bomb factory without major alterations in its operation and production routines. This is something that any spy satellite can detect, and it would draw a lot of attention on the ground as well.

All told, it would be far more discreet, easier, and cheaper to use a small "production reactor" instead: A few kilograms of yellowcake powder would be inserted into what amounts to a nuclear kiln, and bombarded with neutrons for about two months—long enough to form the Pu-239 for a bomb, but not enough time for too much of it to transmute into higher Pu isotopes and spoil the batch.

19.1 NOT AS EASY AS IT SOUNDS

The path to either bomb, uranium or plutonium, is fraught with obstacles that only nation-states have the resources to tackle. For example, to enrich the material for a uranium weapon, about 11,000 tonnes of raw material must be dug from the earth, from which 3.5 tonnes of yellowcake powder is extracted (as you recall from section 4.2, the average rock-to-uranium ratio is 3,000:1). The yellowcake must then be converted to uranium hexafluoride gas (UF_6) and spun in a cascade of centrifuges, thousands of times over. This will (eventually) concentrate about 25 kilos of bomb-grade U-235. [5]

In sharp contrast, a plutonium device like the W-88 warhead [6] only needs about five kilograms of Pu-239 (the exact amount is classified). This weapons-grade plutonium can be transmuted from just five kilograms of uranium yellowcake (derived from about 15 tonnes of raw material), instead of using 3 tonnes (3,000 kilograms) of yellowcake to make one uranium bomb. [7]

Five kilos of uranium yellowcake is a trifling amount (the volume of a pint glass) that could be mined and refined in secret. The purified uranium powder (99.3% U-238) could then be irradiated with neutrons in a production reactor to produce five kilos of weapons-grade plutonium-239. [8] All in all, a lot more bang for your buck—so long as you can fuel the production reactor, which requires the resources of a nation-state to enrich enough uranium to make enough fuel.

Running flat-out for one year, North Korea's 50-megawatt production reactor can make enough Pu-239 for ten bombs (we break down the numbers here: [9]). Chernobyl's RBMK reactors were designed to make more bombs than that, while also supplying electricity to the Soviet people, their secondary function.

Some are concerned that a medical laboratory's microreactor can be subverted for weapons use, but this is entirely unrealistic. A lab's kiln-type production reactor is tiny, usually 1 MW or less. Typically used for experiments or for making medical isotopes, a lab reactor has no need to be any larger. This means a no-goodnik lab rat would have to run the reactor day and night for five years straight to make enough Pu-239 for one bomb.

As you can imagine, this would attract a lot of unwelcome attention from the other lab rats, to say nothing of the professors, the dean, and the Feds—not a plausible scenario. And even if some enterprising no-goodnik built their own reactor at home with a pile of graphite blocks, [10] they would still have to rustle up enough fuel to generate enough neutron flux to make enough plutonium to make a bomb—not a plausible scenario, either.

The neutron bombardment to produce plutonium can be done in a commercial light-water reactor, but as we mentioned each run-time must be less than two months long to avoid excess neutron absorption. It takes a month to shut down a power reactor, and swap out even one fuel assembly, and get the reactor up and running again. This pattern of two months on and one month off would be a dead giveaway, detectable by any spy satellite or human observer. Which means that no one is going to be able to hide making weapons-grade plutonium with a commercial nuclear power plant. It's foolish to even try. [11]

A small production reactor can be used, if the bad guys can get their hands on enough fuel to run the thing. The problem is, there are dozens of different fuel assembly configurations. They would need to get the right design, loaded with pellets enriched to the proper level for their particular reactor, or they would have to build a reactor around the available fuel. A rogue state could do this, but not a grad student.

And then there's the problem of building the weapon itself. Even a relatively simple first-generation implosion-type plutonium device (Trinity and Nagasaki) is considerably more difficult to build than a uranium "gun-type" Hiroshima device. The high-explosive shape charges in a plutonium bomb form a large sphere around a hollow plutonium pit. The shape charges must all go off at precisely the same fraction of an instant, and burn in precisely the same way, to squeeze the pit from all sides to the exact same degree—not something you could throw together in the garage, even with a physics degree. And even if you could, the krytron triggers that set off the shape charges are ultra-precise, high-tech wonders, with nation-states maneuvering behind the scenes to acquire them. Not something you can order online. [12]

So either the bomb is easy to build, and very difficult to arm (uranium); or the bomb is very difficult to build, and a little bit easier to arm (plutonium). Long story short: You pretty much can't get there from here, unless you're already a player.

19.2 DIRTY BOMBS AND OVERACTIVE IMAGINATIONS

A dirty bomb is a conventional (non-nuclear) bomb laced with radioactive material. In theory, the dispersed material will increase the number of casualties expected from a similar-sized "clean" bomb.

In practice, however, the enhanced danger of a dirty bomb would only be proportional to the nuclear fear of the target populace. In other words, a dirty bomb is a terror weapon whose additional effectiveness lies in the victim's psyche, and not in the weapon itself. Without fear and disinformation as a force multiplier, it's just another bomb. As the CDC (Centers for Disease Control and Prevention) explains:

- The main danger from a dirty bomb comes from the explosion, not the radiation.

- The explosion from a dirty bomb can cause serious injuries and property damage.

- Only people who are very close to the blast site would be exposed to enough radiation to cause immediate serious illness. However, the radioactive dust and smoke can spread farther away and could be dangerous to health if people breathe in the dust, eat contaminated food, or drink contaminated water. [13]

The DHS (Department of Homeland Security) puts it like this:

"Most injuries from a dirty bomb would probably occur from the heat, debris, radiological dust, and force of the conventional explosion used to disperse the radioactive material, affecting only individuals close to the site of the explosion. At the low radiation levels expected from an RDD [radiation dispersal device], the immediate health effects from radiation exposure would likely be minimal. Psychological effects from fear of being exposed may be one of the major consequences of a dirty bomb." [14]

DJ LeClear ("The Rad Guy"), one of our favorite nuclear vloggers, weighs in on the topic thusly:

"It's time we deflate the fear factor out of radioactive material dirty bombs. Bombs and bullets are always more harmful than betas and gammas." [15]

The Russian occupation of the Chernobyl and Zaporizhzhia nuclear power plants in Ukraine has raised the specter of Russian troops fashioning used fuel pellets into dirty bombs. Russian propaganda claims that it was the Ukrainians who were planning to make and use dirty bombs, and not them. Either scenario is completely unrealistic, but let's explore this anyway.

For starters, the material for a dirty bomb wouldn't come from a nuclear plant's dry storage casks. Since the casks are bolted shut with the bolts welded in place, and weigh in at 150 tonnes each, they're way too heavy to be carted away and dismantled. This means the work would have to be done on-site, which means it would never happen, since there is more readily available material at a nuclear plant than the spent fuel in a storage cask.

The used fuel pellets for a dirty bomb would almost certainly come from a fuel assembly craned out of a plant's spent-fuel cooling pool. And even though they've cooled off and shrank a bit after their stint in the reactor, the pellets will be stuck in the fuel rods. That's because during fission, fuel pellets expand from the buildup of gaseous fission products, "bambooing" the hollow zirconium rod from the inside.

Even if the pellets could be removed (not bloody likely), using them in a bomb wouldn't do much more than scatter them like shrapnel; the pieces could be found with Geiger counters and retrieved during a clean-up of the bomb site. The only way to make an effective dirty bomb with used fuel is to chop the fuel rods into pieces, crush them, collect the powder, and disperse it with conventional explosives.

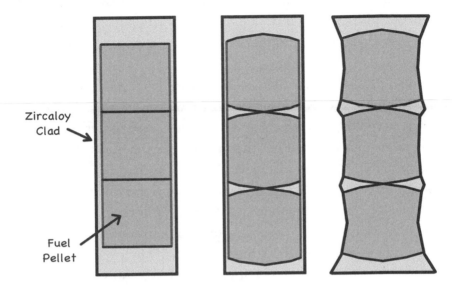

Fig. 79: Fuel Pellet "Bambooing"
*Source: https://www.researchgate.net/figure/Illustration-of-fuel-swelling-and-clad
-deformation-due-to-PCI-a-Fresh-fuel-before_fig7_342124328*

But there's a problem with this approach: Dispersing toxic material reduces its harmful effects—the solution to pollution is dilution. This means you'll need a lot of it to make the idea work. Much like a battlefield round of poison gas, a dirty bomb would have to scatter multiple kilos of fine powder to inflict any significant harm beyond the explosion itself—and that's if the wind cooperates. This involves building, and deploying, an exceptionally heavy bomb for not a whole lot of extra payoff.

Compounding these difficulties, uranium oxide powder has a natural static charge and sticks to *everything*. While this may be a post-explosion advantage for the bad guys, it also makes their bomb-building project a suicide mission, with more of them likely to succumb to excess radiation than their intended victims.

Their workflow could be improved by freezing the chopped-up rods in liquid nitrogen prior to crushing them; but a cryogenic bath, even for inorganic matter, requires a team of specialists. Plus there's all the shielding and air filtration needed for this particular task. All in all, not a plausible option for soldiers in the field, and deadly for those who don't do it right.

In the real world, a dirty bomb made from used fuel pellets would almost certainly be assembled in a weapons factory. The result, however, would be largely useless—a heavy weapon with no tactical or strategic purpose other than inciting terror, and with no real expectation of increased casualties. Aside from that, the geopolitical fallout from using the weapon would be far worse for the bad guys than whatever harm the bomb may inflict on their victims. In the final analysis, there is no advantage for a nation-state—even a rogue nation-state—to build a dirty bomb, especially in a war zone.

Like most nuclear fear scenarios, the dirty bomb scare is highly exaggerated and falls apart upon close inspection.

CHAPTER TWENTY

"Strange Days, Indeed."
– John Lennon

IN THE WAKE OF FUKUSHIMA, Japan shut down every reactor they had and increased their use of fossil fuel. Germany did the same, and now both countries are building even more coal plants; in February 2020, Japan signed up for nearly two dozen more. [1] At the same time, Japan is (finally) restarting at least some of their reactors, but there is no indication they'll be canceling any coal plants as a consequence.

This is unfortunate, because quite aside from all the excess CO_2 and air pollution, coal ash is far more radioactive per kilowatt-hour of energy generated than nuclear waste (Fig. 80). [2] More importantly, coal is chemically dangerous, containing forever toxins like mercury, lead, and cadmium. From the mine to their emissions, to their waste ponds of toxic ash, the coal industry is nowhere near as regulated as nuclear power. If it were, it would be shut down overnight.

What will be the long-term result of this post-Fukushima reversion to fossil fuel, aside from more smog, acid rain, coal ash, and radioactive particles released to the biosphere? The earlier quote (Chapter 8) about coal-induced deaths in Japan and Germany is worth another read. And once again, note that the 28,000 deaths occurred in the first six years after Fukushima. We're up to about twelve years now, so the number has likely doubled to 56,000 premature deaths, equal to nearly twenty 9/11s.

This grim statistic hits home for us. Bruce Boyd, our senior science advisor's father, died from complications due to 9/11 dust, not unlike a premature death from coal pollution. The larger tragedy in all of this is that if the US and the world had built out nuclear power in the 20th century, our priorities in the Middle East would have been much different than they were. Had that been the case, there's a good chance 9/11 would have never even happened. [3]

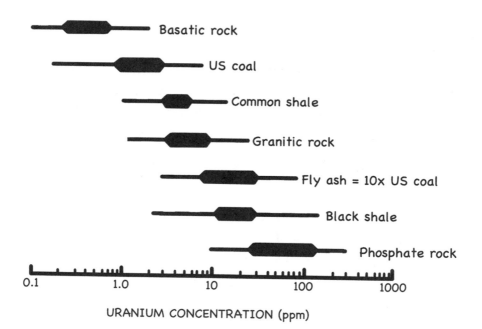

Fig. 80: Uranium Concentrations in Coal And Other Rock
Source: https://pubs.usgs.gov/fs/1997/fs163-97/FS-163-97.html

Japan's foray into utility-scale solar has produced only modest reductions in greenhouse gas emissions. [4] It's becoming clear to them that shutting down their reactors was a really bad idea. So not only are some of them being restarted, but their government is also considering advanced reactor designs. After a decade of media scare stories and government-generated nuclear fear over a casualty-free industrial accident, it seems that Japan is finally starting to rethink their clean-energy future, and good for them. [5] Germany, not so much.

The purpose of Germany's *Energiewende* is to substantially reduce carbon emissions by turning away from fossil fuel and, ever since Fukushima, from nuclear power as well. But after about ten years of trusting the plan, their departure from fossil still hasn't happened to any great degree. In fact, coal-related deaths in Germany have gone *up* since Fukushima, and are now the highest in Europe. [6]

France had their own successful version of *Energiewende* back in the 1970s and '80s, turning away from most of their fossil fuel consumption with a fifteen-year buildout of nuclear power. Since then, they've enjoyed the lowest per-capita carbon footprint in Europe. [7]

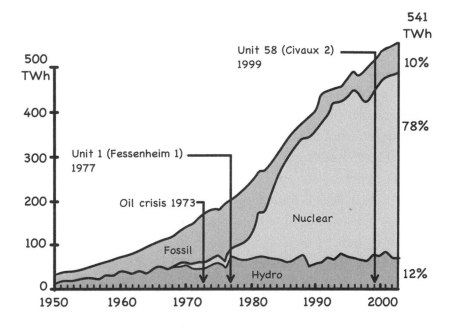

Fig. 81: Nuclear Buildout in France
Source: https://en.wikipedia.org/wiki/Nuclear_power_in_France

Sweden followed suit with their own rapid buildout, while Poland is keen on building nuclear power, and so is Slovakia and the Czech Republic. [8] The world will be in a better place when the rest of us start doing the same. A reactor anywhere means a cleaner planet everywhere.

20.1 DEATH BY A THOUSAND PROTOCOLS

Shutting down the nuclear industry after three accidents would be like shutting down commercial aviation for the same specious reason. When it comes to radiation, we're cautioned that we can never be too careful. But actually, we can: Remember Randy, the little brother in *A Christmas Story* who was so bundled up for winter that he fell over in the snow, and couldn't get up again? [9]

Given their excellent safety record, the nuclear industry should be allowed to rectify their shortcomings and carry on, much like the aircraft industry does, without taking on so much excess baggage that most new reactor projects either don't fly, or take decades to build, or get priced out of the market. Sometimes all three.

To that point: The application process for a reactor license at the NRC is a fantastically expensive, multi-year process. Coincidentally, it's also a major source of funding for the NRC. NuScale, for example, has spent an astonishing $500 million, and two million hours of skilled labor, to generate the information required to submit a 12,000-page Design Certification Application (DCA). [10] This was in spite of the fact that their design is a smaller, simpler, safer light-water reactor based on mature, already-licensed, and well-proven technology. In addition to that, their reactor's core damage frequency (CDF) was estimated by the NRC itself to be 10,000 times safer than NRC requirements. [11] But we never can be too careful, now, can we?

Every revision and resubmission adds to the cost of the application process. And when regulators use ALARA (as low as reasonably achievable) as their guiding principle, the thought process can often go something like this: "Any safety measure that does not price a proposed reactor out of the market is a prudent and sensible improvement, whether it actually increases safety or not. And no one can say we didn't try to make it foolproof, so our butts will be covered."

The way this plays out in practice is that any especially low-cost reactor design (such as a molten-salt reactor) presented for certification can obviously afford another layer of safety and still compete in the marketplace. Such as requiring an airtight containment dome for a reactor that cannot melt down, no matter what, and would have no radioactive steam to release even if it did. Or maybe a third set of emergency backup pumps, or how about another inch of concrete on the superfluous containment dome, just for just-in-case? [12] This keeps prices sky-high and does nothing to improve safety, while stifling innovation and competition. It's no wonder that some people call the NRC the Nuclear Refusal Commission.

The ALARA principle (the "reasonable" part is open to interpretation) makes every application far more difficult and expensive than it has any good reason to be. As we saw in Chapter 18, this was the insidious reality that some anti-nuclear groups have deliberately promulgated to deny sufficient energy for a growing population, in a shortsighted effort to preserve their California dreams.

Even when a nuclear plant does get built, the operation and maintenance protocols are often overbuilt as well. As a professor of nuclear physics once remarked to us: *"You practically need an act of Congress to tighten a bolt on a US reactor."* The Sierra Club will be pleased to hear this—death by a thousand protocols has been their strategy for decades.

Or death by a billion fish larvae. Anti-nukes have been making a fuss about the seawater intake pipes at Diablo Canyon nuclear plant on California's central coast:

"On one side [of the issue] is the plant's substantial, but not overwhelming, effect on nearby marine life. Diablo Canyon's impact on adult fish is trivial . . . barely enough for a busy weekend at a seafood restaurant. 'Entrainment' is another matter; about 1.5 billion fish eggs and larvae pass through the screens into the plant's maw each year." [13]

A whole other fuss was made about the plant's discharge pipe—the back end of this nuclear beast (¡*El Diablo!*), spewing warm, clean seawater, the same water that was gulped into the plant to cool the spent steam from the plant's turbines. This of course warms the discharge water, which slightly raises the temperature of a nearby cove. The bat signal went up, and a proposal was championed by Friends of the Earth to mitigate this alarming non-issue: A pair of cooling towers for a cool $1.62 billion.

A fish farm to replace the lost larvae and adult fish would be substantially cheaper, on the order of $20 million or less to build and operate. This is an expense that Diablo could easily absorb, and a far more attractive option than two new cooling towers. It's quite telling that this wasn't even seriously considered as an alternative solution. Nuclear fear strikes again.

The fact is, Diablo Cove is nowhere near distressed, and the local sea critters have been thriving in the slightly warmer water. [14] Even the whales are happy.

Diablo Whale
Source: John Lindsey https://www.youtube.com/watch?v=NuwjORurP1s

20.2 KEEPING OUR COOL

The US Energy Information Administration (EIA) projects that by 2050 the world will need around 70%–80% more electricity than humanity is using now. [15] And even though "renewable energy" is an appealing catch-phrase, wind and solar are not a practical solution, by themselves, for the challenges we face.

Even if the wind and solar projected to be built by 2050 actually *does* get built (see Fig. 81 below), the increased generation would only cover about 80% of increased global demand. The remaining 20% of increased demand—along with 100% of *existing* demand—would have to come from other sources.

Don't take our word for it. The graph below was compiled from a Stanford University meta-study of forty-seven recent papers, [16] most of which were written by renewable-energy advocates and not by nuclear folks.

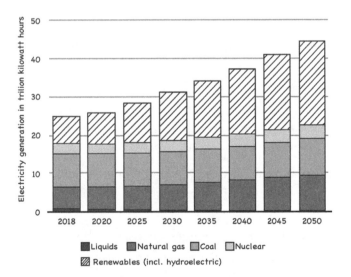

Fig. 82: Projected Electricity Generation Worldwide to 2050
NOTE: This data was compiled from *pro-renewable* energy studies.
Source: *https://www.weforum.org/agenda/2020/02/renewable-energy-future-carbon-emissions/ (World Economic Forum)*

The predicted increase in RE capacity, between now and mid-century, is shown by the top segment of each column. By carefully scaling the graph, we found they're predicting that global fossil-fuel consumption for electric power will actually increase 28% by 2050. (See our supplement on global electricity demand: [17])

This of course would not result in anything close to an all-renewables future. And to be clear, these are numbers from analysts who are favorably disposed to wind and solar, and bullish about its inevitable future.

It gets worse: Because the world is warming, about one-third of increased global electricity demand by midcentury will most likely be for air-conditioning *alone*. [18] This would consume about half of the wind and solar buildout predicted by RE advocates for midcentury. So even with their hoped-for expansion of renewables, the world will fail to make its Carbon Zero goals for 2050 unless we go nuclear as well. And even then it'll be tight. According to the graph, the view of the renewable energy community is that:

- Fossil fuel use will expand 28% by midcentury.

- RE expansion by midcentury will not provide anywhere near the energy humanity will need to achieve Carbon Zero by 2050—or by 2100, for that matter.

We appreciate their candor; perhaps now we can have a productive discussion about building a feasible clean-energy grid in the years ahead.

Renewables do provide sorely needed carbon-free energy, and they can help reduce fossil fuel consumption, but they won't be able to do the heavy lifting all by themselves. Without a ridiculous amount of seasonal storage, and without their own long-distance transmission network—neither of which would be needed by a nuclear buildout of local, independent reactors—a national or even regional all-renewables grid simply will not work. Here's another view of the same data:

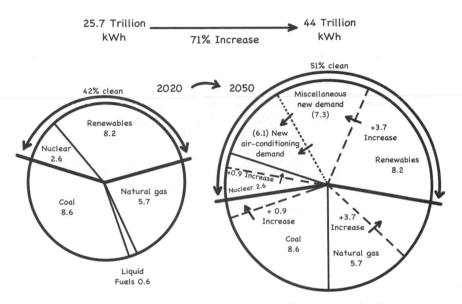

Fig. 83: Projected Electricity Generation 2020 To 2050, Another View
Credit: By the authors

Natural gas is (grudgingly) favored by wind and solar advocates as their backup fuel of choice, at least until the RE industry can build enough batteries or make enough hydrogen to finally kiss methane goodbye. So it's no surprise that Big Gas is positioning themselves as the perfect partner for renewables.

Their intentions are easily traced: The more renewables we have, the more gas-fired power plants we'll need to firm their intermittent output. This is why Exxon Mobil likes to say that gas and renewables go together "like peanut butter and jelly." [19]

British Petroleum has been a bit more up front with their "Beyond Petroleum" greenwashing. Their angle seems to be that if people invest in wind and solar, BP can continue selling gas and oil for when the sun don't blow and the wind don't shine.

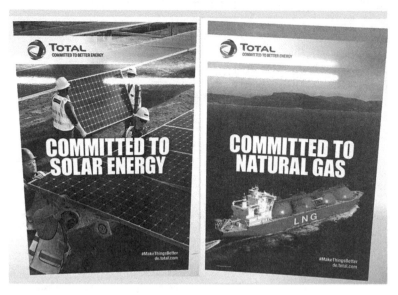

Total Solar and Gas Ad
Source: https://www.forbes.com/sites/michaelshellenberger/2019/03/28/the-dirty-secret-of-renewables-advocates-is-that-they-protect-fossil-fuel-interests-not-the-climate

Gas, Wind & Solar

The Economist 2/23/19

Share of fuel 1990-2030 (% shares of world energy use)		
	1990	2030
Renewables*	0.4	6.3
Nuclear	5.6	6.0
Hydroelectric	6.0	6.8
Coal	27.3	27.7
Natural gas	21.8	25.9
Oil	38.9	27.2

*Renewable energy includes biofuels

Wind turbines are flying high. But how do you keep the lights on when the wind stops blowing? At BP, we see a simple answer: We see cleaner-burning natural gas. It's a perfect partner to renewables.

British Petroleum Gas, Wind, and Solar Ad
Source: https://twitter.com/HenryK_B_/status/1207302792014180352/photo/1

In the absence of nuclear power, grid-scale wind and solar will require a massive buildout of gas plants, and / or thousands of ginormous batteries, and probably more pumped hydro as well (pumping water uphill, to generate electricity later). The problem with natural gas—aside from all the leaks before we burn the stuff, and the CO_2 emissions that come from burning it—is that it comes with an insurmountable problem: It's finite. And if we continue at our current rate of consumption (about 1% per year of known US recoverable reserves), we'll be down to 70% of reserves by 2050 (we break down the numbers here: [20]).

That may not sound like such a big crisis, but a dedicated renewables buildout will dramatically accelerate natural gas depletion, even while the prospects for exporting the stuff at a tidy profit are improving by leaps and bounds. A liquefied natural gas buildout has already been under way in the US gas industry, with near-term projections to double our LNG exports over 2018 levels. [21] Russia's disruption of Europe's gas imports will only accelerate this boom. [22]

This of course means that our domestic rate of methane depletion will inevitably increase as well, and that in turn will put a squeeze on a national RE buildout. Unless and until a fundamental breakthrough in battery technology comes along, any massive buildout of wind and solar will have to rely on gas backup—or suffer a fate worse than global warming and *(gasp!)* rely on nuclear power.

In 2019, the Trump Administration announced that the US would be increasing the export of our finite supply of these "molecules of freedom." [23] Rick Perry's Department of Energy gave it another Orwellian spin, calling it "freedom gas." [24] At some point, Washington will have to develop a national energy policy that doesn't sound like it was written by *The Onion*. [25]

20.3 ENERGY AND CIVILIZATION

Though energy is often viewed through a free-market lens, as if it were just another commodity, it's more fundamental than almost any commodity we use—more fundamental than trade or commerce, or even the market itself. Even a pre-market subsistence economy depended on energy, typically in the form of food to fuel human and animal labor, with wood, peat, and dung for heat, along with candles and a variety of oils (olive, whale, etc.) for a bit of light.

Energy keeps the lights on and keeps us warm and cool, and keeps the machines and computers running so trade and commerce can occur. While more energy can fuel more economic growth within an existing society, denser forms of energy can power the formation of entirely new societies. We saw this in the Industrial Revolution, when a subsistence economy powered by wood was replaced by a market economy powered by coal.

Coal's energy density is twice that of wood, which was the world's first universal fuel other than food. It was wood that took us from the cave to the cottage. Coal (and later petroleum and gas) took us from the cottage to the condo, but now it's time to move on. As we mentioned, the energy density of nuclear fuel, pound for pound, is 100,000 times greater than coal. History and physics suggest that transitioning to a carbon-free society in a matter of a few decades, or even over the course of several decades, will require a denser form of energy than fossil fuel, rather than a more diffuse one like renewables.

Energy drives society, so it's hardly just another commodity, though it is often treated like one. A company called Enron shamefully demonstrated this in 2000 by manipulating energy prices in a newly deregulated market. Electricity costs in California skyrocketed by as much as twenty times. [26] In a similar unregulated free-market fiasco, spot prices soared even higher in the Texas Freeze of 2021. [27]

Energy is absolutely fundamental to the game board on which civilization plays Monopoly, upon this delicately balanced card table we call planet Earth. Because this is so, and because humanity has been busy spoiling its leafy green host, the energy production for ten billion people by midcentury absolutely *must* be decoupled from the whims of Mother Nature, as much as humanely possible. [28]

In our view, it must also at some point be decoupled from the whims of the market. Doing both requires a national energy policy informed by science and backed by sensible regulations. Everybody wants a clean planet; the trick is getting there. And as we pointed out, there is no Planet B.

If the US is serious about building a 100% renewables grid (not something we would advise), we'll also need an equally serious energy policy to restrict the export of LNG, and good luck with that; especially now. The free market is pulling our gas overseas, and will be doing so for some time to come.

The other big problem with using natural gas as a "bridge fuel" to an all-renewables future is that gas is by no means a renewable resource. And when it's gone, it's gone, at least for the next hundred million years. That's about the time it takes for the earth to convert organic matter into coal, petroleum, and gas.

Also keep in mind that we use natural gas for a lot more than heating, cooking, and making electricity. The fertilizer market could not exist in its present state without copious supplies of natural gas. In fact, it's estimated that synthetic nitrogen fertilizer made with natural gas currently feeds between 3 and 3.5 billion people—nearly half the world's population. [29] That being the case, agribusiness (as distinct from agriculture) could be thought of as a global enterprise that uses crops to turn hydrocarbons into carbohydrates. Having to decide between making enough fertilizer to feed ourselves, and providing enough gas backup for wind and solar, should tell us we're on the wrong track.

The growing demand for hydrogen cannot be supplied by existing methods, either—nearly all hydrogen is currently made by the steam reforming of methane, which is the main ingredient of natural gas. [30] And what do they burn to heat the water to make the steam? You guessed it.

The most popular mass production alternative for making hydrogen is electrolyzing potable water. Waste water and seawater can also be used, but they must be cleaned and desalinated first, requiring even more energy. And in order to be even vaguely cost-effective, the high-volume electrolysis of water requires a large amount of electricity and a steady supply thereof. Since the energy density of wind and solar is so low, while its intermittency is so high, using either one of them to electrolyze water requires substantial battery backup. Using them to prepare waste water or seawater for electrolysis would be even more expensive.

A fleet of dedicated reactors for powering mass hydrogen production would be a much cheaper, and far more effective, approach. And when molten-salt reactors and high-temperature gas reactors [31] come into commercial use, they could directly contribute their high heat to the process, reducing costs even more.

There's another thing to consider: If we build out wind and solar instead of nuclear, renewables will have to be overbuilt by a factor of three to five times to compensate for their low average capacity, in addition to being backed up with massive amounts of energy storage (and/or natural gas). Solar and wind have capacity factors of about 25% to 40% respectively, compared to nuclear's 90+%. But with enough overbuild, backup, and storage, wind and solar could, in theory, make enough hydrogen to take at least some of the burden off an all-renewables grid. At least, that's the plan.

Making hydrogen with excess renewable energy and stashing it away for a rainy day may sound like a good idea, but hydrogen is a poor energy storage medium with an energy throughput of about 15%–30%. In other words, that's all the energy you get back when you use electricity to isolate hydrogen, and use the hydrogen to power an onboard fuel cell to make electricity to propel a vehicle.

Hydrogen's shortcomings only compound when the natively weak performance of wind and solar is used to make it. Even worse, we show in *Roadmap to Nowhere* that Mark Jacobson's 2021 proposal for a 100% RE grid for the US by midcentury would require more than *one million times* the volume of hydrogen our country is producing today. [32] Not a plausible scenario.

A national 100% renewable-energy grid may sound like a wonderful jobs program, but in our view the labor and resources saved by building reactors would be better spent on everything else we have to fix or upgrade in this country, such as repairing 45,000 bridges. [33] We have a lot more to do over the next few decades than just greening the grid.

Visualize ten billion people in the grip of a global heat wave (this is unfortunately becoming far too easy to imagine). We'll need more clean energy than we have ever generated before, and we'll be needing it for a *lot* more than air conditioning. Trying to meet this increased demand by passively harvesting transient energy, and storing enough surplus to transmit over long distances on short notice, whenever the clouds come up or the wind dies down, is not the best use of our available technology and resources, or the best use of our common sense. To be perfectly blunt:

> If you want energy, *make it*.
> Don't just harvest the stuff.

A practical solution to the issues we face already exists, and it's already been proven at scale. The only obstacle is nuclear fear.

CHAPTER TWENTY-ONE

Make Energy, Not War

SHUTTING DOWN THE NUCLEAR POWER INDUSTRY won't stop the bad guys from getting a bomb, [1] but it will prevent us from dealing with air pollution, acid rain, and the twin monsters of global warming and ocean acidification.

There is nothing in commercial nuclear power that is essential for weapons development other than enrichment technology, which is already understood the world over. While reactors and weapons do share the same basic physics, they do not rely on each other for their existence. [2]

The best way to eradicate nuclear weapons is to convert them into clean electricity, and the best way to reduce the desire for nuclear weapons is to achieve the clean energy abundance that makes for prosperous, stable, and peaceful societies. [3] At this decisive moment in history, that comes down to building as many reactors as fast as we can.

In the age of global warming, issues like energy poverty and water scarcity will aggravate international tensions, and could lead to armed conflict before midcentury. Food scarcity and mass migration will surely follow. The unfolding reality of a warming planet, with a dying ocean and a burgeoning human population, will subsume every other issue in the decades to come. It truly is the story of the century.

And let's get real: The popular fantasy of migrating off-planet to save humanity (or least, the super-rich humans) is just that—a fantasy. At the risk of repeating ourselves: There is no planet B, Elton John was right about Mars, the moon isn't any better, and there is no earth-like planet that anyone could possibly get to without warp drive. Earth is it.

Colonizing Mars to make a fresh start for humanity is like moving to Antarctica because we don't want to clean our room. If the Moguls that Be are really serious about going, they should develop the reactors they'll need and test them here at home (might be a nice little side business, fellas). Even if we just send robots and drones, they'll need a reactor to do any serious work. [4] Whatever the Moguls ultimately decide to do, let's hope they keep one thing in mind:

> *"If you have the power to turn another planet into Earth,*
> *then you have the power to turn Earth back into Earth."*
> —NEIL DeGRASSE TYSON

Sustainable health, prosperity, and peace absolutely depend on energy abundance. And energy always comes first: With enough energy, you can do what you need or build what you need to procure your other resources. Moving to another world won't alter the equation, and achieving these laudible goals off-planet would cost a *lot* more time, effort, and energy than doing it here at home. Our primary focus, for at least the next few decades, should be on building a sustainable ecosystem for the home planet—from there, we can rule the galaxy! Bill Gates is on board with this idea (the first part). [5] We hope he inspires others.

21.1 RELAX – THE GENIE'S BEEN OUT OF THE BOTTLE FOR SEVENTY YEARS

Once again—shutting down commercial nuclear power won't do *bupkis* to stop the bad guys. They don't need a fuel enrichment facility or a nuclear power plant to conceal their nefarious schemes.

This is why banning nuclear power to prevent nuclear weapons makes about as much sense as banning petroleum to prevent napalm. There are lots of good reasons to not use fossil fuel, and an international treaty to ban napalm goes without saying. But nobody would seriously consider banning petroleum as a way to rid the world of napalm. In the same way:

We could shut down
every commercial reactor
and its supporting infrastructure
and fail to block a path
to the bomb.

Furthermore, commercial nuclear power is *anti*-proliferative. That's because clean and abundant energy goes a long way toward resolving social ills, which in turn reduces international tensions. Simple basics like clean water, clean heat, electric stoves, refrigeration, washing machines, and reliable electricity can transform the health, stability, and prospects of entire countries.

Even better, a nation does not need a domestic nuclear industry to benefit from nuclear power. A seaworthy nuclear barge, towed to a harbor or docked on a river, could jump-start a country's consumption of clean energy. Beyond running their own electric grid, no special skills would be needed by the locals.

Russia is already doing this along their eastern Arctic coast. [6] Expect many more of these plug-and-play nuclear plants in the years to come, on both land and sea. ThorCon Power of Florida is planning to go one better than the Russians with their ThorCon Isle, a Generation-IV molten-salt reactor that will be water-mobile, meltdown proof, and cheaper than a coal plant. Seaborg is doing something similar with their MSR Power Barges. [7]

Commercial power reactors can also consume "downblended" weapons material as fuel through a megatons-to-megawatts program, in which the material's high fissile content is reduced to reactor-fuel levels by adding depleted uranium. When the Soviet Union collapsed, we gave Ukraine about $8 billion for about 12,000 nuclear weapons. (Such a deal!)

ThorCon Nuclear Isle – Towable MSR Platforms
Credit: ThorCon (CC-BY-SA-4.0)

Through the rest of the 1990s, about one-third of carbon-free US electricity came from hundreds of tonnes of reactor fuel made by downblending the material from dismantled Soviet weapons. [8] Say it again, loud and proud:

MAKE ENERGY, NOT WAR!

Right on. All that banning nuclear energy would do (should we ever be so foolish) is make it nearly impossible to respond to smoggy skies, a warming climate, rising seas, and a growing population. While wind and solar are indeed helpful in reducing carbon fuel consumption, they have not shown how they can serve—by themselves, or even largely by themselves—as a comprehensive long-term solution for powering civilization and addressing climate change.

21.2 THE JEVONS PARADOX

A fellow named Jevons famously observed that improved energy efficiency does not necessarily reduce overall energy consumption, because the more energy we make available through increased efficiency the more ways we find to use it. [9] We mass-produce energy-saving TVs, and now we can afford to have one in every room. So now we must generate even more power to manufacture and run multiple TVs. [10]

And don't expect the Internet of Things (IoT) to resolve the paradox. With the advent of 5-G, a tsunami of data may gobble up as much as one quarter of all global electricity by the end of the 2020s. [11] One recent example: In its heyday, bitcoin mining was consuming more energy than the entire country of Switzerland. [12] Large data centers present the same challenges. And you thought we were going to have challenges running all those extra air-conditioners and charging our cars. Energizing the Cloud will soon be just as energy-intensive.

Few people have a firm grasp on the enormous amount of energy we actually use, to say nothing of the colossal amount of energy we'll need in the years ahead. This is entirely understandable—big numbers are difficult for anyone to visualize, even people who work with them for a living, and sometimes visualizations don't help: We're wowed by the factoid that one trillion dollars is a stack of $100 bills piled 630 miles high, even though we can't visualize anything 630 miles high. (If it's any help, that's more than ten times higher than Richard Branson's suborbital space flight.)

Another drawback of our cognitive limitations is that it's been all too easy to convince ourselves of improbable mega-schemes, such as the fantasy that renewables alone are enough to power the country. Gosh, all it would take is 9.7 billion full-size solar panels and 394,000 jumbo wind turbines (see *Roadmap to Nowhere*). Further complicating matters are the social theorists who contend that humanity could— and should—go on an energy diet and reduce our consumption of resources in general.

This is yet another fantasy, quite aside from Jevons Paradox. The harsh reality is that billions of people will burn whatever they have to in order to stay warm, cook their food, and make a bit of light—even animal dung. It's also true that one ill-advised policy change, or one unfortunate change of administration, can under-mine whatever conservation measures individual citizens may practice.

This criticism isn't leveled at individuals, rich or poor (well, maybe the uber-rich). Energy is a *national* issue, and pollution knows no boundaries, so effective energy conservation is much more than a matter of personal responsibility. Whether we like it or not, we are in fact all in this together. And even if personal conservation could be enforced, it would take several generations to change our Jevonsian tendencies—if it could be done at all. [13] Social engineering has been tried before, often with disappointing results.

To that point, a national campaign to change America's mind (and Washington's mind) about nuclear power will have to be more far-reaching than billboards, catchy jingles, and radio ads. Even so, a campaign to reframe nuclear energy would have much better prospects for beneficial change than keeping hope alive for a fundamental breakthrough in energy storage. Or the fond hope of wiring together tens of thousands of RE farms to form an inter-dependent, self-supporting, fuel-free national harvesting grid.

The public has already seen that nuclear can generate reliable power for years on end; they've just been snookered into thinking it's expensive and dangerous, to the point where it can seem politically incorrect to entertain the idea of building more reactors, let alone keeping the ones we have.

With the challenge to preserve life as we know it (or to preserve as much as we can), now is not the time to go on an energy diet. That would be like Eisenhower restricting the troops' caloric intake in the run-up to D-Day to turn them into lean, mean, fighting machines. Quite the opposite: The more clean energy we can reliably generate, the better prepared for the future we'll be. A battle for survival requires energy to spare. We can all go on a diet when the war is over, and take a vacation to Mars.

Overly dramatic? About one in five people live in South Asia. Consider the consequences of energy scarcity when millions of climate refugees from low-lying Bangladesh or Pakistan migrate next door to India, a country that is struggling to feed itself even now. [14] Or consider Miami Beach. It could be ankle deep by midcentury—the city's highest natural elevation is only six feet above sea level. They can't all start living on houseboats or in high-rises. Millions will have to relocate, and wherever they go the infrastructure will have to expand to accommodate them.

In every corner of the world, clean energy, fresh water, sewage treatment, and public health services are the best hope of preventing a manageable humanitarian crisis from becoming an unmitigated catastrophe. Deadly heatwaves have now become annual events. Summers in the Middle East are approaching Death Valley temperatures, and there is every reason to believe these trends will continue. This is especially true since we now know that the oceans, which have been acting thus far as a heat sink, are warming about 40% faster than previously thought. [15] A warming ocean also means an expanding ocean, which means more coastal flooding. Even now, Miami's king tides are routinely flooding the streets. [16]

For all these reasons and more, we're going to need all the carbon-free energy we can get. As the IEA (International Energy Agency) points out, net-zero emissions isn't just a nice idea, it's an existential imperative. [17] The 2021 IPCC report doesn't pull any punches, either: *"Global surface temperature will continue to increase until at least mid-century under all emissions scenarios considered."* [18] And Net Zero is just the start, the point where global emissions finally level off. From there, we'll need to go Carbon Zero, which means just what it says—zero carbon emissions. And then we'll need even *more* clean energy to go Carbon Negative, by actively removing excess carbon from the environment. (Some people are calling this Climate Positive, which we quite like.)

This final target entails lowering our atmospheric CO_2 to pre-industrial levels, and restoring the pH balance of our oceans as well. Hoping that renewables alone will perform on this scale is simply not a sensible plan for long-term survival. On the other hand, all the carbon-free fuel we need to get the job done is available for the taking, dissolved in the same oceans we're trying to save.

21.3 "COOL HEAD MAIN T'ING." — OLD HAWAIIAN SAYING

An objective comparison of renewables and nuclear shows that if an exclusionary, 100% choice has to be made (which is Mark Jacobson's idea, not ours), nuclear would be a much better choice for powering the nation and the world—if we must pick only one.

We're not pressing for 100% nuclear, if only because so much RE infrastructure is already in place. And given the momentum of the renewables movement, large-scale wind and solar will continue to be deployed well into the future. That's fine, so long as it isn't relied upon for more than it can deliver. Purists on either side of the nuclear vs. renewables debate should face the fact that energy is a mixed bag and always has been. That's the world we live in.

After the public gets over the false hope of 100% renewables, and puts their nuclear fears in perspective, we can all have a more productive discussion. Here are some key points to keep in mind as you study clean energy issues:

- Fuel is energy storage.

- Renewables are fuel-free systems, and require some form of storage.

- Wind, water, and sunlight are not fuel; they are ambient natural phenomena.

- With carbon-free fuel, we won't have to rely on fuel-free systems.

Without full grid connectivity, and a fortune in dirt-cheap storage, a renewables-heavy grid will not work as advertised. Sorry, but it's true. For these reasons and more, a comprehensive energy solution for the nation and the world must include carbon-free fuel, and not rely solely on fuel-free passive energy harvesting.

Fuel is stable, portable stuff you can use to make energy, when and where you need it, and nuclear fuel is incredibly dense stuff. So much so that every nuclear power plant has about two years of energy storage, on-site and ready to use—the fuel inside the reactor's core.

The ability to make energy on demand—day and night, for two years straight if need be—renders the pivotal issues of mass energy storage and seasonal storage moot. Local reactors would also reduce the need for an HVDC transmission network in direct proportion to the number of reactors installed. With local nuclear power, long-distance transmission corridors can be reduced, and in many cases eliminated.

To be fair, uranium and plutonium aren't the only carbon-free fuels we have. Hydrogen gas can also be used as a fuel—but only after we invest a *lot* more energy making it than what we could ever get out of using it in a fuel cell, or burning it in a combustion engine. This is why hydrogen is more properly regarded as an energy *carrier* or a storage medium, rather than a fuel.

Storing and piping hydrogen does have its drawbacks. Blending it with natural gas seems to work well, but as the world moves away from methane, the increasingly pure hydrogen in our pipelines will leak and cause embrittlement to the valves and welds. [19] Chilling hydrogen to a liquid mitigates the problem, but that takes additional energy, expensive equipment, and insulated pipes. Converting it into ammonia is another option for storage and transport, where the ammonia is converted back to hydrogen before use. But that costs energy as well. [20]

Still, hydrogen can serve as a form of storage—if there's enough off-peak energy to electrolyze enough water to keep the hydrogen tanks topped up. And of course, it there's enough energy to purify the water beforehand. Efficient electrolysis, done at scale, requires a steady supply of electric energy, necessitating either battery backup or reliable power from a carbon-free source like nuclear or hydro.

Renewables are being touted as an elegant way to achieve a distributed energy system of local, independent micro-grids, free from the clutches of Big Energy, the vagaries of weather, and other inconveniences. But as we examine the issues involved, it becomes increasingly clear that small modular reactors (SMRs) are the kind of energy systems that could actually make this vision a reality.

A small modular reactor could provide a community with true all-weather energy independence. In addition to 24/7 reliable clean energy, local reactors could also provide desalinated water, local heating, synthetic fuel, hydrogen production, and the electricity needed for plasma-arc waste disposal. Nuclear power is as independent and distributed as clean energy gets.

Small-town America
Population 5,000 / Transportation hub / Light industry
Hospital / Water desalination
100% independent
All-weather carbon-free power

You Can't Do This with Renewables
Credit: Shutterstock

These are some of the fundamental differences that must be appreciated to properly evaluate any grid-scale renewable energy system. In our view, an RE grid, whether regional or national, is not something that most people would seriously consider if they were fully aware of RE's limitations and the advantages of nuclear power.

Fear-based protocols, established in response to the concerns of a misinformed populace, have resulted in construction delays and abandoned projects. Reviving the struggling US nuclear industry will not be an easy process, and any difficulties along the way will surely prompt even more anti-nuclear sentiment. The 100% renewables meme has only increased the spin of this vicious circle. The purpose of our books is to help counteract the spin.

On a level playing field, with standardized designs, a solid supply chain, and sensible regulations, reactors can be completed on time and on budget to the highest international standards. It's already happening in Asia and the Middle East. Even large reactors are now routinely completed in about five years, with much lower construction costs than here in the US. This includes South Korea's APR-1400, a Generation-III light-water reactor that has already been approved by our Nuclear Regulatory Commission for construction and operation in the US. [21]

South Korea is building APR-1400s in-country for $2300 /kW (per kilowatt of installed capacity). This is about the same price of China's in-country Hualong One Gen-III+ reactors. South Korea recently completed four APR-1400s in the United Arab Emirates for $4250 /kW, on-time and on-budget. They've also been awarded a contract to build six APR-1400s in Poland for $31 billion, or $3700 per installed kW. (For cost and build times, see: [22]). At the same time, Westinghouse has been awarded a contract to build six AP-1000s in Poland for $30 billion. [23] As we explain in *Roadmap to Nowhere*, there is good reason to think that a domestic US buildout of APR-1400s, in volume production, with US wages and material, would cost about $3,000/kW. You will note this is one-third more than what the South Koreans are spending on nuclear in their own country. So we think it's an entirely reasonable—indeed, a conservative—"should-cost" estimate for a US nuclear buildout. With molten-salt reactors, the price could easily approach $2,000 /kW.

Reactors are not fantastically expensive, nor do they take forever to build, once a company gets up to speed building the same reactor over and over again—with a complete set of blueprints, a seasoned labor force, and a robust supply chain. While nuclear physics is fairly esoteric, building a nuclear power plant is about as complex as building a cargo ship. [24] During World War II, the US built

3,000 Liberty ships, without the aid of computers, the internet, or cell phones. [25] There is no reason in the world why we can't build 3,000 reactors. And if we can't figure out how to build them on-time and on-budget, let's hire the South Koreans to come over here and show us how it's done, until we get the hang of it and start cranking them out like sedans.

Asia is doing nuclear power right; the rest of the world needs to catch up.

Credit: Micah Brown

CHAPTER TWENTY-TWO

We Will Now Begin Boarding

LEARNING TO MANAGE OUR NUCLEAR FEAR, like we manage our fear of flying, will go a long way toward ensuring our energy abundance and security in the years ahead. We humans must learn to overcome our fears in order to save ourselves and other living things from the mess we've created in the process of powering ourselves to modernity. It's time to take the next step with a denser and cleaner fuel. As with any form of dense energy, caution is always warranted, but fear is the mind-killer.

Although airline crashes are attention-grabbing disasters, they don't dissuade us from flying. We want to fly, or believe we must, so we swallow our fears and get on board. We comfort ourselves with reassuring statistics and the realization that everything in life is a balance of risk versus benefit.

There is a widely-held assumption that renewables can be a comprehensive solution to the climate crisis. Unfortunately, this sentiment does not conform to reality. The fact is, we cannot have a reliable 100% national RE grid, or anything close to it, without a fundamental breakthrough in storage, coupled with a nationwide HVDC (high voltage direct current) overlay. Even the National Renewables Energy Lab (NREL) doesn't think we can unify the grid by midcentury. [1]

And gambling our future on hoped-for technical developments to come along and save our butts would be folly. We need to strategize and prioritize based on the mature and emerging technologies that can be scaled up and put to practical use within our existing energy infrastructure. Current nuclear technology, imperfect though it may be, can do the job, all on its own if need be. And it's worth repeating that there are two years' worth of energy storage in the core of every reactor—it's called "fuel."

The only thing seriously lacking in nuclear energy is a seat at the table in every serious discussion on clean energy. Humanity already has the technology to save the world from itself. We're just not using it anywhere near as much as we need to. If you get just one takeaway from this book, we hope it's that.

Although it is entirely feasible with today's technology, a 100% nuclear grid will not happen before midcentury. Having the technology is one thing, having the work force and supply chain to do the job is another. Besides which, the world has already invested over $4 trillion in renewable systems; [2] it would serve no good purpose to replace them before the end of their service lives. By the same token, old reactors should either be refurbished or replaced with new reactors, rather than renewables or gas. And when renewables wear out, they should be replaced with reactors as well.

> *"If I were young today, given my bent for science, I would choose nuclear engineering, not space sciences. We must find a way for people to have abundant, affordable energy with a small environmental footprint."*
> —JAMES HANSEN, "FIRE ON PLANET EARTH"

22.1 ONE STEP FORWARD AND TWO STEPS BACK

Replacing a reliable carbon-free system with an unreliable carbon-free system, backed by a methane-leaking and carbon-emitting network of natural gas plants, is not our idea of progress. Shutting down nuclear power in an effort to clean up the planet is placing fear above facts, and hoping that physics will conform to wishful thinking. We need to do better than that.

A case in point: The Indian Point nuclear plant in New York State has been phased out in favor of renewables. [3] The plant once produced more than 80% of downstate New York's clean energy, but the sales pitch by greenwashing politicians and idealists like the Riverkeeper organization [4] was that no new gas plants would be needed.

Their rosy prediction was flatly contradicted by the 2017 deactivation assessment conducted by the New York Independent System Operator (NYISO). These are the people who have to manage whatever energy grid that NY state policy dictates, and try to keep the lights on. Their conclusion, based on more than a century of hands-on experience, was that the three large gas plants which were about to start construction at the time (the same new gas plants the antis said would not be needed) would in fact be needed to maintain system reliability after Indian Point closed. [5]

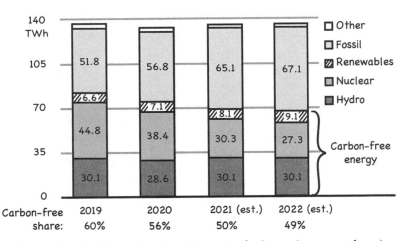

Fig. 84: New York State Electricity Generation by Source (in terawatt-hours)
Sources: NYISO Power Trends (2020), Gold Book (2020, for rooftop solar), Real-Time Fuel Mix (for 2020), Interconnection Queue (for grid solar & wind, probability adjusted)

In 2019, the two reactors at Indian Point provided 16,100 GWh of electricity, about 25% of the NYC metro region's demand. The reactors were shut down in 2020 and 2021. To replace the lost energy, wind and solar were projected to expand by about 13%, while fossil fuels would slightly increase from 65% to 67%. Meanwhile, New York opened the Cricket Valley and CPV gas plants. [6]

New York City will also have to import hydro power from Quebec on new transmission lines. [7] The same thing will likely happen in California if Diablo Canyon is ever replaced by gas-backed renewables, which the antis plan to do in 2030. We'll see about that.

Replacing a reactor with anything other than a new reactor is a clear example of nuclear fear masquerading as environmental stewardship. The sensible thing to do is to replace carbon power plants (coal and gas) with carbon-free power plants. It makes no sense to replace carbon-free power plants like San Onofre with fuel-free passive energy-harvesting systems, backed by copious volumes of natural gas.

The Green New Deal will succeed only if all useful forms of clean energy are enlisted. Even so, the first iteration of the bill focused exclusively on "renewables," which everyone knows is the code word for wind and solar, even though nuclear power is far more renewable than either one. In our view, any GND that is formally enacted must include a strong element of nuclear power: a Green Nuclear Deal. Otherwise, the job will not get done by 2050. Or 2100, for that matter. [8]

Belying the stereotype that only conservatives are pro-nuclear, key Washington Democrats favor the technology as well. This includes former Secretary of State John Kerry, President Biden's Special Envoy for Climate. [9] Here's hoping they can make common cause on this vital issue with their colleagues across the aisle. Like other industries, nuclear has its fair share of support from conservatives and libertarians, while other pronukers are card-carrying socialists and proud to tell you. The scientists, engineers, and advocates of nuclear energy are all over the political map; the issue that brings us together is clean energy, not politics. [10]

Renewables, distributed energy production, and underground thermal energy storage (UTES) are all fine ideas, so long as their limitations are respected, but it's a mistake to expect any technology to perform outside its wheelhouse. And as we mentioned, wind and solar, when they're buffered with backup and storage, have even lower EROIs than when they aren't buffered. [11]

With an EROI less than 7:1, renewables will not be able to sustain a modern society. Charles Hall, the noted ecologist who developed the EROI concept, recommends an EROI of at least 10 to 12 to power a society like the US. [12] Even so, Jacobson's new US Roadmap has reduced its proposed national CSP fleet to 8 GW, down from 185 GW in his original 2015 proposal. This seems odd to us. As we saw in Chapter 3, concentrated solar power is the only buffered renewable that can poke its head above the 7:1 threshold.

For your convenience, here's the EROI graph again:

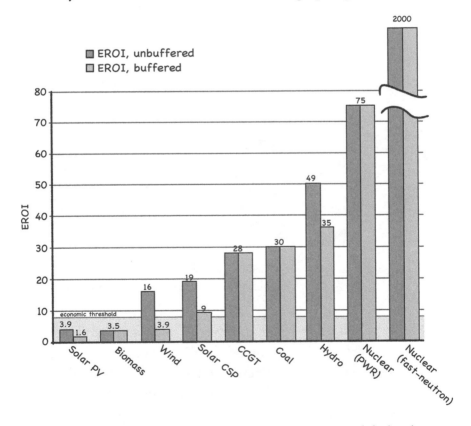

Wind and solar manufacturing are energy-intense global industries with their own supply chains. The mining, refining, manufacturing and delivery of wind and solar equipment consume piles of energy, very little of which is produced by renewables themselves. [13] Wind and solar are often touted in terms of how many households they can power. No disrespect to homemakers, but there is a *lot* of work that gets done in this world aside from running households. Most electricity is consumed by commerce and industry, and a significant portion of that energy must be available on demand, with no excuses. Our lives and fortunes have operated on that universal assumption for more than a century, and this expectation is not expected to change any time soon.

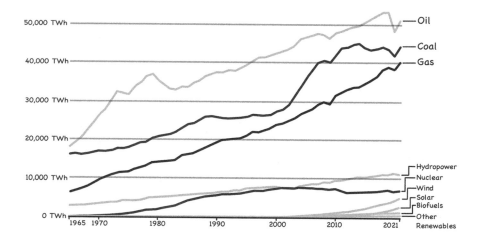

Fig. 85: Global Energy Consumption by Source (1970–2021)
Source: https://ourworldindata.org/grapher/primary-sub-energy-source

Modern civilization is not an off-grid hideaway, and should not be powered as if it were. Living off-grid is all well and good, but if the grid didn't exist, then neither would the developed societies that build and ship the panels and turbines that enable true believers and deep ecologists to live at a clean and green arm's length from the rest of us. Green living is a nice idea, but just like renewables, it doesn't work in isolation. Even Thoreau realized that he needed the rest of us to make his axe so he could harmonize with nature on Walden Pond (we should all be so lucky).

22.2 FINAL THOUGHTS

An abundance of nuclear power is the best chance humanity has to achieve the deep decarbonization required for a sustainable future. Keep in mind, we're talking about primary energy and not just the electricity sector: transportation, industrial process heat, space heating, desalination, and synthetic fuels for air transport. Even ocean transport can be remedied by nuclear power.

About 30,000 1-GW nuclear power plants (30 TW) could supply all the primary energy the planet is expected to demand by 2050. (See our supplement on global primary energy in 2050. [14]) And don't forget, Carbon Zero is just the start.

By our estimate, an additional fleet of 3,000 one-gigawatt reactors—equal to 10% of the global midcentury fleet—may be needed to power an effective CO_2-removal effort. The two-phase, multi-generational task is to de-acidify the oceans and bring atmospheric CO_2 down to 350 ppm, then lower the ppm even further to the pre-industrial ideal of 280 ppm. These and other long-term future projects will be explored in *Power to the Planet*.

Granted, that's a lot of manufacturing and construction, but let's put things in perspective: The number of container ships in the global merchant fleet totals about 5,400 vessels, with about 900 new ships of various sizes being built each year, all around the world. There is no good reason why a global fleet of power reactors, both land-based and water-mobile, cannot be built with the same shipyard efficiency. Think Liberty ships, not superyachts. [15]

With the daunting challenges we face, why in the world would people even *attempt* to generate this insane amount of power using wind and solar if we don't absolutely, positively have to? And with a buildout of nuclear, we don't.

There is a school of environmental thought that says the world must power down. That we must learn to adjust our needs and schedules to whatever energy we can harvest from transient natural phenomena, use what we need, and (hopefully) store the rest for a rainy day. Hence the strategy of wind and solar "farms" passively harvesting the bounty of a capricious Mother Nature.

It's a romantic notion, but like most romance, it's nowhere near as easy or workable as it's made out to be—especially in a world of eight billion humans, many of whom lack sufficient energy as it is. We don't need to power down; what the world needs to do is *power up*, especially with so many nations striving for industrialization and a piece of the pie. Because if cheap, clean, sustainable energy isn't available, they'll just use something else. Nearly any person would do the same. This is one good reason why we shouldn't categorically demonize all fossil fuel. Burning propane for indoor cooking would be a giant step up from burning wood, coal, or dung. This is why we say that the world should transition to carbon-free fuel as fast as *humanely* possible.

While we're increasing our global supply of primary energy, we will also need to decouple our energy production from the natural world as much as we possibly can. [16] The idea is to reduce our impact on the environment while sustainably providing the basics of modern life for ten billion people by midcentury. In other words, we need to reduce our footprint without reducing ourselves. This will require an extremely dense source of carbon-free energy, with a minimal footprint and a small, tightly-managed waste stream.

A renewables future would take an entirely different approach: Jacobson's 100% renewables proposal for the US recommends a nationwide system of over 50,000 passive energy-harvesting farms, connected by a network of new high-voltage transmission corridors to (hopefully) form a self-supporting, fuel-free RE grid in harmony with nature's ebb and flow. That's the plan. We get it; we just don't think it'll get us where we need to go, especially in the time we need to get there. And, it seems like an enormous waste of resources, land, and labor that could be better allocated to other pressing needs.

Decoupling our energy systems from nature—as much as possible, and as fast as possible—is a vitally important task. Especially since we'll be living in a world of ever-less-predictable weather, with ever-less-abundant resources, while an ever-greater portion of the world's population climbs the economic ladder. And they'll all be wanting refrigerators, air conditioners, and personal transportation.

Regardless of anyone's preferred technology, the ultimate goal on which all environmentalists can agree is to restore the planet as closely as possible to pre-industrial levels of atmospheric and oceanic CO_2. And maybe after that, we will finally be living in harmony with nature. But as things stand right now, this planet's ecosphere cannot long sustain the population we already have, much less the one we will almost certainly have by midcentury. And as gas reserves dwindle, so will the global breadbasket.

Even if human-induced global warming didn't exist (it does), emissions must still be brought to a halt. Yes, combustion got us this far, but it's taken a grievous toll in the process. We're already seeing pieces of the world as we know it calving from the iceberg of civilization, adrift in a warming sea. The pollution alone is killing us, and it's killing the oceans too. Climate change just supercharges the dilemma.

But as daunting as all this sounds, none of it means we have to re-engineer civilization and subsist on transient energy harvested from natural sources. Humanity is way past the point of waiting for whatever Nature sends our way. We took the first steps out of that idyllic state when we transitioned from hunting and gathering to farming and livestock. It's way too late for a do-over. And now here we are, all eight billion of us and counting, living in our built environment—architect slang for not-the-natural environment. Turning back would be cataclysmic, and probably require a global disaster to spark the process. We'd rather not.

On the other hand, keeping this show on the road will require our nation and the rest of the world to decarbonize. Because even though most people are good at heart, they will keep burning carbon fuel to get the energy they need, unless there's a better solution—which, we are happy to report, there is.

Nuclear power plants can now be built as cheap, or cheaper, than coal plants. And when long-term operation (LTO) costs are taken into account (see the graph below), nuclear power is cheaper than the lowest price range the IEA (International Energy Agency) projects for global utility-scale solar:

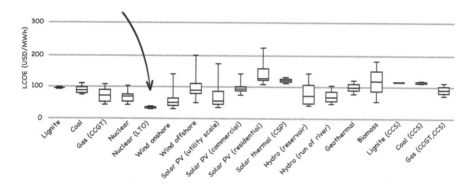

Fig. 86: Levelized Cost of Energy by Technology, at 7% Discount Rate
Check out the interactive graph at: https://www.iea.org/reports
/projected-costs-of-generating-electricity-2020

While the deep ecology of living in sync with nature may appeal to some, most of humanity is surging in the opposite direction. Half of humanity already live in cities, and two-thirds will be urbanized by midcentury. Increased population density requires increased energy density, with power available on demand regardless of weather, season, or time of day. All the greenwashing in the world won't change that fact.

The best way to be in harmony with nature is to let it be. Industrializing nature with a wholesale shift to diffuse and intermittent harvesting systems would be a giant leap backward for humanity and the natural world. And though we as individuals can and should conserve energy, we as a society need all the clean energy we can get. Then we'll need even more to clean up the atmosphere and oceans.

The benefits of nuclear power clearly outweigh the risks—it isn't even close. The sooner we start having an informed conversation on the subject, the better chance we'll have of responding to climate change. But that chance may never come if we're too afraid to board the proverbial plane.

The only thing we have to fear is fear itself.

Endnotes

CHAPTER ONE: The Only Thing We Have to Fear

1. Wikipedia – Aviation Accidents and incidents
   https://en.wikipedia.org/wiki/Aviation_accidents_and_incidents
   Section "Bureau of Aircraft Accidents Archives" [B3A]. The
   sum from 1970 to 2019 is about 86,000 deaths.

2. Wikipedia – Banqiao Dam Failure
   https://en.wikipedia.org/wiki/1975_Banqiao_Dam_failure
   YouTube – Banqiao Dam Disaster
   https://www.youtube.com/watch?v=ctPLXim9WG8

3. HydroQuebec.com – Exports to New York
   https://www.hydroquebec.com/clean-energy-provider/markets
   /new-york.html
   See para. 7
   The Conversation.com – Resistance to hydropower evaporating
   as science takes center stage
   https://theconversation.com/resistance-to-hydropower-is
   -evaporating-as-science-takes-centre-stage-77807
   See para. 4

4. Mercury News – Death toll in San Bruno pipeline explosion
   climbs to eight
   https://www.mercurynews.com/2010/09/28/death-toll-in-san
   -bruno-pipeline-explosion-climbs-to-eight/

5. Wikipedia – List of natural gas and oil production accidents
   https://en.wikipedia.org/wiki/List_of_natural_gas_and_oil
   _production_accidents_in_the_United_States

6. World Nuclear.org – Three Mile Island Accident
   https://www.world-nuclear.org/information-library/safety-and
   -security/safety-of-plants/three-mile-island-accident.aspx
   From section "Health Impacts of Accident":
   "Indeed, more than a dozen major, independent health studies
   of the accident showed no evidence of any abnormal number
   of cancers around TMI years after the accident. The only
   detectable effect was psychological stress during and shortly
   after the accident ... The studies found that the radiation
   releases during the accident were minimal, well below any levels
   that have been associated with health effects from radiation
   exposure. The average radiation dose to people living within
   10 miles of the plant was 0.08 millisieverts, with no more than
   1 millisievert to any single individual. *The level of 0.08 mSv is
   about equal to a chest X-ray, and 1 mSv is about a third of the
   average background level of radiation received by US residents in
   a year.*" [emphasis ours]

7. Timothy Maloney.net – Coal death reduction spreadsheet
   http://www.timothymaloney.net/Preventable_Coal_Deaths
   _files/Coal%20death%20reduction%20-%20spreadsheet%20
   ref.pdf

8. Wikipedia – Chernobyl Disaster
   https://en.wikipedia.org/wiki/Chernobyl_disaster
   UNSCEAR – Assessments of the radiation effects from
   Chernobyl accident
   https://www.unscear.org/unscear/en/areas-of-work/chernobyl
   .html

9. IAEA – Present and future environmental impact of the Chernobyl
   accident
   https://www-pub.iaea.org/MTCD/Publications/PDF/te_1240
   _prn.pdf
   Frame 12

10. UNSCEAR 2008 – Sources and effects of ionizing radiation
    Annex D
    http://www.unscear.org/docs/reports/2008/11-80076_Report
    _2008_Annex_D.pdf
    Frame 8. Also see pg. 58 Article 49, Fig. VII.

There were 134 first responders who experienced Acute Radiation Syndrome (ARS). Following the 28 deaths that occurred in the immediate aftermath of the accident, twenty of the responders died from various causes by year 2004. Eighty-six first responders were still living.

11. IAEA – Chernobyl looking back to go forward
   https://www-pub.iaea.org/MTCD/publications/PDF/Pub1312 _web.pdf Frame 17

12. UNSCEAR 2008 – Sources and effects of ionizing radiation Annex D
   http://www.unscear.org/docs/reports/2008/11-80076_Report _2008_Annex_D.pdf
   Frame 18 Fig. VII "Outcome for Patients with ARS." Scale the light blue segments termed "later deaths." The total is about 20.

13. UNSCEAR 2008 – Sources and effects of ionizing radiation Annex D
   http://www.unscear.org/docs/reports/2008/11-80076_Report _2008_Annex_D.pdf
   Pgs. 182, 183, esp. para. D251 re: doses below 100 mSv

14. Vice.com – Why Hundreds of Thousands of Women Ended Their Pregnancies After Chernobyl
   https://www.vice.com/en/article/ywyqzv/why-hundreds-of -thousands-of-women-ended-their-pregnancies-after-chernobyl
   AP News – Greek Doctors Estimate 2,500 Abortions Last Year Over Chernobyl Fears
   https://apnews.com/article/0bd15dfe73811f65233dc774598f9130

15. NCBI – Health consequences after Chernobyl disaster are less than feared
   https://www.ncbi.nlm.nih.gov/pmc/articles/PMC1200614/
   "The report considered that the mental health impact of Chernobyl was the largest public health problem. People living in affected areas tended to have negative self-assessments of their health, believed they had shortened life expectancy, and tended to lack initiative and depend on help from the state. Persistent myths and misperceptions about the threat of radiation have resulted in "paralyzing fatalism" among residents, the report warned, noting that "Poverty, lifestyle

diseases now rampant in the former Soviet Union, and mental health problems pose a far greater threat to local communities than does radiation exposure."

16. UNEP. Org – No immediate health risks from Fukushima nuclear accident says UN expert science panel
https://www.unep.org/news-and-stories/press-release/no
-immediate-health-risks-fukushima-nuclear-accident-says-un
-expert

17. Japan Times – Fukushima stress deaths tops 3/11 poll
https://www.japantimes.co.jp/news/2014/02/20/national/post-quake-illnesses-kill-more-in-fukushima-than-2011-disaster/#.XnF7ehd7nMU
Our World in Data. org – What was the death toll from Chernobyl and Fukushima?
https://ourworldindata.org/what-was-the-death-toll-from
-chernobyl-and-fukushima

18. World Nuclear.org – Fukushima radiation exposure
http://www.world-nuclear.org/information-library/safety
-and-security/safety-of-plants/appendices/fukushima-radiation
-exposure.aspx
Section "Radiation Effects" para.1

19. WebMD – Radiation doses from CT scans
https://www.webmd.com/cancer/radiation-doses-ct-scans

20. NY Times – The Evacuation zones around the Fukushima Daiichi nuclear plant
https://archive.nytimes.com/www.nytimes.com/interactive
/2011/03/16/world/asia/japan-nuclear-evaculation-zone.html
Note the panicked wording of the graphic. The article was published less than a week after the accident.
Medium.com – For The First Time, World Learns Truth About Risk Of Nuclear
https://medium.com/generation-atomic/for-the-first-time-world
-learns-truth-about-risk-of-nuclear-6b7e97d435df

21. Blogs.Nature.com – Fire in Fukushima unit 4 may have been caused by unit 3
https://blogs.nature.com/news/2011/05/exclusive_fire_in
_fukushima_un.html

22. Slate – After the Fukushima disaster, a U.S. mistake undermined the Japanese government
https://slate.com/technology/2013/09/nrc-response-to-fukushima-a-mistake-turned-public-opinion-against-japan.html

23. CSPAN – Japan nuclear plant crisis
https://www.c-span.org/video/?298535-1/japan-nuclear-plant-crisis
Forbes – Former U.S. Nuclear Safety Regulator Withheld Vital Aid During Fukushima Crisis
https://www.forbes.com/sites/michaelshellenberger/2019/01/22/former-u-s-nuclear-safety-regulator-withheld-vital-aid-during-fukushima-crisis-says-top-official/
4thgeneration.energy – I read former NRC chairman's book of lies so you don't have to
https://4thgeneration.energy/i-read-former-nrc-chairmans-book-of-lies-so-you-dont-have-to-part-i/
Wikipedia – Gregory Jaczko
https://en.wikipedia.org/wiki/Gregory_Jaczko
See second to last para. After his term at the NRC, Jaczko left government to start a for-profit wind consulting company

24. Atomic Insights – Spent Fuel Pools Protect The Public. Don't Believe Skeptics
https://atomicinsights.com/spent-fuel-pools-protect-public-dont-believe-skeptics/
NRC.gov – Technical Study of Spent Fuel Pool Accident Risk at Decommissioning Nuclear Power Plants
https://www.nrc.gov/docs/ML0104/ML010430066.pdf Frame 27

25. NIRS.org – National Diet of Japan official report of the Fukushima nuclear accident
https://www.nirs.org/wp-content/uploads/fukushima/naiic_report.pdf
Frame 9

26. Japan Times – Fukushima plant site originally was a hill safe from tsunami
https://www.japantimes.co.jp/news/2011/07/13/national/fukushima-plant-site-originally-was-a-hill-safe-from-tsunami/
Royal Society Publishing – The Fukushima accident was preventable

https://royalsocietypublishing.org/doi/10.1098/rsta.2014.0379

27. NY Times – Tsunami warnings, written in stone
https://www.nytimes.com/2011/04/21/world/asia/21stones.html

28. Bulletin of Atomic Scientists – Onagawa: The Japanese nuclear power plant that didn't melt down on 3/11
https://thebulletin.org/2014/03/onagawa-the-japanese-nuclear -power-plant-that-didnt-melt-down-on-3-11/
Actinide Age – When Your Local Nuclear Plant is the Safest Place in the World
https://actinideage.wordpress.com/2016/09/17/when-your-local -nuclear-plant-is-the-safest-place-in-the-world/

29. IAEA.org – The Westinghouse AP-1000
https://inis.iaea.org/collection/NCLCollectionStore/_Public /42/026/42026956.pdf

30. World Nuclear.org – Fukushima Daiichi accident
https://www.world-nuclear.org/information-library/safety-and -security/safety-of-plants/fukushima-accident.aspx
Search for "sequence of evacuation orders." In that section, see map titled: "State of Reconstruction of Fukushima Prefecture" (updated 2020)

31. CS Monitor – Japan wants Fukushima evacuees to go home. They're not so sure
https://www.csmonitor.com/World/Asia-Pacific/2018/0221 /Japan-wants-Fukushima-evacuees-to-go-home.-They-re-not -so-sure

32. Washington Post – Eight years after Fukushima's meltdown, the land is recovering, but public trust is not
https://www.washingtonpost.com/world/asia_pacific/eight -years-after-fukushimas-meltdown-the-land-is-recovering-but -public-trust-has-not/2019/02/19/0bb29756-255d-11e9-b5b4 -1d18dfb7b084_story.html
Asahi.com – New residents near Fukushima nuclear plant can get 2 million yen
https://www.asahi.com/ajw/articles/14031389
(About $20,000 USD)
France24 – Fukushima evacuees resist return as Olympics near
https://www.youtube.com/watch?v=oEQeeqtuXZc

33. *Ibid* # 23 above. See third link
     Heritage.org – Mr. Jaczko went to Washington and look
     what happened
        https://www.heritage.org/nuclear-energy/commentary/mr-jaczko
        -went-washington-and-look-what-happened
        Para. 3
34. Dropbox – NRC letter about Jaczko
        https://www.dropbox.com/s/23ulwm5kmlpcinh/NRC_letter
        _about_Jaczko_13-X-2011.png
35. JCER – Accident Cleanup Costs Rising to 35-80 Trillion Yen
     in 40 Years
        https://www.jcer.or.jp/english/accident-cleanup-costs-rising
        -to-35-80-trillion-yen-in-40-years
36. French Academy of Sciences – Dose–effect relationship and
     estimation of the carcinogenic effects of low doses of ionizing
     radiation
        https://www.redjournal.org/article/S0360-3016(05)01135-1
        /fulltext
37. World Nuclear.org – Fukushima: radiation exposure
        https://www.world-nuclear.org/information-library/safety
        -and-security/safety-of-plants/appendices/fukushima-radiation
        -exposure.aspx
     "No harmful health effects were found in 195,345 residents
     living in the vicinity of the plant who were screened by the end
     of May 2011. All the 1,080 children tested for thyroid gland
     exposure showed results within safe limits, according to the
     report submitted to IAEA in June. By December, government
     health checks of some 1700 residents who were evacuated from
     three municipalities showed that two-thirds received an external
     radiation dose within the normal international limit of 1 mSv,
     98% were below 5 mSv/yr, and ten people were exposed to more
     than 10 mSv."
38. PNAS – Radiation dose rates now and in the future for residents
     neighboring restricted areas of the Fukushima Daiichi Nuclear
     Power Plant
        https://www.pnas.org/doi/10.1073/pnas.1315684111

"Measurements were taken of the full-body contamination from cesium exposure of 9498 residents who had returned to the town and stayed there between September 2011 and March 2012. The study found that two-thirds of the residents had no detectable levels of cesium. Of the rest, only one appeared to have received an equivalent dose more than 1 mSv, and that was 1.07 mSv. The current ambient dose rate in the town is about 3 mSv/yr from external sources, well within the government's 20 mSv/yr limit for returnees."

39. HPS.org – Radiation risk in perspective
https://hps.org/documents/radiationrisk.pdf

40. NY Times – Japan to Pay Cancer Bills for Fukushima Worker
https://www.nytimes.com/2015/10/21/world/asia/japan-cancer-fukushima-nuclear-plant-compensation.html
Unseen Japan.com – Fukushima Workers Battle Leukemia – and Bureaucracy
https://unseenjapan.com/fukushima-workers-leukemia-bureaucracy/

41. ANS.org – No, The Cancer Death Was Probably Not From Fukushima
https://www.ans.org/news/article-2074/no-the-cancer-death-was-probably-not-from-fukushima/
Forbes – Top Scientist Says Japan's Decision To Financially Reward Fukushima Worker Is Not Based On Science
https://www.forbes.com/sites/michaelshellenberger/2018/09/06/top-scientist-says-japans-decision-to-financially-reward-fukushima-worker-is-not-based-on-science/

42. NCBI – Lessons from Fukushima: Latest Findings of Thyroid Cancer After the Fukushima Nuclear Power Plant Accident
https://www.ncbi.nlm.nih.gov/pmc/articles/PMC5770131/

43. Ednocrineweb.com – What is thyroid cancer?
https://www.endocrineweb.com/conditions/thyroid-cancer
Section "Thyroid Cancer Prognosis"
JPANDS.org – Thyroid Cancer Following Childhood Low Dose Radiation Exposure: Fallacies in a Pooled Analysis
https://www.jpands.org/vol22no4/cuttler.pdf

CHAPTER TWO: Zoomies! We're Doomed!

(Note: For the proper pronunciation of "doomed," see this short video:)
YouTube – We're Doomed!!
https://www.youtube.com/watch?v=nNzkIgGpsAw

1. Wikipedia – Cosmic ray
   https://en.wikipedia.org/wiki/Cosmic_ray
   Space.com – Death rays from space: How bad are they?
   https://www.space.com/7193-death-rays-space-bad.html
   NCBI – Basic Principles of Radiation Biology
   https://www.ncbi.nlm.nih.gov/books/NBK232435/
   UN library.org – Sources and effects of ionizing radiation
   (UNSCEAR 2000)
   https://www.un-ilibrary.org/content/books/9789210582483
   /read Pg. 84 / frame 86
2. Metadata Berkeley.edu – Decay chains
   https://metadata.berkeley.edu/nuclear-forensics/Decay%20
   Chains.html
3. YouTube – Radiation how much is too much?
   https://www.youtube.com/watch?v=niFizj29h5c
4. Healthline – What Does Potassium Do for Your Body?
   https://www.healthline.com/nutrition/what-does-potassium-do
5. Phyast.pitt.edu – Test of the linear no threshold theory of radiation
   carcinogenesis for inhaled radon decay products
   http://www.phyast.pitt.edu/~blc/LNT-1995.PDF
6. NCBI – Toxicological profile for uranium health effects
   https://www.ncbi.nlm.nih.gov/books/NBK158798/
   Para. 8: "Ingested uranium is less toxic than inhaled
   uranium, which may be partly attributable to the relatively low
   gastrointestinal absorption of uranium compounds."
7. See the supplement "Sieverts and Grays" in the back of this book.
8. Scientific American – Why is the earth's core so hot?
   https://www.scientificamerican.com/article/why-is-the-earths
   -core-so/
9. Wikipedia – Geothermal gradient
   https://en.wikipedia.org/wiki/Geothermal_gradient

10. Wikipedia – Magnetosphere
    https://simple.wikipedia.org/wiki/Magnetosphere#/media
    /File:Structure_of_the_magnetosphere-en.svg
    BBC.om – Why Mars doesn't have a molten core: Nasa probe
    determines Mars' internal structure
    https://www.bbc.com/news/science-environment-57935742

11. Science Daily – Wave-particle duality
    https://www.sciencedaily.com/terms/wave-particle_duality.htm

12. NASA.gov – What are gamma rays?
    https://heasarc.gsfc.nasa.gov/docs/cgro/epo/vu/overview
    /whatare/whatare.html

13. Wikipedia – Rosalind Franklin
    https://en.wikipedia.org/wiki/Rosalind_Franklin

14. Nature – A brief history of the DNA repair field
    https://www.nature.com/articles/cr2007113

15. NY Times – Nobel Prize in Chemistry Awarded to Tomas
    Lindahl, Paul Modrich and Aziz Sancar for DNA Studies
    https://www.nytimes.com/2015/10/08/science/tomas-lindahl
    -paul-modrich-aziz-sancarn-nobel-chemistry.html
    YouTube – Why we shouldn't fear radiation
    https://www.youtube.com/watch?v=agzrhxZD5kc
    Ars Technica – Systems that keep the genome safe earn three
    Nobels in chemistry
    https://arstechnica.com/science/2015/10/systems-that-keep-the
    -genome-safe-earn-three-nobels-in-chemistry/

16. Wikipedia – Great Oxidation Event
    https://en.m.wikipedia.org/wiki/Great_Oxidation_Event
    Amazon – Oxygen: The molecule that made the world
    https://www.amazon.com/Oxygen-molecule-Oxford
    -Landmark-Science/dp/0198784937
    Nature – Rising levels of atmospheric oxygen and evolution of Nrf2
    https://www.nature.com/articles/srep27740

17. Blogs.VOA.news – Earth's Atmosphere Got Its Oxygen in Periodic
    Bursts Over a 100 Million Year Period
    https://blogs.voanews.com/science-world/2015/11/20/earths
    -atmosphere-got-its-oxygen-in-periodic-bursts-over-a-100-million
    -year-period/

Universe Today – Radioactive Hot Spots on Earth's Beaches May Have Sparked Life

https://www.universetoday.com/12425/radioactive-hot-spots-on-earths-beaches-may-have-sparked-life/

Ibid. #16 above. See third link

18. NCBI – Oxidative stress, inflammation, and cancer: How are they linked?

https://www.ncbi.nlm.nih.gov/pmc/articles/PMC2990475/

See abstract

News.USC.edu – Oxygen is a blessing, right? Well, it can also be a curse

https://news.usc.edu/87372/oxygen-is-a-blessing-right-well-it-can-also-be-a-curse/

19. Science Daily – Dynamic duo takes out cellular trash

https://www.sciencedaily.com/releases/2014/09/140907180642.htm Para. 4

20. WebMD – Understanding cancer - the basics

https://www.webmd.com/cancer/guide/understanding-cancer-basics#1

Space Daily – Low-level radiation exposure less harmful to health than other modern lifestyle risks

https://www.spacedaily.com/reports/Low_level_radiation_exposure_less_harmful_to_health_than_other_modern_lifestyle_risks_999.html

21. NCBI – Greetings: 50 years of Atomic Bomb Casualty Commission–Radiation Effects Research Foundation studies

https://www.ncbi.nlm.nih.gov/pmc/articles/PMC33856/ Para. 6

22. Bio One.org – Solid Cancer Incidence among the Life Span Study of Atomic Bomb Survivors: 1958–2009

https://bioone.org/journals/radiation-research/volume-187/issue-5/RR14492.1/Solid-Cancer-Incidence-among-the-Life-Span-Study-of-Atomic/10.1667/RR14492.1.full

See table 3. The cancer rate for the exposed population can be derived from Rows 1 and 2.

There were 105,444 persons (73,401 + 32,043) in the two cities. They experienced 22,538 cancers (16,387+ 6,151) over the 50+-year course of the study. Their cancer rate was 22,538 ÷ 105,444 = 21.4%.

The non-exposed cohort is documented in Row 12, labeled NIC. There were 5,222 cancers among 25,239 subjects, for a cancer rate of 5,222 ÷ 25,239 = 20.7%.

Therefore, the radiation-exposed population experienced a less than 1% greater rate of cancer compared to the unexposed population (21.4 – 20.7 = 0.7%).

23. NCBI – The radiation hypersensitivity of cells at mitosis
https://pubmed.ncbi.nlm.nih.gov/12556342/
24. Kahn Academy – DNA proofreading and repair
https://www.khanacademy.org/science/high-school-biology /hs-molecular-genetics/hs-discovery-and-structure-of-dna/a /dna-proofreading-and-repair See introduction
25. You Tube – The myths of nuclear energy
https://www.youtube.com/watch?v=c1QmB5bW_WQ
Ibid. #20 Chapter One second link

CHAPTER THREE: A Half-Life Well Lived

1. Wikipedia – Exxon Valdez oil spill
https://en.wikipedia.org/wiki/Exxon_Valdez_oil_spill
2. Greenbiz.com – It's time to trash recycling
https://www.greenbiz.com/article/its-time-trash-recycling
Reuters – The recycling myth big oil's solution for plastic waste littered with failure
https://www.reuters.com/investigates/special-report/environment -plastic-oil-recycling/
3. Wikipedia – Electric arc furnace
https://en.wikipedia.org/wiki/Electric_arc_furnace
Amazon – Prescription for the planet
https://www.amazon.com/Prescription-Planet-Painless-Remedy -Environmental/dp/1419655825
See Chapter 7
4. Wikipedia – Radioisotope thermoelectric generator
https://en.wikipedia.org/wiki/Radioisotope_thermoelectric _generator
5. Wikipedia – Voyager 2

https://en.wikipedia.org/wiki/Voyager_2

6.   MDPI – What is the Minimum EROI that a Sustainable Society
      Must Have?
      https://www.mdpi.com/1996-1073/2/1/25

7.   Forbes – EROI a Tool To Predict The Best Energy Mix
      https://www.forbes.com/sites/jamesconca/2015/02/11
      /eroi-a-tool-to-predict-the-best-energy-mix/?sh=49fdb197a027
      Festkoerper – Energy intensities, EROIs, and energy payback
      times of electricity generating power plants
      https://festkoerper-kernphysik.de/Weissbach_EROI_preprint.pdf

8.   National Review – Dr. Charles Hall on sustainable energy sources
      https://therationalview.podbean.com/e/dr-charles-hall-on
      -sustainable-energy-sources/

CHAPTER FOUR: A Fate Worse than Global Warming

1.   Monbiot.com – Atomised: The Fukushima crisis should not spell
      the end of nuclear power
      https://www.monbiot.com/2011/03/16/atomised/

2.   Wikipedia – SL-1
      https://en.wikipedia.org/wiki/SL-1

3.   Inis.iaea.org – Significant incidents in nuclear fuel cycle facilities
      https://inis.iaea.org/collection/NCLCollectionStore/_Public
      /27/060/27060437.pdf

4.   World Nuclear.org – World Nuclear Performance Report 2018
      https://world-nuclear.org/getmedia/b392d1cd-f7d2-4d54-9355
      -9a65f71a3419/performance-report.pdf.aspx
      Section 2.1 Global Highlights, pg. 10 fig. 1
      Tim sez: With a careful scaling of the "Nuclear Electricity
      Production" bar graph, the purple segments for North America
      show that total historical US and Canadian nuclear electric
      production is about 30,000 TWh, 30 trillion kWh.
      World-Nuclear.org – Nuclear power in Canada
      https://www.world-nuclear.org/information-library/country
      -profiles/countries-a-f/canada-nuclear-power.aspx
      Working with the historical Canada reactor chart and other
      data in this link indicates that total historical Canadian nuclear
      electric production is about 3.3 trillion kWh. Therefore, US

historical total is about 27 trillion kWh: 2,700,000 megawatt-hours or 2,700 terawatt-hours.

5. Forbes – Which Industry Offers The Safest Jobs In America: Nuclear Or Logging?
   https://www.forbes.com/sites/jamesconca/2019/12/13/the-safest-and-the-most-dangerous-jobs-in-america--nuclear-and-logging/?sh=26a8292c455b

6. Nature – Nuclear power plants cleared of leukemia risk
   https://www.nature.com/articles/news.2011.275
   The clue that led to the actual cause of the contaminated soil was that somebody finally noticed the elevated leukemia rates near some of the proposed sites for future reactors. Further investigation revealed that these empty lots had been contaminated by previous industries.

7. Washington Post – In the past five years, at least six Americans have been shot by dogs
   https://www.washingtonpost.com/news/wonk/wp/2015/10/27/a-dog-shoots-a-person-almost-every-year-in-america/

8. The Guardian – 'Ignored for 70 years': human rights group to investigate uranium contamination on Navajo Nation
   https://www.theguardian.com/environment/2021/oct/27/human-rights-group-uranium-contamination-navajo-nation

9. NPR – For the Navajo Nation, uranium mining's deadly legacy lingers
   https://www.npr.org/sections/health-shots/2016/04/10/473547227/for-the-navajo-nation-uranium-minings-deadly-legacy-lingers

10. World Nuclear.org – Uranium mining overview
    https://www.world-nuclear.org/information-library/nuclear-fuel-cycle/mining-of-uranium/uranium-mining-overview.aspx

11. NCBI – Life Cycle Greenhouse Gas Emissions from Uranium Mining and Milling in Canada
    https://pubmed.ncbi.nlm.nih.gov/27471915/ Abstract, esp. lines 12 and 13

12. UNECE.org – Carbon Neutrality in the UNECE Region: Integrated Life-cycle Assessment of Electricity Sources

https://unece.org/sites/default/files/2022-04/LCA_3_FINAL%20March%202022.pdf

Fig. 1 Frame B

13. Nowtricity.com – Current emissions in Germany
https://www.nowtricity.com/country/germany/

14. Energy.gov – What's the lifespan for a nuclear reactor? Much longer than you might think
https://www.energy.gov/ne/articles/whats-lifespan-nuclear-reactor-much-longer-you-might-think

Utility Drive – How long can a nuclear plant run? Regulators consider 100 years
https://www.utilitydive.com/news/how-long-can-a-nuclear-plant-run-regulators-consider-100-years/597294/

Scientific American – How long can a nuclear reactor last?
https://www.scientificamerican.com/article/nuclear-power-plant-aging-reactor-replacement-/

15. NEI magazine – Renewal by annealing
https://www.neimagazine.com/features/featurerenewal-by-annealing-7171272

NRC.gov – Section 50.66 Requirements for thermal annealing of the reactor pressure vessel
https://www.nrc.gov/reading-rm/doc-collections/cfr/part050/part050-0066.html

16. World Nuclear News – Rosatom launches annealing technology for VVER-1000 units
https://www.world-nuclear-news.org/Articles/Rosatom-launches-annealing-technology-for-VVER-100

17. EPA.gov – TENORM uranium mining residuals
https://www.epa.gov/radiation/tenorm-uranium-mining-residuals

18. Forbes – Nuclear Power - Where's The Uranium Coming From?
https://www.forbes.com/sites/jamesconca/2019/05/28/nuclear-power-wheres-the-uranium-coming-from/?sh=2e17e3e37b9f

19. *Ibid.* # 17 above
Wikipedia – In situ leaching
https://en.m.wikipedia.org/wiki/In_situ_leach

20. See Part One of the mining supplement at the back of this book.

21. See Part Two of the mining supplement at the back of this book.

22. World Nuclear.org – Environmental aspects of uranium mining
     https://world-nuclear.org/information-library/nuclear-fuel-cycle/
     mining-of-uranium/environmental-aspects-of-uranium-mining.aspx

23. HPS.org – Radiation from granite countertops information sheet
     https://hps.org/documents/Radiation_granite_countertops.pdf
     An EPA (Environmental Protection Agency) survey of
     abandoned U mines in the Ambrosia Lake region of New
     Mexico found that 85% of the 22,000-acre site had levels of 5
     to 15 μR/hr. See frame 11:
     EPA.gov – Aerial radiological surveys Ambrosia Lake uranium
     mines
     https://www.epa.gov/sites/default/files/2015-05/documents
     /nm_grants_aerial-radiological-survey-of-the-ambrosia-lake
     -uranium-mines.pdf

CHAPTER FIVE: Waste Not, Want Not

1. Schlissel technical.com – Coal-fired power plant construction costs
     https://schlissel-technical.com/docs/reports_35.pdf
     See introduction. As far back in 2006, the price of a new coal
     plant in Asia was $3,500 per installed kW
     Science council – The cost of nuclear power
     https://www.thesciencecouncil.com/index.php/advisors/active
     -advisers/dr-barry-brook/175-the-cost-of-nuclear-power In 2021,
     reactors in Asia were being built anywhere from $1,300 to $3600
     /kW
     Robert Hargraves – Thorium Energy Cheaper than Coal @
     TEAC 12
     https://www.youtube.com/watch?v=ayIyiVua8cY

2. You Tube – Our friend the atom
     https://www.youtube.com/watch?v=pkwadgJORFM

3. World nuclear.org – Plutonium
     https://world-nuclear.org/information-library/nuclear-fuel-cycle
     /fuel-recycling/plutonium.aspx
     In theory, Pu-239 has a 78% rate of fission in slow spectrum.
     But in practice, only about 55% of the Pu-239 that forms in a
     fuel pellet ever gets hit with a neutron. So the resulting rate of

Pu-239 fission in a light-water reactor is only 43% (0.78 × 0.55 = 0.43).

4. Powermag.com – Rapid advancement for fast reactors
https://www.powermag.com/rapid-advancements-for-fast
-reactors/

5. Nwtrb.gov – Management and Disposal of U.S. Department of Energy Spent Nuclear Fuel
https://www.nwtrb.gov/docs/default-source/reports/nwtrb
-mngmntanddisposal-dec2017-508a.pdf?sfvrsn=12

Tim sez: Cumulative SNF was 72,000 tonnes in 2017 and is going up at a rate of about 2,500 tonnes a year. Round that up to the generally accepted value of 80,000 tonnes SNF in the US in 2020.

At a burn rate of about 1.05 GWe-years / tonne, that comes to about 84,000 GWe-yr. Multiply by 8.76 TWh per GWe-yr for about 740,000 TWh available from SNF.

Divide by 4178 TWh / yr to obtain 177 years of US electricity from SNF.

6. NRC.gov – Background information on depleted uranium
https://www.nrc.gov/waste/llw-disposal/llw-pa/uw-streams
/bg-info-du.html Para 1

7. YouTube – OPG's deep geological repository
https://www.youtube.com/watch?v=m1CFxyWYUZw
The Drive.com – Wait, this mysterious heavily-armored blue train caboose belongs to the navy?
https://www.thedrive.com/the-war-zone/39654/wait-this
-mysterious-heavily-armored-blue-train-caboose-belongs-to
-the-navy
YouTube – Nuclear fuel cask test - 1978
https://www.youtube.com/watch?v=Bu1YFshFuI4
Facebook – Zeilony Atom – Train hits container for radioactive waste
https://www.facebook.com/watch/?v=242266033894867
YouTube – The real bad stuff (high-level waste)
https://www.youtube.com/watch?v=KnxksKmJa6U

8. See the supplement "reserves of Depleted Uranium" in the back of this book.

9. All great quotes – James Hansen quotes
https://www.allgreatquotes.com/quote-197420/

10. Jack Devanney Substack – Taking to people about nuclear power
https://jackdevanney.substack.com/p/talking-to-people-about
-nuclear-power?utm_source=substack&utm_medium=email
After 600 years:
https://worksinprogress.substack.com/p/a-tale-of-two-particles
After 1,000 years:
https://world-nuclear.org/information-library/nuclear-fuel-cycle
/nuclear-wastes/radioactive-wastes-myths-and-realities.aspx
See bullet point #4
Jack Devanney Substack - Deep geologic hubris
https://jackdevanney.substack.com/p/deep-geologic-hubris?utm
_source=substack&utm_medium=email

CHAPTER SIX: Back to the Future

1. Wikipedia – Experimental breeder reactor Two
https://en.wikipedia.org/wiki/Experimental_Breeder_Reactor_II

2. YouTube – Argonne explains nuclear recycling in 4 minutes
https://www.youtube.com/watch?v=MlMDDhQ9-pE

3. YouTube – Integral fast reactor (IFR)
https://www.youtube.com/watch?v=tgFYLVcXSGw
From the 8:30 mark
Vimeo.com – Making a Contribution: The Story of EBR-II
(Full Version)
https://vimeo.com/35261457

4. YouTube – Thorium remix EBR-2 test
https://www.youtube.com/watch?v=hUJ_FV5xeFc

5. Hackaday.com – Coal to nuclear transition to decarbonize the grid
https://hackaday.com/2022/09/22/coal-to-nuclear-transition
-to-decarbonize-the-grid/
INL.gov – Investigating Benefits and Challenges of Converting
Retiring Coal Plants into Nuclear Plants
https://fuelcycleoptions.inl.gov/SiteAssets/SitePages/Home
/C2N2022Report.pdf
A DoE report on coal-to-nuclear

6. Terrapower.com – molten chloride fast reactor technology

https://www.terrapower.com/our-work/molten-chloride-fast
-reactor-technology/

7. Springer.com – Synthesis of americium trichloride via chlorination of americium oxide using zirconium tetrachloride in LiCl–KCl molten salt
https://link.springer.com/article/10.1007/s10967-022-08527-3

8. YouTube – Recycling used nuclear fuel - Orano la Hague - English
https://www.youtube.com/watch?v=V0UJSlKIy8g

9. France's emissions are 1/8th of Germany's.
For France's electricity:
5.13 t /person × (0.3 cm / 2.9 cm) = *0.53 t /person:*
Worldometer.info – France CO_2 emissions
https://www.worldometers.info/co2-emissions/france-co2
-emissions/
The light blue electricity segment = 1.2 cm. Total bar height = 2.7 cm
For Germany's electricity:
9.44 t /person × (1.2 cm / 2.7 cm) = *4.20 t /person:*
Worldometer.info – Germany CO_2 emissions
https://www.worldometers.info/co2-emissions/germany-co2
-emissions/
The light blue electricity segment = 0.3 cm. Total bar height = 2.9 cm
Comparing the two countries:
4.20 t /person Germany ÷ 0.53 t /person France = *7.9 × greater in Germany.* To track the numbers in real time go to:
Electricitymaps.com
https://app.electricitymaps.com/map
Apple store – Gridwatch
https://apps.apple.com/us/app/gridwatch/id1553702520
m.Akpure.com – Gridwatch
https://m.apkpure.com/gridwatch-beta/com.umich.gridwatch

10. Global construction review – China to unveil world's first waterless molten salt reactor, suitable for deserts
https://www.globalconstructionreview.com/china-unveil-worlds
-first-waterless-molten-salt-re/

Powermag.com – China approves commissioning of thorium-powered reactor

   https://www.powermag.com/china-approves-commissioning -of-thorium-powered-reactor/

11.   TMSR.com – Why China's 600 fte MSR program wants to cooperate with Delft TU and NRG in Petten

   https://www.thmsr.com/wp-content/uploads/2020/05/20171205 -Report-MenG-SINAP-researchers-anonymized-def-.pdf

12.   US thorium reserves are about 200,000 tonnes:

USGS.gov – Thorium deposits of the United States energy resources for the future?

   https://pubs.usgs.gov/circ/1336/pdf/C1336.pdf

   See table 2, pg. 3 / frame 9

Ibid. #13 above – Energy content per tonne of thorium is about the same as for uranium.

   By proportion, (6.4 million tonnes Th ÷ 4.5 billion tonnes U) × (240,000 years PRI NRG from Uranium) = about 340 years of global primary energy from Thorium.

Uranium from rare earth:

World nuclear.org – Uranium from rare earth deposits

   https://world-nuclear.org/information-library/nuclear-fuel-cycle /uranium-resources/uranium-from-rare-earths-deposits.aspx

13.   World nuclear.org – Uranium from phosphates

   https://www.world-nuclear.org/information-library/nuclear -fuel-cycle/uranium-resources/uranium-from-phosphates.aspx

14.   LiveScience – Facts about thorium

   https://www.livescience.com/39686-facts-about-thorium.html

World nuclear.org – Thorium

   https://world-nuclear.org/information-library/current-and -future-generation/thorium.aspx

Wikipedia – Thorium on the moon

   https://en.wikipedia.org/wiki/Compton%E2%80%93 Belkovich_Thorium_Anomaly

JPL.NASA.gov – Map of Martian thorium at mid-latitudes

   https://www.jpl.nasa.gov/images/pia04257-map-of-martian -thorium-at-mid-latitudes

15.   CleanCore energy – ANEEL: Thorium-Based Reactor Fuel Could Support A New Wave Of Nuclear Power
       https://cleancore.energy/news/blog-post-title-one-69b2e
       YouTube – Thorium + HALEU = clean core energy Mark Nelson
       https://www.youtube.com/watch?v=nAUDuaqpVW8

16.   Mike sez:

The uranium / plutonium fuel cycle plays out along these lines. (Note: The numbers here are approximate—no process is 100% efficient):

U-235 has an 85% chance of fission in slow spectrum. The unfissioned 15% will absorb neutrons and eventually become plutonium-239, which has a 40% rate of fission in slow spectrum. The remaining 60% means that 9% of the original U-235 fuel mass will continue absorbing neutrons (0.6 × 0.15 = 0.09), and become even larger, long-lived transuranics.

The thorium / uranium fuel cycle starts at a lower number on the Periodic Table:

World-Nuclear.org – fuel recycling
       https://www.world-nuclear.org/information-library/nuclear
       -fuel-cycle/fuel-recycling/plutonium.aspx

Thorium-232 absorbs a neutron and becomes U-233, which has a 92% rate of fission in slow spectrum. The remaining 8% absorbs neutrons and becomes U-235, of which 85% will fission. The remaining 15% is 1.2% of the original thorium mass (0.15 × .08 = 0.012). These newly-minted U-235s will absorb neutrons and become Pu-239, of which 40% will fission. The remaining 60% is 0.7% of the original thorium mass (0.6 × 0.012 = 0.007).

So while 9% of U-235 fuel used in slow spectrum will ultimately become long-lived TRUs, only 0.7% of Th-232 will do the same. That's about 13 times less long-lived waste, by using the smallest, lightest atom that fissions: U-233. And while highly fissile U-233 is almost non-existent in nature, it can be bred quite easily in a reactor from fertile Th-232, which is abundant, requires no fuel processing, and is hardly radioactive at all.

17.   IAEA.org – Uranium and REE recovery from Florida phosphates looking back and going forward

https://inis.iaea.org/collection/NCLCollectionStore/_Public
/48/039/48039440.pdf

18.   Amazon – Superfuel by Richard Martin
https://www.amazon.com/SuperFuel-Thorium-Energy-Source
-Future/dp/113727834X

19.   IER – Big Wind's Dirty Little Secret: Toxic Lakes and Radioactive
Waste
https://www.instituteforenergyresearch.org/renewable/wind
/big-winds-dirty-little-secret-rare-earth-minerals/

20.   Forbes – Uranium Seawater Extraction Makes Nuclear Power
Completely Renewable
https://www.forbes.com/sites/jamesconca/2016/07/01/uranium
-seawater-extraction-makes-nuclear-power-completely-renewable
/?sh=d5671a2159ae
YouTube – Scientists Extract Uranium Powder from Seawater
with Yarn
https://www.youtube.com/watch?v=Y6g5pj9QfMg
Ars Technica – Attention knitters: Researchers harvest
uranium from the sea with a yarn "net"
https://arstechnica.com/science/2018/06/attention-knitters
-researchers-harvest-uranium-from-the-sea-with-a-yarn-net
China's method of uranium extraction from seawater:
Research Gate – Selective extraction of uranium from seawater
with biofouling-resistant polymeric peptide
https://www.researchgate.net/publication/350825165_Selective
_extraction_of_uranium_from_seawater_with_biofouling
-resistant_polymeric_peptide

21.   India's method of uranium extraction from seawater:
IndianExpress.com – IISER Pune scientists successfully
extract over 95% uranium from seawater
https://indianexpress.com/article/cities/pune/iiser-pune-scientists
-successfully-extract-over-95-uranium-from-seawater-8096491/
India Today – Indian scientists extract record uranium from
seawater that could power nuclear plants
https://www.indiatoday.in/science/story/indian-researchers
-extract-record-uranium-from-seawater-that-powers-nuclear
-plants-1986196-2022-08-10

22. CNA.ca – There's uranium in seawater and it's renewable
https://cna.ca/2016/07/27/theres-uranium-seawater-renewable/
Source paper (2011):
Large.stanford.edu – Amidoxime Uranium Extraction From Seawater
http://large.stanford.edu/courses/2011/ph241/chan1/

23. Terrestrial energy – Integral molten salt reactor: carbon-free, low-cost, high-impact
https://www.terrestrialenergy.com/technology/

24. World Nuclear.org – Uranium in Namibia
https://world-nuclear.org/information-library/country-profiles /countries-g-n/namibia.aspx Search for "China"
English.news.cn – Chinese-invested Namibian uranium mine launches sustainable development report
https://english.news.cn/20220901/19c3f090afc34c80b85cf3f 944cb839f/c.html

25. Tim sez:
Assume that fissioning one tonne of uranium produces one gigawatt-year of electric energy, or 1 GWe-yr. Therefore, the ocean's 4.5 billion tonnes of uranium can produce 4.5 billion GWe-yr. This can be expressed as 4.5E18 watt-years.
Multiply 4.5E18 W-years × 8760 h /yr equals approx. 40E21 watt-hours of total ocean-sourced nuclear energy.
Global annual primary energy (PRI NRG) consumption is about 170,000 TWh. That is, 170,000E12 Wh /yr.
Duration of ocean-sourced uranium equals total ocean-sourced energy divided by PRI NRG annual consumption. Therefore, 40E21 Wh divided by 170,000E12 Wh /yr = 240,000 years of primary energy for the entire planet.

26. Mike sez:
For the last several years, the yellowcake that gets made into fuel pellets has been averaging around $40 a kilo. See chart in the section "Plant Operating Costs":
World nuclear.org – Economics of nuclear power
https://world-nuclear.org/information-library/economic-aspects /economics-of-nuclear-power.aspx

Given the paltry ore grade of the world's top uranium mines (see Section 3.1), prices for uranium are sure to rise with a buildout of nuclear power. Meanwhile, the near-term projections for ocean-sourced uranium are $200 – $300 a kilo, or roughly five to eight times the current mined price:
Nature.com – Selective extraction of uranium from seawater with biofouling-resistant polymeric peptide
    https://www.nature.com/articles/s41893-021-00709-3
The price gap will of course narrow as terrestrial uranium becomes more scarce. Also consider that, unlike carbon power plants (biomass, coal, and natural gas), reactors use very little fuel. And what little they do use only accounts for about a third of the cost of a complete fuel assembly:
World-nuclear.org – Economic of nuclear power
    https://world-nuclear.org/information-library/economic-aspects
/economics-of-nuclear-power.aspx
    See "plant operating costs"
Consumer prices in Germany have gone up by far more than that, and they're nowhere near generating 100% renewable energy. Aside from all of that, a price drop is assured by the fact that most countries have direct access to the sea, or have an ally who does. With open and equal access to a limitless source of uranium, prices should drop to well below $200 a kilo.

The new extraction method developed in India and cited in endnote #17 above could slash these prices to a fraction of the estimated $200 per kilo. But even if it doesn't, $5 a month would be a small premium to pay for access to an endless supply of reliable clean energy, with a mining footprint that's about 5% of solar and 15% of wind.

27. EIA.gov – EIA projects nearly 50% increase in world energy usage by 2050, led by growth in Asia
        https://www.eia.gov/todayinenergy/detail.php?id=41433

CHAPTER SEVEN: To Be Perfectly Blunt

1. Percent of US energy supplied by RE (renewable energy):
    A) For primary energy (i.e., all forms of energy): EIA.gov – US energy facts explained

https://www.eia.gov/energyexplained/us-energy-facts/
See first figure.

B) For electricity only: Electricity EIA.gov – Electricity explained:
Electricity in the United States
https://www.eia.gov/energyexplained/electricity/electricity-in
-the-us.php See first figure.

2. Bloomberg – Wind and Solar Power Have Become Amazingly
Affordable
https://www.bloomberg.com/opinion/articles/2019-11-07
/wind-and-solar-power-have-become-amazingly-affordable

3. Greentech media – Beyond Declining Battery Prices: 6 Ways to
Evaluate Energy Storage in 2021
https://www.greentechmedia.com/articles/read/beyond
-declining-battery-prices-six-ways-to-evaluate-energy-storage-in
-2021

4. Spectrum IEEE.org – How Inexpensive Must Energy Storage
Be for Utilities to Switch to 100 Percent Renewables?
https://spectrum.ieee.org/what-energy-storage-would-have
-to-cost-for-a-renewable-grid

5. Energy storage news – Long-duration storage mystery revealed:
Form Energy discloses details of 'multi-day battery' tech (iron-
air batteries)
https://www.energy-storage.news/long-duration-storage-mystery
-revealed-form-energy-discloses-details-of-multi-day-battery-tech/

6. Wikipedia – Energy in California
https://en.wikipedia.org/wiki/Energy_in_California
See graphic. Note that nearly half of CA electricity comes
from natural gas.
Ethree.com – Long-run resource adequacy under deep
decarbonization pathways for California
https://www.ethree.com/wp-content/uploads/2019/06/E3
_Long_Run_Resource_Adequacy_CA_Deep-Decarbonization
_Final.pdf
Pg. 57 "Key Findings"
LA Times – California faces a crossroads on the path to 100%
clean energy

https://www.latimes.com/environment/story/2019-12-12
/california-clean-energy-gas-plants

7. MIT news – Simple solar-powered water desalination
https://news.mit.edu/2020/passive-solar-powered-wate
-desalination-0207

8. EIA.gov – Average frequency and duration of electric distribution
outages vary by states
https://www.eia.gov/todayinenergy/detail.php
See graph on right, "All Utility Types." Average 2 hours with
4 hours.

Energy.gov – Maintaining reliability in the modern power system
https://www.energy.gov/sites/prod/files/2017/01/f34
/Maintaining%20Reliability%20in%20the%20Modern%20
Power%20System.pdf
Pg. 4 / frame 8

9. By concentrating utility solar farms in sunny locales, recent US
annual capacity factors have risen to 25%. See:
EIA.gov – Electric power monthly table 6.07.B "Capacity
Factors for Utility Scale Generators Primarily Using Non-
Fossil Fuels"
https://www.eia.gov/electricity/monthly/epm_table_grapher
.php?t=epmt_6_07_b
In 2020, US annual average capacity factor for wind was
35.4%. See:
Statista – Solar photovoltaic capacity factors in the United
States between 2014 and 2017, by select state
https://www.statista.com/statistics/1019796/solar-pv-capacity
-factors-us-by-state/
For technical increase in panel efficiency (as distinct from
annual capacity factor):
Spectrum.IEEE.org – Perovskite Solar Out-Benches Rivals in 2021
https://spectrum.ieee.org/oxford-pv-sets-new-record-for
-perovskite-solar-cells

10. Greentech media – Why Moore's Law Doesn't Apply to Clean
Energy Technologies
https://www.greentechmedia.com/articles/read/why-moores-law
-doesnt-apply-to-clean-technologies

11. PV magazine – Post bankruptcy, Crescent Dunes CSP plant owner wants project back online by year's end
https://pv-magazine-usa.com/2020/08/03/post-bankruptcy-and-doe-loan-owner-of-crescent-dunes-wants-csp-plant-online-by-years-end/

12. Wikipedia – Crescent dunes solar energy project
https://en.wikipedia.org/wiki/Crescent_Dunes_Solar_Energy_Project

13. Popular mechanics – The $1 Billion Solar Plant Is an Obsolete, Expensive Flop
https://www.popularmechanics.com/technology/a30472835/crescent-dunes-solar-plant

14. Timothymaloney.net – Hydrogen by electrolysis
https://timothymaloney.net/Hydrogen_by_Electrolysis.html
See infographics at bottom of page for hydrogen efficiencies

CHAPTER EIGHT: How Clean Should Clean Energy Be?

1. BBC – The dystopian lake filled by the world's tech lust
https://www.bbc.com/future/article/20150402-the-worst-place-on-earth

2. YouTube – Dependence on China for Military Hardware
https://www.youtube.com/watch?v=Dih30mUexrA

3. Mackinac.org – Bright panels dark secrets the problem of solar waste
https://www.mackinac.org/blog/2022/bright-panels-dark-secrets-the-problem-of-solar-waste
Science direct – Silicon tetrachloride
https://www.sciencedirect.com/topics/earth-and-planetary-sciences/silicon-tetrachloride

4. Grist.org – Going 100% renewable power means a lot of dirty mining
https://grist.org/article/report-going-100-renewable-power-means-a-lot-of-dirty-mining/
Mother Jones – Solar's death panels
https://www.motherjones.com/environment/2011/02/solar-panels-desert-tortoise-mojave/

Deseret.com – The dark side of 'green energy' and its threat to the nation's environment
    https://www.deseret.com/utah/2021/1/30/22249311/why-green-energy-isnt-so-green-and-poses-harm-to-the-environment-hazardous-waste-utah-china-solar

5. UNSCEAR.org – Sources, effects and risks of ionizing radiation (2016)
    https://www.unscear.org/docs/publications/2016/UNSCEAR_2016_Report-CORR.pdf
    See page 13, para. 48

6. Navalpost.com – How safe are US nuclear powered warships?
    https://navalpost.com/how-safe-the-u-s-nuclear-powered-warships

7. Colorado Sun – The wind turbines on his Colorado farm are 20 years old. Who's going to take them down?
    https://coloradosun.com/2022/10/23/logan-county-wind-energy-regulation

8. IER.org – Will solar power be at fault for the next environmental crisis?
    https://www.instituteforenergyresearch.org/uncategorized/will-solar-power-fault-next-environmental-crisis
    Bloomberg – Wind Turbine Blades Can't Be Recycled, So They're Piling Up in Landfills
    https://www.bloomberg.com/news/features/2020-02-05/wind-turbine-blades-can-t-be-recycled-so-they-re-piling-up-in-landfills

9. World Nuclear.org –Decommissioning nuclear facilities
    https://www.world-nuclear.org/information-library/nuclear-fuel-cycle/nuclear-wastes/decommissioning-nuclear-facilities.aspx
    Decommisioning fee ranges from 0.1 cent to 0.2 cent per kWh.
    A 1000-MWavg reactor produces about 8.8 billion kWh per year. With a decommission fee of 0.1 cent /kWh, annual fee = $8.8 million. After 40-year lifetime, decommission fund = $8.8M × 40 = $350 million.
    With a decommision fee of 0.2 cent /kWh, annual fee = $17.6 million. Decommission fund = $700 million.

10. KPBS.org – Counting customer costs for San Onofre closure: $10.4 billion
https://www.kpbs.org/news/2015/aug/03/counting-customer
-costs-san-onofre-closure- 95-bill/
See para. 15, and the fourth table

11. NY Times – California to ban the sale of new gasoline cars
https://www.nytimes.com/2022/08/24/climate/california-gas
-cars-emissions.html
One week later, California asks citizens not to charge EVs during heat wave:
NY Times – Amid Heat Wave, California Asks Electric Vehicle Owners to Limit Charging
https://www.nytimes.com/2022/09/01/us/california-heat-wave
-flex-alert-ac-ev-charging.html
Days later, the California Assembly votes 67-3 to keep Diablo open:
Cal Matters.org – Diablo Canyou: Nuke plant a step closer to staying open longer
https://calmatters.org/environment/2022/09/diablo-canyon
-legislature-california/
The folly of trying to replace Diablo with renewables:
KQED.org – Why Plans to Replace Diablo Canyon With 100 Percent Clean Energy Could Fall Short
https://www.kqed.org/science/801694/why-plans-to-replace-
diablo-canyon-with-100-percent-clean-energy-could-fall-short

12. Unherd.com – Greta Thunberg kills off the anti-nuclear campaign
https://unherd.com/thepost/greta-thunberg-kills-off-the-anti
-nuclear-campaign/

13. Tim sez:

Estimates vary, but the general consensus (see links below) is that the US loses about 12,000 people a year to premature death due to coal emissions. If we had continued building out nuclear after TMI, instead of coal, and if we had achieved a fleet of 50% nuclear by 2019, forty years after TMI, we would have had roughly one quarter the coal deaths per year over those forty years. That's 120,000 lives. We lower this back-of-the-envelope calculation to a conservative estimate of 100,000 shortened by building out coal instead of nuclear.

Spectrum.IEEE.org – Coal pollution fatalities
https://spectrum.ieee.org/coal-pollution-fatalities
See para. 2: 10,000 deaths per year in US due to coal electricity generation

The Guardian – 'Invisible killer': fossil fuels caused 8.7m deaths globally in 2018
https://www.theguardian.com/environment/2021/feb/09 /fossil-fuels-pollution-deaths-research
See first graphic: 13,100 deaths /yr

Washington.edu – Emissions from electricity generation lead to disproportionate number of premature deaths for some racial groups
https://www.washington.edu/news/2019/11/20/electricity -generation-emissions-premature-deaths/ See para. 2: 16,000 deaths per year

14.   Knoema.com – Japan CO_2 emissions per capita
https://knoema.com/atlas/Japan/CO2-emissions-per-capita

15.   Statista – Annual carbon dioxide emissions in Germany from 1990 to 2021
https://www.statista.com/statistics/449701/co2-emissions -germany/
Note that from 2011 to 2018, there has only been a 4.6% decrease in Germany's CO_2 emissions.

16.   News.climate.columbia.edu – How Energy Choices After Fukushima Impacted Human Health and the Environment
https://news.climate.columbia.edu/2019/06/17/post-fukushima -energy-japan-germany
Para. 5
Grist.org – The cost of Germany turning off nuclear power: Thousands of lives
https://grist.org/energy/the-cost-of-germany-going-off-nuclear -power-thousands-of-lives

17.   Forbes – Natural gas and the new deathprint for energy
https://www.forbes.com/sites/jamesconca/2018/01/25 /natural-gas-and-the-new-deathprint-for-energy

Forbes – How Deadly Is Your Kilowatt? We Rank The Killer Energy Sources

https://www.forbes.com/sites/jamesconca/2012/06/10/energys -deathprint-a-price-always-paid

18. Nacleanenergy.com – Wind turbine worker safety the human factor

https://www.nacleanenergy.com/articles/32210/wind-turbine -worker-safety-the-human-factor

Firetrace.com – Five high profile wind turbine fires

https://www.firetrace.com/fire-protection-blog/wind-turbine -death

Forbes – Forget Eagle Deaths, Wind Turbines Kill Humans

https://www.forbes.com/sites/jamesconca/2013/09/29/forget -eagle-deaths-wind-turbines-kill-humans

19. WHO Int news room –Radiation: The Chernobyl accident

https://www.who.int/news-room/questions-and-answers/item /radiation-the-chernobyl-accident

Para 5: "Mention has often been made of WHO's 1959 agreement with the IAEA. This is a standard agreement similar to agreements it has with other UN agencies as a means of setting out respective areas of work. This agreement has never once been used to stop or restrict WHO's work."

Para 7: "The agreement between WHO and IAEA does not affect the impartial and independent exercise by WHO of its statutory responsibilities, nor does it subordinate one Organization to the other."

Rational Wiki – WHO / IAEA conspiracy

https://rationalwiki.org/wiki/WHO-IAEA_conspiracy

20. Forbes – Nuclear Power Continues To Break Records In Safety And Generation

https://www.forbes.com/sites/jamesconca/2021/03/25/nuclear -power-continues-to-break-records-in-safety-and-generation

CHAPTER NINE: How It All Began

1. PubMed NCBI – Muller's Nobel lecture on dose-response for ionizing radiation: ideology or science?

https://pubmed.ncbi.nlm.nih.gov/21717110/

2. Science.org – Attack on radiation geneticists triggers furor
   https://www.science.org/content/article/attack-radiation
   -geneticists-triggers-furor Para. 12
   Physics Today – Low-dose radiation exposure should not be feared
   https://physicstoday.scitation.org/doi/10.1063/PT.3.3037

3. Atomic Insights – On the origins of the linear no-threshold
   (LNT) dogma by means of untruths, artful dodges and blind faith
   https://atomicinsights.com/wp-content/uploads/LNT-and
   -NAS-Environ.-Res.-1.pdf

   "Detailed documentation indicates that actions taken in support
   of this policy revolution were ideologically driven and deliberately
   and deceptively misleading; that scientific records were artfully
   misrepresented; and that people and organizations in positions
   of public trust failed to perform the duties expected of them . . .
   The impact of these deceptions has been substantial and, to this
   day, they significantly affect and dominate regulatory policies
   and risk assessment practices."

4. Babel.hathitrust.org – The biological effects of atomic radiation
   a report to the public
   https://babel.hathitrust.org/cgi/pt
   This is the 1956 BEAR-I report. See especially page 3: "Effects
   on Humans"

5. *Ibid.* #3 above. See frame 9, pg. 440. Also see:
   Springer.com – The Genetics Panel of the NAS BEAR I
   Committee (1956) epistolary evidence suggests self-interest
   may have prompted an exaggeration of radiation risks that
   led to the adoption of the LNT cancer risk assessment model
   https://link.springer.com/article/10.1007/s00204-014-1306-7
   Theodosius Dobzhansky wrote "Mankind Evolving," a textbook
   used in college genetics courses around the world:
   Amazon.com – Mankind Evolving Dobzhansky
   https://www.amazon.com/Mankind-Evolving-Evolution
   -Silliman-Memorial/dp/0300000707
   Milislav Demerec was a director of the department of Genetics
   at Carnegie University, and a prominent *Drosophila* researcher
   (fruit flies used in genetics work):

Wikipedia – Milislav Demerec https://en.wikipedia.org/wiki/Milislav_Demerec

6. Peh-med.biomedcentral.com – Was Muller's 1946 Nobel Prize research for radiation-induced gene mutations peer-reviewed? https://peh-med.biomedcentral.com/articles/10.1186/s13010-018-0060-5

7. Genome News Network – Hermann J. Muller (1890–1967) demonstrates that X rays can induce mutations http://www.genomenewsnetwork.org/resources/timeline/1927_Muller.php

8. Thermofisher.com – Mass spectrometry https://www.thermofisher.com/us/en/home/industrial/mass-spectrometry.html

9. Tim sez:

    2700E-3 Sv ÷ 3.5 min × 60 min /hr × 8760 hr /yr = 405E3 Sv /yr.

    Average background dose is about 3 mSv per year, or 3E-3 Sv /yr.
    Ratio of the two doses = 405E3 Sv /yr ÷ 3E-3 Sv /yr = 135E6.
    That is, the Muller minimum dose rate is 135 million times greater than the average background dose rate.

10. Radiologyinfo.org – Radiation dose https://www.radiologyinfo.org/en/info/safety-xray

11. New scientist.com – Stem cell timeline: The history of a medical sensation https://www.newscientist.com/article/dn24970-stem-cell-timeline-the-history-of-a-medical-sensation/

12. Goodreads.com – Paracelsus https://www.goodreads.com/author/quotes/121121.Paracelsus
    WebMD – Hyponatremia https://www.webmd.com/a-to-z-guides/what-is-hyponatremia#1

13. May Clinic.org – Hyponatremia https://www.mayoclinic.org/diseases-conditions/hyponatremia/symptoms-causes/syc-20373711
    CBS News.com – Woman dies after water drinking contest https://www.cbsnews.com/news/woman-dies-after-water-drinking-contest/

14. Healthline.com – What You Need to Know About Taking Too Much Aspirin
    https://www.healthline.com/health/aspirin-overdose#overdose
    Under "Toxic Amounts" heading: The lethal dose for aspirin is 500 mg/kg of body weight. The average adult weighs about 70 kg (154 lbs.) So their lethal dose would be 35,000 mg. A 100-ct. bottle of 500 mg pills contains 50,000 mgs.

15. The Wire.in – Has Science Let Radiation Scare Us to Death?
    https://thewire.in/the-sciences/has-science-let-radiation-scare-us-to-death

16. HPS.org – Radiation risk in perspective
    https://hps.org/documents/radiationrisk.pdf

17. Pub.IAEA.org – Speech on behalf of the United Nations Scientific Committee on the Effects of Atomic Radiation (UNSCEAR) to the Fukushima Ministerial Conference on Nuclear Safety, 15 – 17 December 2012, Fukushima, Japan
    https://www-pub.iaea.org/iaeameetings/Fukushima/UNSCEAR_Statement.pdf Para. 4

18. Scientific American – Fukushima Residents Return Despite Radiation
    https://www.scientificamerican.com/article/fukushima-residents-return-despite-radiation/

19. Forbes – Absurd radiation limits are a trillion dollar waste
    https://www.forbes.com/sites/jamesconca/2014/07/13/absurd-radiation-limits-are-a-trillion-dollar-waste/?sh=1299dab53d60

20. Wikipedia – List of causes of death by rate
    https://en.wikipedia.org/wiki/List_of_causes_of_death_by_rate#Developed_vs._developing_economies
    See table: "Leading causes of death by age groups, US – 2015." Note that in age group 45-54, 43,000 cancer deaths out of 136,000 deaths (top ten causes) = 32%. In age group 55-64, 116,000 cancer deaths out of 288,000 deaths = 40%. In age group 65-plus, 596,000 cancer deaths out of 1,514,000 deaths = 39%. Cancer causes 22% of deaths in US (2017):
    Healthline – What Are the 12 Leading Causes of Death in the United States?
    https://www.healthline.com/health/leading-causes-of-death

Cancer causes 16% of deaths globally (2019):
BBC.com – What do the people of the world die from?
   https://www.bbc.com/news/health-47371078
Cancer.org – Global cancer facts and figures 4th edition
   https://www.cancer.org/content/dam/cancer-org/research
   /cancer-facts-and-statistics/global-cancer-facts-and-figures/global
   -cancer-facts-and-figures-4th-edition.pdf

CHAPTER TEN: There Is No Safe Dose of BS

1. The German Eye – Radiation fraud must be exposed
      https://thegermanyeye.com/radiation-fraud-must-be-exposed
   -3711
   Science Direct – The linear No-Threshold (LNT) dose
   response model: A comprehensive assessment of its historical
   and scientific foundations
      https://www.sciencedirect.com/science/article/pii/S0009279
   718311177
   Rockefeller Foundation influence over NAS and BEAR-I
   Science Direct – The Muller-Neel dispute and the fate of
   cancer risk assessment
      https://www.sciencedirect.com/science/article/abs/pii/S00139
   35120308562
      The BEAR Committee ignored bomb survivor evidence from
   Japan. From the abstract: "The National Academy of Sciences
   (NAS) Atomic Bomb Casualty Commission (ABCC) Human
   Genetic Study" (i.e., The Neel and Schull, 1956a report) *showed
   an absence of genetic damage* in offspring of atomic bomb survivors
   in support of a threshold model, but [the Neel-Shull report] *was
   not considered for evaluation* by the NAS Biological Effects of
   Atomic Radiation (BEAR) I Genetics Panel." [emphasis ours]
2. HPS.org – The History of the Linear No-Threshold (LNT) Model
      http://hps.org/hpspublications/historylnt/episodeguide.html
      See episode 11.
      (*Note:* These video interviews give the full story of LNT, which
   we summarize in the companion book to this volume, *The LNT
   Report*.)
3. *Ibid.* #5 in Chapter 9, second reference. See table 1

4. *Nuclear Fear* by Spencer Weart
   https://www.amazon.com/Nuclear-Fear-Spencer-R-Weart/dp/0674628365
   See Part Three "New Hopes and Horrors 1953 – 1963"

5. NCBI – DNA damage and repair in human reproductive cells
   https://www.ncbi.nlm.nih.gov/pmc/articles/PMC6337641/

6. Science.org – Leukemia and ionizing radiation (Lewis, 1957)
   https://www.science.org/doi/10.1126/science.125.3255.965

7. Core.AC.UK – Edward Lewis and radioactive fallout the impact of CalTech biologists on the debate over nuclear weapons testing in the 1950s and 60s
   https://core.ac.uk/download/pdf/11810195.pdf
   See frame 50 / pg. 44

8. Books.google.com – The nature of radioactive fallout and its effects on man. Hearings before the special subcommittee on radiation of the joint committee on atomic energy, Congress of the United States
   https://books.google.com/books/about/The_Nature_of_Radioactive_Fallout_and_It.html (see pg. 2000)

9. *Ibid.* #7 above. Unna, W., All Radiation Held Perilous: Nation's Top Geneticists Unanimous In Opinion, Fallout Produced Now Will Shorten Lives in Future, Congress is Told, in Washington Post. 1957: Washington, D.C. p. A1 and A6. Courtesy of the Caltech Archives.

10. NCBI – Estimating risk of low radiation doses - a critical review of the BEIR VII report and its use of the linear no-threshold (LNT) hypothesis
    https://pubmed.ncbi.nlm.nih.gov/25329961/ See para. 5

11. New Scientist – Timeline the evolution of life
    https://www.newscientist.com/article/dn17453-timeline-the-evolution-of-life/
    NCBI – Hormesis defined https://www.ncbi.nlm.nih.gov/pmc/articles/PMC2248601/
    See section: "The Concept of Hormesis" para. 3

12. ICRP.org home page – https://www.icrp.org/

13. Spg.FAS.org – A brief history of radiation
    https://sgp.fas.org/othergov/doe/lanl/pubs/00326631.pdf

14. NRC.gov – Lethal dose
    https://www.nrc.gov/reading-rm/basic-ref/glossary/lethal-dose-ld
    .html
15. Sagepub – Application of low doses of ionizing radiation in
    medical therapies
    https://journals.sagepub.com/doi/10.1177/1559325819895739
16. NCBI – Health impacts of low-dose ionizing radiation: current
    scientific debates and regulatory issues
    https://www.ncbi.nlm.nih.gov/pmc/articles/PMC6149023/
    See "Medical Occupational Exposure"
17. UNSCEAR.org – Report of the united nations scientific
    committee on the effects of atomic radiation (1958)
    https://www.unscear.org/docs/publications/1958/UNSCEAR
    _1958_Report.pdf
    Annex G, Mammalian Somatic Effects, page 170, frame 172:
    (e) Linearity has been assumed primarily for purposes of
    simplicity . . .
    (f) . . . There may or may not be a threshold dose. The two
    possibilities of threshold and no-threshold have been retained
    because of the very great differences they engender.

CHAPTER ELEVEN: How Ed Lewis Went Off the Rails

1. Research Gate – Leukemia and ionizing radiation revisited
   (Cuttler & Welch 2015)
   https://www.researchgate.net/publication/338819152_Leukemia
   _and_Ionizing_Radiation_Revisited
2. *Ibid.* #6 Chapter 10
3. RERF – Life-Span Study (LSS)
   https://www.rerf.or.jp/en/programs/research_activities_e/outline
   _e/proglss-en/
   NCBI – Japanese legacy cohorts: the life span study atomic bomb
   survivor cohort and survivors' offspring
   https://www.ncbi.nlm.nih.gov/pmc/articles/PMC5865006/
4. MDPI – Overview of Biological, Epidemiological, and Clinical
   Evidence of Radiation Hormesis
   https://www.mdpi.com/1422-0067/19/8/2387/htm

Pub Med – Biological stress response terminology: Integrating the concepts of adaptive response and preconditioning stress within a hormetic dose-response framework
https://pubmed.ncbi.nlm.nih.gov/17459441/
NCBI – Hormesis defined
https://www.ncbi.nlm.nih.gov/pmc/articles/PMC2248601/
Neuroscience News – Low doses of radiation may improve quality of life for those with severe Alzheimer's
https://neurosciencenews.com/radiation-treatment-alzheimers-18334/

5. YouTube – What is hormesis?
https://www.youtube.com/watch?v=jOfcpsXpFgA
PubMed. NIH.gov – Hormesis: the dose response for the 21st century
https://pubmed.ncbi.nlm.nih.gov/31330228

6. Research Gate – Documented optimum and threshold for ionising radiation
https://www.researchgate.net/publication/239664703_Documented_optimum_and_threshold_for_ionising_radiation

7. Explore Claremore History
https://exploreclaremorehistory.wordpress.com/2017/01/27/the-radium-wells-bath-house-claremore-oklahoma/

8. UNSCEAR – 1958 report Annex FGHI
https://www.unscear.org/docs/publications/1958/UNSCEAR_1958_Annex-F-G-H-I.pdf
See pg. 165 / frame 41

CHAPTER TWELVE: Who Framed Nuclear Power?

1. Journal of Nuclear Medicine – Eliminating use of the linear no-threshold assumption in medical imaging
https://jnm.snmjournals.org/content/58/6/1014.1.short?rss=1
Pub Med – Death of the ALARA radiation protection principle as used in the medical sector
https://pubmed.ncbi.nlm.nih.gov/32425724/
From para. 1 of the abstract:
"ALARA is the acronym for As Low As Reasonably Achievable. It is a radiation protection concept borne from the Linear

No-Threshold hypothesis. There are no valid data today supporting the use of LNT in the low-dose range . . ."
Sage Pub – It Is time to move beyond the linear no-threshold theory for low-dose radiation protection
  https://journals.sagepub.com/doi/pdf/10.1177/15593258 18779651
Dr. Ed Calabrese has assembled a complete history of LNT:
HPS.org – History of LNT – Dr. Edward Calabrese Documentary
  https://hps.org/publicinformation/ate/cat78.html
We highly recommend this series of video interviews:
Health Physics Society interviews of Ed Calabrese: The History of the Linear No-Threshold (LNT) Model Episode Guide
  http://hps.org/hpspublications/historylnt/episodeguide.html
For further reading on LNT and low-dose radiation, see the following papers:
NCBI – Remedy for radiation fear – discard the politicized science
  https://www.ncbi.nlm.nih.gov/pmc/articles/PMC4036393/
NCBI – The BEIR VII estimates of low-dose radiation health risks are based on faulty assumptions and data analyses: a call for reassessment
  https://www.ncbi.nlm.nih.gov/pubmed/29475999
MIT – A new look at prolonged radiation exposure
  http://news.mit.edu/2012/prolonged-radiation-exposure-0515
Science Direct – The LNT model for cancer induction is not supported by radiobiological data
  https://www.sciencedirect.com/science/article/pii/S00092797 18311013?dgcid=author
Pub Med – Subjecting Radiologic Imaging to the Linear No-Threshold Hypothesis: A Non Sequitur of Non-Trivial Proportion
  https://pubmed.ncbi.nlm.nih.gov/27493264/
Pub Med – Dose optimization to minimize radiation risk for children undergoing CT and nuclear medicine imaging is misguided and detrimental
  https://www.ncbi.nlm.nih.gov/pubmed/28490467
Pub Med – Are we approaching the end of the linear no-threshold era?
  https://www.ncbi.nlm.nih.gov/pubmed/30262515

2. National Academies Press – BEIR VII: Health risks from exposure to low levels of ionizing radiation
    https://www.nap.edu/resource/11340/beir_vii_final.pdf
    Para. 1
    HPS.org – Radiation risk in perspective https://hps.org /documents/radiationrisk.pdf
3. Babel.hathitrust.org – The biological effects of atomic radiation A report to the public
    https://babel.hathitrust.org/cgi/pt
    Pg. 17
4. ANS.org – The Muller-Neel dispute and the fate of cancer risk assessment
    http://local.ans.org/ne/wp-content/uploads/2020/10/Muller -Neel-History-Published-Version.pdf
    See intro. Also see pg. 4, para. 6
5. NCBI – Heath effects of exposure to low levels of ionizing radiation
    https://www.ncbi.nlm.nih.gov/books/NBK218704/
6. Radiation Answers.org – Radiation and Me
    https://www.radiationanswers.org/radiation-and-me/effects -of-radiation.html
    NIH.gov – On the risk to low doses (<100 mSv) of ionizing radiation during medical imaging procedures
    https://www.ncbi.nlm.nih.gov/pmc/articles/PMC3683301/
    Paragraph A86 of Report 103 of the International Commission on Radiological Protection (ICRP) states that "There is, however, general agreement that epidemiological methods used for the estimation of cancer risk do not have the power to directly reveal cancer risks in the dose range up to around 100 mSv."
    Actually, Report 103 says there is more than general agreement—it's "unanimous." Here is report 103 (it's in French. Google translation is below):
    https://inis.iaea.org/collection/NCLCollectionStore/_Public /49/042/49042116.pdf?r=1
    Inis.iaea.org – Recommandations 2007 de la commission internationale de protection radioloique
    Frame 196, Section A.4, from paragraph A86:

"The epidemiological methods used to estimate the risk of cancer, however, are **unanimous** as to the fact that they do not have the power sufficient to directly reveal cancer risks in the range of doses up to approximately 100 mSv."
Physics Today – Low-dose radiation exposure should not be feared
https://physicstoday.scitation.org/doi/10.1063/PT.3.3037

7. ABC News – Could alleged Botox deaths scare patients?
https://abcnews.go.com/Health/Wellness/botox-led-daughters-death-mother-claims/story
Para. 23
Huffpost – More men are getting Botox than ever here's how 'Brotox' is different
https://www.huffpost.com/entry/botox-for-men_l_62e3dc22e4b00fd8d83eb7fd

8. Pub Med – A nationwide study of mortality associated with hospital admission due to severe gastrointestinal events and those associated with nonsteroidal anti-inflammatory drug use
https://pubmed.ncbi.nlm.nih.gov/16086703/

9. Mayo Clinic – Thalidomide: Research advances in cancer and other conditions
https://www.mayoclinic.org/diseases-conditions/cancer/in-depth/thalidomide/art-20046534

10. Wikipedia – Enantiopure drug
https://en.wikipedia.org/wiki/Enantiopure_drug
ACS.org – Thalidomide
https://www.acs.org/content/acs/en/molecule-of-the-week/archive/t/thalidomide.html

11. Atomic Insights – Robert F. Kennedy Jr. tells the Colorado Oil and Gas Association that Wind and Solar Plants are Gas Plants
https://atomicinsights.com/robert-f-kennedy-jr-tells-the-colorado-oil-and-gas-association-that-wind-and-solar-plants-are-gas-plants/

12. PE.com – Ivanpah solar plant, built to limit greenhouse gases, is burning more natural gas
https://www.pe.com/2017/01/23/ivanpah-solar-plant-built-to-limit-greenhouse-gases-is-burning-more-natural-gas/

13. World Nuclear.org – Chernobyl accident 1986

https://www.world-nuclear.org/information-library/safety-and
-security/safety-of-plants/chernobyl-accident.aspx

14. Medium.com – George Lakoff's framing 101
https://medium.com/@ennuid/george-lakoffs-framing
-101-7b88e9c91dac

15. The Guardian – Ocean temperatures hit record high as rate of
heating accelerates
https://www.theguardian.com/environment/2020/jan/13
/ocean-temperatures-hit-record-high-as-rate-of-heating-accelerates
CNN.com – The Pacific Ocean is so acidic that it's dissolving
Dungeness crabs' shells
https://www.cnn.com/2020/01/27/us/pacific-ocean-acidification
-crabs-dissolving-shells-scn-trnd/index.html
Washington Post – Global warming to push billions outside
climate range that has sustained society for 6,000 years,
study finds
https://www.washingtonpost.com/weather/2020/05/04/human
-climate-niche/
The Guardian – One billion people will live in insufferable
heat within 50 years
https://www.theguardian.com/environment/2020/may/05/one
-billion-people-will-live-in-insufferable-heat-within-50-years-study

16. Napalm Biography – Napalm in brief
http://www.napalmbiography.com/faq

17. Brave New Climate – An open letter to environmentalists on
nuclear energy
https://bravenewclimate.com/2014/12/15/an-open-letter-to
-environmentalists-on-nuclear-energy/

CHAPTER THIRTEEN: LNT vs. the 100 mSv Threshold

1. UNSCEAR.org – 2013 report
https://www.unscear.org/docs/reports/2013/13-85418_Report
_2013_Annex_A.pdf
Frame 17 / pg. 10, in Section 3, para. 39:
"The doses to the general public, both those incurred during the
first year and estimated for their lifetimes, are generally low or
very low. No discernible increased incidence of radiation-related

health effects are expected among exposed members of the public or their descendants. The most important health effect is on mental and social well-being, and the fear and stigma related to the perceived risk of exposure to ionizing radiation."

Also see frame 254 / pg. 247, para. E8:

"The Committee has evaluated the uncertainty associated with its estimates of risk due to radiation exposure [U14]. For a hypothetical group of male radiation workers in the United Kingdom, the Committee estimated the average additional lifetime risk of solid cancer due to a whole-body dose of 100 mSv to be about 1%, and the 95% subjective confidence interval on this value to be from a factor of 2.5 lower to a factor of 2 higher. These estimates are similar to those made by the United States Committee on the Biological Effects of Ionizing Radiation [N22] and imply that for a population incurring an acute exposure of 100 mSv, the lifetime risk of cancer would increase from about 41% to about 42%. For 10 mSv, the theoretical increase would be from about 41% to 41.1%."

NIH.gov – Death of the ALARA radiation protection principle
   https://www.ncbi.nlm.nih.gov/pmc/articles/PMC7218317/

"There are no valid data today supporting the use of LNT in the low-dose range, so dose as a surrogate for risk in radiological imaging is not appropriate, and therefore, the use of the ALARA concept is obsolete."

2. YouTube – Chernobyl Episode 1 - 1:23:45 - A nuclear engineer reacts
   https://www.youtube.com/watch
History vs. Hollywood – Chernobyl 2019
   http://www.historyvshollywood.com/reelfaces/chernobyl/
Forbes – Top UCLA Doctor Denounces HBO's "Chernobyl" As Wrong And "Dangerous"
   https://www.forbes.com/sites/michaelshellenberger/2019/06/11/top-ucla-doctor-denounces-depiction-of-radiation-in-hbos-chernobyl-as-wrong-and-dangerous/
Forbes – Chernobyl truth drowns in dramatized movie
   https://www.forbes.com/sites/jamesconca/2019/04/25/chernobyl-truth-drowns-in-dramatized-movie/

3. Wayback Machine – Hermann J. Muller 1946 Nobel lecture
https://web.archive.org/web/20220407083711/https://www
.nobelprize.org/prizes/medicine/1946/muller/lecture/
Para. 10

4. You Tube - Helen Caldicott: Fukushima's Ongoing Impact
https://www.youtube.com/watch?v=0-d-_uOypQo
See especially 0:48 to 1:47. The map she uses is falsified.
See Chapter 17.)
YouTube – Helen Caldicott "Thorium Documentary"
https://www.youtube.com/watch?v=Qaptvhky8IQ&t=128s

5. TimothyMaloney.net – IEER and Dr. Arjun Makhijani are
wrong about two things
http://www.timothymaloney.net/IEER_and_Dr._Arjun
_Makhijani_are_wrong_about_two_things.html pl
Refer to the sixth axis from the top of Fig. 2, labeled 0.0001
Sv to 1 mSv. The NRC's cleanup criterion for industrial
site decommissioning for unrestricted future use is given as
250 microSieverts (0.25 mSv).

So according to the Nuclear Regulatory Commission, the
regulated maximum radiation level for decommissioned nuclear
power facilities is *less than one-tenth* of the 3 mSv natural average
background radiation in the United States.

6. Kahn Academy – Cell cycle checkpoints
https://www.khanacademy.org/science/ap-biology/cell
-communication-and-cell-cycle/regulation-of-cell-cycle/a
/cell-cycle-checkpoints-article
Nature.com – DNA Damage & Repair: Mechanisms for
Maintaining DNA Integrity
https://www.nature.com/scitable/topicpage/dna-damage
-repair-mechanisms-for-maintaining-dna-344
NCBI – Molecular biology of the cell, 4th edition
https://www.ncbi.nlm.nih.gov/books/NBK21054/
Chapter 5

7. YouTube – Brazil 2012 sunbathing on radioactive beaches
https://www.youtube.com/watch?v=RvgAx1yIKjg
Wikipedia – Guarapari
https://en.wikipedia.org/wiki/Guarapari

See "Radioactivity"

Semantic Scholar – Modelling Natural Radioactivity in Sand Beaches of Guarapari, Espírito Santo State, Brazil
https://pdfs.semanticscholar.org/8168/71ec74982c760be32b0 e2f3f1dc6a7ddd305.pdf

8. Forbes – In Germany And Austria, Visits To Radon Health Spas Are Covered By Health Insurance
https://www.forbes.com/sites/geoffreykabat/2019/02/02 /in-germany-and-austria-visits-to-radon-health-spas-are-covered -by-health-insurance/
Para. 3

9. MIT News – A new look at prolonged radiation exposure
http://news.mit.edu/2012/prolonged-radiation-exposure-0515

10. Research Gate – Leukemia and ionizing radiation revisited
https://www.researchgate.net/publication/338819152 _Leukemia_and_Ionizing_Radiation_Revisited (evidence of 700 mSv threshold)
Research Gate – What would become of nuclear risk if governments changed their regulations
https://www.researchgate.net/publication/338819152_Leukemia _and_Ionizing_Radiation_Revisited
(Evidence of possible 1,100 mSv threshold.)
InderScience Online – Documented optimum and threshold for ionising radiation
https://www.inderscienceonline.com/doi/abs/10.1504/IJNUCL .2007.014806

11. NCBI – Death of the ALARA radiation protection principle as used in the medical sector
https://www.ncbi.nlm.nih.gov/pmc/articles/PMC7218317/
See abstract
"There are no valid data today supporting the use of LNT in the low-dose range, so dose as a surrogate for risk in radiological imaging is not appropriate, and therefore, the use of the ALARA concept is obsolete."

CHAPTER FOURTEEN: LNT vs. Common Sense

1. HPS.org – Calabrese interviews

http://hps.org/hpspublications/historylnt/episodeguide.html
Episode 17 @ 2:41

2. *Ibid.* #1 above. Episode 17 @ 17:30

3. Springer.com – How the US National Academy of Sciences misled the world community on cancer risk assessment: new findings challenge historical foundations of the linear dose response
https://link.springer.com/article/10.1007/s00204-013-1105-6
Pub Med – LNTgate: How scientific misconduct by the U.S. NAS led to governments adopting LNT for cancer risk assessment
https://pubmed.ncbi.nlm.nih.gov/27131569/
Atomic Insights – Edward Calabrese challenges Science Magazine to right a 59 year-old case of scientific misconduct
https://atomicinsights.com/edward-calabrese-challenges-science-magazine-to-right-a-59-year-old-case-of-scientific-misconduct/

4. Science.org – Genetic effects of atomic radiation BEAR-I June 29, 1956
https://www.science.org/doi/10.1126/science.123.3209.1157
(This is the BEAR-I Genetics Panel technical report)

5. NAS.org – Societal Threats from Ideologically Driven Science
https://www.nas.org/academic-questions/30/4/societal_threats_from_ideologically_driven_science

6. National Academies Press – The future of low dose radiation research in the United States (2019)
https://nap.nationalacademies.org/catalog/25578/the-future-of-low-dose-radiation-research-in-the-united-states
See Pg. 9:

Human health effects at low doses of radiation are expected to be small but the uncertainties associated with current best estimates of risks are considerable. Nevertheless, exposures at low doses are of primary interest for setting standards to protect individuals against the adverse effects of ionizing radiation. The main health effect associated with exposure to low doses of ionizing radiation is cancer. *Non-cancer effects such as hereditary, cardiovascular, and central nervous system effects are receiving more attention than they did previously, but the occurrence and possible mechanisms at low doses of radiation remain uncertain.*

7. NAP.Edu – Health Risks from Exposure to Low Levels of Ionizing Radiation: BEIR VII Phase 2
    https://www.nap.edu/download/11340
    Pg. 7

8. The Linear No-Threshold Model of Low-Dose Radiogenic Cancer: A Failed Fiction
    https://pubmed.ncbi.nlm.nih.gov/30792613/
    See abstract for quote.

9. PBS.org – How risky is flying?
    https://www.pbs.org/wgbh/nova/planecrash/risky.html
    BTW, the same author also wrote:
    Radiation effects.org – The dangers of radiophobia
    https://radiationeffects.org/wp-content/uploads/2016/11/The-dangers-of-radiophobia.pdf

10. News.USC.edu – Oxygen is a blessing, right? Well, it can also be a curse
    https://news.usc.edu/87372/oxygen-is-a-blessing-right-well-it-can-also-be-a-curse/

11. Amazon – The Rise of Nuclear Fear (Weart)
    https://www.amazon.com/Rise-Nuclear-Fear-Spencer-Weartebook/dp/B0087GZIBA/ref=sr_1_fkmrnull_1

12. ANS Nuclear Café – What did we learn from Three Mile Island?
    http://ansnuclearcafe.org/2014/03/25/what-did-we-learn-from-three-mile-island/#sthash.9XY8NCTc.dpbs

13. Pub.IAEA.org – Safety of nuclear power plants: Design
    https://www-pub.iaea.org/MTCD/Publications/PDF/Pub1715web-46541668.pdf

14. Healthy Children.org – Vaccine safety: examine the evidence
    https://www.healthychildren.org/English/safety-prevention/immunizations/Pages/Vaccine-Studies-Examine-the-Evidence.aspx

15. NHK.org – Contaminated soil piles up in vast Fukushima cleanup project
    https://www3.nhk.or.jp/nhkworld/en/news/backstories/1942/

16. Scientific American – Fukushima residents return despite radiation
    https://www.scientificamerican.com/article/fukushima-residents-return-despite-radiation/

17. PNAS.org – Radiation dose rates now and in the future for residents neighboring restricted areas of the Fukushima Daiichi Nuclear Power Plant
   https://www.pnas.org/content/111/10/E914
   Para. 1

18. NRC.gov – Assessment of Variations in Radiation Exposure in the United States
   https://www.nrc.gov/docs/ML1224/ML12240A227.pdf
   Top of pg. 8

19. Forbes – It Sounds Crazy, But Fukushima, Chernobyl, And Three Mile Island Show Why Nuclear Is Inherently Safe
   https://www.forbes.com/sites/michaelshellenberger/2019/03/11/it-sounds-crazy-but-fukushima-chernobyl-and-three-mile-island-show-why-nuclear-is-inherently-safe/
   See section "Anxiety, Displacement and Panic" para. 13

20. Sci Tech Daily – Fukushima Soil Decontamination: Lessons learned
   https://scitechdaily.com/fukushima-soil-decontamination-lessons-learned/

21. Forbes – After five years, what is the cost of Fukushima?
   https://www.forbes.com/sites/jamesconca/2016/03/10/after-five-years-what-is-the-cost-of-fukushima/
   Para. 3
   JCER.or.jp – Accident Cleanup Costs Rising to 35–80 Trillion Yen in 40 Years
   https://www.jcer.or.jp/jcer_download_log.php

22. *Ibid.* #21 above, second link. See first two pages

23. Forbes – Japan's expert panel agrees that dumping radioactive water into the ocean is best
   https://www.forbes.com/sites/jamesconca/2020/02/01/japans-expert-panel-agrees-that-dumping-radioactive-water-into-the-ocean-is-best/
   World Nuclear.org – Discharge to sea will have minimal impact, TEPCO says
   https://www.world-nuclear-news.org/Articles/Discharge-to-sea-will-have-minimal-impact-Tepco-sa

24. Reuters.com – Japan to release Fukushima water into ocean from Aug 24
     https://www.reuters.com/world/asia-pacific/japan-release-fukushima-water-into-ocean-starting-aug-24-2023-08-22/
     Generation Atomic calculator (1.6 MBq per liter):
     Glide.page – Radiation Dose Calculator
     https://radiation-dose-calculator.glide.page/dl/welcome
     (Also see our supplement in endnote #25 below.)
     ALPS treatment (filtering for radionuclides):
     Plus.Reuters.com – Treatment and release of ALPS water
     https://plus.reuters.com/the-treatment-and-release-of-alps-treated-water/p/1
     Discharge of tank water:
     Wiki – Discharge of radioactive water at Fukushima
     https://en.wikipedia.org/wiki/Discharge_of_radioactive_water_of_the_Fukushima_Daiichi_Nuclear_Power_Plant
     Cumulative radiation of tank water:
     Science alert.com – Expert explains why radioactive water should be released
     https://www.sciencealert.com/an-expert-explains-why-the-radioactive-water-stored-at-fukushima-should-be-released
     https://www.nra.go.jp/data/000418886.pdf
     (See section 4 "Discharge Limits")

25. See the supplement "Dietary Potassium vs. Fuku Water."

26. *Ibid.* #24, third link
     Use the Calculator in #24 to compare air travel and tritiated water
     148E15 ÷ 860E14 = 172 days.

27. NY Times – The Jane Fonda effect
     https://www.nytimes.com/2007/09/16/magazine/16wwln-freakonomics-t.html

28. Public Accountability.org – Jerry Brown's ties to the oil and gas industry
     https://public-accountability.org/report/jerry-browns-ties-to-the-oil-and-gas-industry/
     Environmental Progress.org – Jerry Brown's secret war on clean energy

http://environmentalprogress.org/big-news/2018/1/11/jerry
-browns-secret-war-on-clean-energy

29. NY Times – Japan crisis could rekindle US antinuclear movement
https://www.nytimes.com/2011/03/19/science/earth/19antinuke
.html

30. SEHN.org – Precautionary principle
https://www.sehn.org/precautionary-principle-understanding
-science-in-regulation
ACSH.org – Precautionary principle: we must ban driving
to whole foods
https://www.acsh.org/news/2019/03/18/precautionary-principle
-we-must-ban-driving-whole-foods-13889

31. IRP-CDN – Nuclear is for life: a cultural revolution
https://irp-cdn.multiscreensite.com/1d1b01d4/files/uploaded
/N4Lch9.pdf
Chapter 9 pages 211-212 "Popular Culture and the Precautionary
Principle"

32. NCBI – The Healthy Worker Effect and Nuclear Industry Workers
https://www.ncbi.nlm.nih.gov/pmc/articles/PMC2889508/

33. Cloudfront.net – Nuclear shipyard worker study (1980–1988): a
large cohort exposed to low-dose-rate gamma radiation
https://d3pcsg2wjq9izr.cloudfront.net/files/6471/articles/6407
/f118361025791241.pdf

34. NCBI – The healthy worker effect and nuclear industry workers
https://www.ncbi.nlm.nih.gov/pmc/articles/PMC2889508/
See introduction, para. 2

35. IOPscience.org – Greater than 99% consensus on human caused
climate change in the peer-reviewed scientific literature
https://iopscience.iop.org/article/10.1088/1748-9326/ac2966
The Guardian – Why we care about the 97% expert consensus
on human-caused global warming
https://www.theguardian.com/environment/climate-consensus
-97-per-cent/2014/jun/24/why-we-care-about-global
-warming-consensus
From the Guardian article:
"Three distinct studies using four different methods have
independently shown that the expert consensus on human-caused

global warming is 97 ± 1%. The result is the same whether we ask the experts' opinions, look at their public reports and statements, or examine their peer-reviewed science. Even studies that quibble about the precise percentage have accidentally reinforced the 97 ± 1% consensus."

36. NCBI – Death of the ALARA Radiation Protection Principle as Used in the Medical Sector
https://www.ncbi.nlm.nih.gov/pmc/articles/PMC7218317
See para. 1 of the abstract: "ALARA is the acronym for As Low As Reasonably Achievable. It is a radiation protection concept borne from the Linear No-Threshold hypothesis. There are no valid data today supporting the use of LNT in the low-dose range . . ."

CHAPTER FIFTEEN: That Depends on What You Mean by "Reasonable"

1. CDC.gov – ALARA - as low as reasonably achievable
https://www.cdc.gov/nceh/radiation/alara.html

2. HPS.org – Answer to Question #10031 Submitted to "Ask the Experts"
https://hps.org/publicinformation/ate/q10031.html
Ibid. #36 in Chapter 14

3. OEHHA.ca.gov – Public health goals for chemicals in drinking water - Titium (March 2006)
https://oehha.ca.gov/media/downloads/water/chemicals/phg/phgtritium030306.pdf
Top of pg. 3 / frame 10
Journals.law.com – What is the Health Risk of 740 Bq L–1 of Tritium? A Perspective
https://journals.lww.com/health-physics/Citation/2013/01000/Commentary___What_is_the_Health_Risk_of_740_Bq_L_1.15.aspx

4. *Ibid.* #5 in Chapter 13. See Fig.1 titled "Tritium Activity, Biological Effects, and Regulations"

5. Pub.IAEA.org – Fukushima accident technical volume 5/5 post-accident recovery

https://www-pub.iaea.org/MTCD/Publications/PDF
/AdditionalVolumes/P1710/Pub1710-TV5-Web.pdf
Page 16 / frame 27

6. NRC.org – NRC Decommissioning Rule from 10 CFR 20 NRC radiation standards: Subpart E-Radiological Criteria for License Termination

https://www.nrc.gov/docs/ML0824/ML082480281.pdf
See pgs. 1-2:

"A site will be considered acceptable for unrestricted use if the residual radioactivity that is distinguishable from background radiation results in a TEDE to an average member of the critical group that does not exceed 25 mrem (0.25 mSv) per year, including that from groundwater sources of drinking water, and the residual radioactivity has been reduced to levels that are as low as reasonably achievable (ALARA)."

7. NRC.gov – Backgrounder on Tritium, Radiation Protection Limits, and Drinking Water Standards

https://www.nrc.gov/reading-rm/doc-collections/fact-sheets
/tritium-radiation-fs.html
See 5th section:

"How does the radiation dose from nuclear power-related tritium compare to the dose a person receives from natural background radioactivity or from medical procedures?"

"As an example, drinking water for a year from a well with 1,600 picocuries per liter of tritium (comparable to levels identified in a drinking water well after a significant tritiated water spill at a nuclear facility) would lead to a radiation dose (using EPA assumptions) of 0.3 millirem (mrem). That dose is:

- **at least 2,000 to 5,000 times lower** than the dose from a medical procedure involving a full-body CT scan (e.g., 500 to 1,500 mrem from a CT scan)
- **1,000 times lower** than the approximate 300 mrem dose from natural background radiation
- **50 times lower** than the dose from natural radioactivity (potassium) in your body (e.g., 15 mrem from potassium)

- **12 times lower** than the dose from a round-trip cross-country airplane flight (e.g., 4 mrem from Washington, D.C., to Los Angeles and back)"

8. EU-ALARA.net – Optimization of radiation protection ALARA: a practical guidebook
   https://www.eu-alara.net/images/stories/EANdocuments/Publications/EAN_ALARA_Book_oct_2019V2.pdf
   See pg./ frame 59

9. See the supplement "Iodine-129" at the back of this book.

10. Wikipedia – Ramsar Iran radioactivity
    https://en.wikipedia.org/wiki/Ramsar,_Iran#Radioactivity

11. YouTube – No more radiophobia
    https://www.youtube.com/watch?v=JpcUCo0ebNA

12. Brave New Climate – What we can learn from Kerala
    https://bravenewclimate.com/2015/01/24/what-can-we-learn-from-kerala/ Para. 5
    NCBI – Background radiation and cancer incidence in Kerala, India-Karanagappally cohort study
    https://www.ncbi.nlm.nih.gov/pubmed/19066487
    Eurasia Review – High level natural radiation areas in kerala: no evidence of adverse health effects
    https://www.eurasiareview.com/09042022-high-level-natural-radiation-areas-in-kerala-no-evidence-of-adverse-health-effects-analysis/

13. *Ibid*. endnote 12 above.
    See Brave New Climate article, para.6.
    Journals.lww.com – Radioactivity in the Diet of Population of the Kerala Coast Including Monazite Bearing High Radiation Areas
    https://journals.lww.com/health-physics/Abstract/1970/10000/Radioactivity_in_the_Diet_of_Population_of_the.8.aspx

14. *Ibid*. endnote 12 above.
    See Brave New Climate article, para. 4
    Times of India – Kerala has the highest lifespan of 74.9 years
    https://timesofindia.indiatimes.com/life-style/health-fitness/health-news/kerala-has-the-highest-lifespan-of-74-9-years/articleshow/55087367.cms

15. *Ibid* footnote #13 above (Brave New Climate article) para. 7
    Ibid. # 12 above, second link

16. Atomic Insights – San Onofre steam generators – honest error driven by search for perfection
    https://atomicinsights.com/san-onofre-steam-generators-hones-error-driven-by-search-for-perfection/
    Wikipedia – San Onofre nuclear generating station
    https://en.wikipedia.org/wiki/San_Onofre_Nuclear_Generating_Station
    (Safe operation at 70% power)

17. Forbes – Megadroughts and desalination another pressing need for nuclear power
    https://www.forbes.com/sites/jamesconca/2019/07/14/megadroughts-and-desalination-another-pressing-need-for-nuclear-power/

18. Forbes – Had They Bet On Nuclear, Not Renewables, Germany & California Would Already Have 100% Clean Power
    https://www.forbes.com/sites/michaelshellenberger/2018/09/11/had-they-bet-on-nuclear-not-renewables-germany-california-would-already-have-100-clean-power/
    Clean Energy Wire – How much does Germany's energy transition cost?
    https://www.cleanenergywire.org/factsheets/how-much-does-germanys-energy-transition-cost
    See last para. In June 2018, the Euro was equal to $1.17 USD. 520 billion Euros × 1.17 = about $600 billion.
    Statista – Annual carbon dioxide emissions in Germany from 1990 to 2021
    https://www.statista.com/statistics/449701/co2-emissions-germany/

19. Tim sez:
    Germany's electricity production for 2019 = 612 TWh.
    A reactor with 1,100 MW capacity averages about 1,000 MW through the year, at a 90% capacity factor. Its electric energy production is:
    1,000 MW × 8760 hours /yr = almost 8.8 TWh /yr.
    612 TWh /y ÷ 8.8 TWh /y per reactor = 69 reactors.

Assume $4500 per installed kW of capacity. (Note: This is the cost per kilowatt of the four South Korean APR-1400 reactors recently completed in the UAE.)

Then total combined cost: 69 reactors × 1,100 MW /reactor × 1,000 kW /MW × $4500 /kW = **$342 billion.**

20. Vox.com – Europe's renewable energy policy is built on burning American trees
    https://www.vox.com/science-and-health/2019/3/4/18216045/renewable-energy-wood-pellets-biomass

21. WBUR.org – This Vermont Town Took A Big Hit When Its Nuclear Plant Closed
    https://www.wbur.org/news/2019/04/23/vermont-yankee-vernon-lessons
    Wikipedia – Vermont Yankee Nuclear Power Plant
    https://en.wikipedia.org/wiki/Vermont_Yankee_Nuclear_Power_Plant

22. Robert Stone Productions – Pandora's Promise
    https://robertstoneproductions.com/project/pandoras-promise/

23. AP.Org – PART II: AP IMPACT: Tritium leaks found at many nuke sites
    https://www.ap.org/press-releases/2012/part-ii-ap-impact-tritium-leaks-found-at-many-nuke-sites
    See "East Coast Issues" para. 5

24. Generation Atomic radiation calculator:
    Glide page – Radiation dose calculator
    https://radiation-dose-calculator.glide.page/dl/welcome

CHAPTER SIXTEEN: The Forbidden Zone

1. YouTube – What's it like inside Chernobyl? Is it safe?
   https://www.youtube.com/watch?v=yLoR7btSxtg
   From the 10:20 mark

2. BBC.com – Chernobyl: Why radiation levels spiked at nuclear plant
   https://www.bbc.com/news/science-environment-60528828
   See eighth sentence

3. UNSCEAR.org – UNSCEAR 2008 report Annex D (corrected)

https://www.unscear.org/docs/publications/2008/UNSCEAR
_2008_Annex-D-CORR.pdf

Pg. 58 / frame 18, Fig. VII "Outcome for patients with ARS." Of the 134 patients with doses greater than 800 mSv, 86 were still living in 2004. 86 ÷ 134 = 64%, or ~ 2/3.

4.   UNSCEAR.org – Evaluation of data on thyroid cancer in regions affected by the Chernobyl accident

https://www.unscear.org/docs/publications/2017/Chernobyl
_WP_2017.pdf

Professor Gerry Thomas, Chair in Molecular Pathology at Imperial College London and the Chernobyl Tissue bank, emailed this explanation to our publisher:

"The number cited [in the above UNSCEAR paper] gives fifteen [deaths in children from thyroid cancer]. These were not instantaneous deaths—thyroid cancer takes many years to kill, and had some of these children had access to the best medical care some would undoubtedly have survived. Some of the very early cases (within 10 years of the accident) presented with lung metastases. As with any cancer, once the primary tumor spreads it becomes more difficult to treat. Overall, and over a period of 50 years we would expect 1% of cases to prove fatal—however, this estimate uses historical data, and as we are getting better at treating even advanced cancer with targeted therapies even 1% may be an overestimate."

This is a lecture by professor Thomas on radiation and human health:

https://www.youtube.com/watch?v=agzrhxZD5kc&t=6s

This is a panel discussion in which she participates:

https://www.youtube.com/watch?v=PZUvoeIArDM

Also see:

IAEA.org – STI/PUB/1312 Chernobyl looking back to go forward

https://www-pub.iaea.org/MTCD/publications/PDF/Pub1312
_web.pdf

Frame 17, page 4

5.   Treehugger.com – Ghost Towns of Chernobyl Becoming Wonderland for Wolves

https://www.treehugger.com/ghost-towns-chernobyl-wonderland
-wolves-4856269

American Scientist.org – Growing up with Chernobyl

https://www.americanscientist.org/article/growing-up-with-chernobyl

6. Phys.org – Defying radiation, elderly residents cling on in Chernobyl
    https://phys.org/news/2016-04-defying-elderly-residents-chernobyl.html
    See para. 16
    Wikipedia – Samosely
    https://en.wikipedia.org/wiki/Samosely
    PRI.org – 30 years after Chernobyl, these Ukrainian babushkas are still living on toxic land
    https://www.pri.org/stories/2016-04-26/30-years-after-chernobyl-these-ukrainian-babushkas-are-still-living-their-toxic

7. NCBI – Activities concentration of radio cesium in wild mushroom collected in Ukraine 30 years after the Chernobyl power plant accident
    https://www.ncbi.nlm.nih.gov/pmc/articles/PMC5757420/
    See abstract

8. WHO.int – Chernobyl: the true scale of the accident
    https://www.who.int/news/item/05-09-2005-chernobyl-the-true-scale-of-the-accident

9. NCBI – Observations on the Chernobyl Disaster and LNT (Jaworowski)
    https://www.ncbi.nlm.nih.gov/pmc/articles/PMC2889503/
    See Section: "Effects of Chernobyl Fallout on the Population", para. 4. Also search for "deficit of cancers" and "Bryansk"

10. UN library.org – UNSCEAR 2000b
    https://www.un-ilibrary.org/content/books/9789210582490/read
    See pg. 517

11. Live Science – Chernobyl radiation effects
    https://www.livescience.com/chernobyl-radiation-effects.html
    Para.3

12. PETA.org – NASA grounds monkey radiation experiments
    https://www.peta.org/blog/nasa-grounds-monkey-radiation-experiments/
    Wikipedia – Ionizing radiation in spaceflight

https://en.wikipedia.org/wiki/Effects_of_ionizing_radiation
_in_spaceflight

13. NCBI – Health Impacts of Low-Dose Ionizing Radiation: Current
     Scientific Debates and Regulatory Issues
     https://www.ncbi.nlm.nih.gov/pmc/articles/PMC6149023/
     See abstract

14. World Nuclear.org – Demystifying radiation - the nemesis of
     nuclear energy?
     http://www.world-nuclear-news.org/Articles/Demystifying
     -radiation-the-nemesis-of-nuclear-ener
     See section "Radiation and the United Nations"

15. The Bulletin.org – Public opinion on nuclear energy: what
     influences it
     https://thebulletin.org/2016/04/public-opinion-on-nuclear
     -energy-what-influences-it/

16. IOP Science – What have we learned from a questionnaire survey
     of citizens and doctors both inside and outside Fukushima?:
     survey comparison between 2011 and 2013
     https://iopscience.iop.org/article/10.1088/0952-4746/35/1
     /N1/pdf
     Frame 7, Fig. 2

CHAPTER SEVENTEEN: What Are the Odds?

1. PBS.org – How risky is flying?
    https://www.pbs.org/wgbh/nova/planecrash/risky.html

2. Environmental Progress – Sierra Club
    http://environmentalprogress.org/sierra-club
    Para. 3
    Wellock, T. R. 1998. *Critical Masses: Opposition to Nuclear
    Power in California, 1958–1978.* Univ of Wisconsin Press.

3. DailyCal.org – 14 things that kill more people than sharks
    https://www.dailycal.org/2017/06/30/14-things-kill-people
    -sharks/ See #3

4. NASA.org – Kharecha and Hansen 2013
    https://pubs.giss.nasa.gov/abs/kh05000e.html
    Scientific American – Nuclear power may have saved 1.8 million
    lives otherwise lost to fossil fuels, may save up to 7 million more

https://blogs.scientificamerican.com/the-curious-wavefunction/nuclear-power-may-have-saved-1-8-million-lives-otherwise-lost-to-fossil-fuels-may-save-up-to-7-million-more/

Pubs.ACS – Prevented Mortality and Greenhouse Gas Emissions from Historical and Projected Nuclear Power
https://pubs.acs.org/doi/abs/10.1021/es3051197

Next Big Future – Anti-nuclear policies increased global carbon by 18% and added 9.5 million air pollution deaths
https://www.nextbigfuture.com/2018/05/anti-nuclear-policies-increased-global-carbon-by-18-and-added-9-5-million-air-pollution-deaths.html

5. Reuters.com – Aspirin risk compares to driving cars, study finds
https://www.reuters.com/article/us-risks/aspirin-risk-compares-to-driving-cars-study-finds-idUSN0737156120070509

This article refers to aspirin deaths being at 10 deaths per 100,000. The reference on aspirin mortality cited in endnote 14 of Chapter 11 refers to aspirin and other NSAIDs (ibuprophen, etc.), at 15 deaths per 100,000. The recent US census says there are about 54 million men 50 years or older. Taking the conservative estimate of 10 per 100,000, that comes to 5,400 deaths per year from aspirin.

6. SeattlePI.com – Someone drowns in a tub nearly every day in America
https://www.seattlepi.com/national/article/Someone-drowns-in-a-tub-nearly-every-day-in-1201018.php

7. NCBI – Fatalities in high school and college football players
https://www.ncbi.nlm.nih.gov/pubmed/23477766

8. Washington Post – In the past five years, at least six Americans have been shot by dogs
https://www.washingtonpost.com/news/wonk/wp/2015/10/27/a-dog-shoots-a-person-almost-every-year-in-america/

People.com – Oklahoma Woman Shot by a Puppy After Yellow Labrador Gets Spooked by a Passing Train
https://people.com/pets/oklahoma-woman-accidentally-shot-by-dog/

Psychology Today – Sit! Stay! Shoot! Do We Need Gun Control for Dogs?

https://www.psychologytoday.com/us/blog/canine-corner /201302/sit-stay-shoot-do-we-need-gun-control-dogs

9. Colorado Sun – Colorado has the highest per-capita rate of skin cancer, thanks to sunshine and high elevation
    https://coloradosun.com/2019/02/04/colorado-skin-cancer-rates/ See para. 2.
    Although Colorado's skin cancer rates (from sunshine, not granite) are the highest in the nation, it is among "the top five states with the lowest percentage of people diagnosed with cancer."

10. PBS.org – Facts about nuclear radiation everyday exposures
    https://www.pbs.org/wgbh/pages/frontline/shows/reaction /interact/facts.html
    See Grand Central Station Employee: 120 mrem /year = 1.2 mSv /yr

11. See the supplement "Savannah Thorium" at the back of this book.

12. Momtastic – Hot spots earth's 5 most naturally radioactive places
    https://www.momtastic.com/webecoist/2013/01/22/hot-spots -earths-5-most-naturally-radioactive-places/2/ See para. 3 and 4 for Guarapari, Brazil and Ramsar, Iran
    NCBI – Human exposure to high natural background radiation: what can it teach us about radiation risks?
    https://www.ncbi.nlm.nih.gov/pmc/articles/PMC4030667/
    See Section 2, Subsections titled India and Iran
    For doses in the Chernobyl exclusion zone see:
    Chernobyl Gallery.com – Radiation Levels
    http://www.chernobylgallery.com/chernobyl-disaster /radiation-levels/
    Note that most sites in the exclusion zone measure less than 1 microSievert /hr, which totals less than 9 mSv /year. This is less radioactive than Kerala, India or Ramsar, Iran. Even some locations near Denver measure up to 12 mSv /yr.
    Wired – The Eerie Repopulation of the Fukushima Exclusion Zone
    https://www.wired.com/story/eerie-repopulation-fukushima -exclusion-zone/ Para. 4

13. IAEA.org – Power reactor information system (PRIS)

https://pris.iaea.org/pris/countrystatistics/reactordetails
.aspx?current=710

14. Zero Hedge.com – Fukushima Radiation Has Contaminated
The Entire Pacific Ocean (And It's Going To Get Worse)
https://www.zerohedge.com/news/2016-10-02/fukushima
-radiation-has-contaminated-entire-pacific-ocean-and-its-going
-get-worse

15. The Guardian – Back in the water: Fukushima no-go zone gets
first surf shop since disaster (2019)
https://www.theguardian.com/world/2019/mar/08/back-in-the
-water-fukushima-no-go-zone-gets-first-surf-shop-since-disaster
Statesman Journal – Fukushima radiation has reached U.S. shores
https://www.statesmanjournal.com/story/tech/science
/environment/2016/12/07/fukushima-radiation-has-reached-us
-shores/95045692/

16. Snopes – Fukushima nuclear fallout map
https://www.snopes.com/fact-check/nuclear-fallout-map/

17. Re: Helen Caldicott: Fukushima's Ongoing Impact
https://www.youtube.com/watch?v=0-d-_uOypQo See
especially 0:48 to 1:47

18. Our radioactive ocean.org
https://ourradioactiveocean.org/

Go to "How we analyze samples and report data." Click on
the graphic next to the heading. Note the entry 12,000 Bq /m3,
due to marine potassium alone.

LBL.gov – Radioactivity in the natural environment
https://www2.lbl.gov/abc/wallchart/chapters/15/3.html Table
15-1, section "Oceans"

Activity from K-40 is 12 Bq /Liter. Multiply by 1000 liters / m^3
for 12,000 Bq /m^3.

Refer to Figures 1 and 2. The light blue dots in Figure 1 are
samples with less than 2 Bq/m^3. Likewise for the far-left data
points in Figure 2. These points represent legacy cesium-137
from before Fukushima.

In 2017, samples were in the range 4 to 7 Bq, as shown in
Figure 2. They correspond to the orange dots in Figure 1. So in

California waters, cesium-137 radioactivity increased by about 4 Bq /m³

19. Timothy Maloney.net – Tritium water leak
http://www.timothymaloney.net/Tritium_water_leak.html
See section "Proper diet containing trace amounts of potassium"

20. Science Daily – Coastal fog linked to high levels of mercury found in mountain lions, study finds
https://www.sciencedaily.com/releases/2019/11/191126121138.htm

21. LA Times – Mountain lions in coastal regions have high mercury levels
https://www.latimes.com/california/story/2019-12-02/mountain-lions-in-coastal-regions-have-high-mercury-levels

22. See the supplement "Fuku Fish" at the back of this book.
Oceana.org – Worried about Fukushima radiation in seafood?
Turns out bananas are more radioactive than fish
https://oceana.org/blog/worried-about-fukushima-radiation-seafood-turns-out-bananas-are-more-radioactive-fish

23. *Ibid.* #19 above. Tim sez:
Dietary potassium in a 70-kilogram human has radiation of 4400 Bq, which emits 6900 MeV /sec.
Using the conversion factor 1 joule = 6.24E12 MeV from footnote 16 above:
5900 MeV /s ÷ 6.24E12 MeV /J = 0.946E-9 J /s (internal dose).
0.946E-9 J /s ÷ 70 kg = 13.5E-12 J /kg per second = 13.5E-12 sieverts per second.
13.5E-12 Sv /s × 3600 s /hr × 24 hr × 365 = 0.43E-3 Sv /yr, or 0.43 mSv /yr.

24. Forbes – Germans Boared With Chernobyl Radiation
https://www.forbcs.com/sitcs/jamcsconca/2014/09/05/germans-boared-with-chernobyl-radiation/ - 5df126b143b0

25. Gov.UK – Ionising radiation: dose comparisons
https://www.gov.uk/government/publications/ionising-radiation-dose-comparisons/ionising-radiation-dose-comparisons

26. See the twin supplements "Energy Density of LEU" and "Energy Density of Natural Gas" at the back of this book.

27. US History.com – Eisenhower

https://www.u-s-history.com/pages/h1814.html

28. National Park express – The Atomic Age in Las Vegas, Nevada!
https://www.nationalparkexpress.com/the-atomic-age-in
-las-vegas-nevada/

CHAPTER EIGHTEEN: Contain Yourself

1. World Nuclear.org – Three mile island accident
https://www.world-nuclear.org/information-library/safety-and
-security/safety-of-plants/three-mile-island-accident.aspx

"Indeed, more than a dozen major, independent health studies of the accident showed no evidence of any abnormal number of cancers around TMI years after the accident. The only detectable effect was psychological stress during and shortly after the accident.

"The studies found that the radiation releases during the accident were minimal, well below any levels that have been associated with health effects from radiation exposure. The average radiation dose to people living within 10 miles of the plant was 0.08 millisieverts, with no more than 1 millisievert to any single individual. The level of 0.08 mSv is *about equal to a chest X-ray*, and 1 mSv is about a third of the average background level of radiation received by U.S. residents in a year." [emphasis added]

2. How nuclear fear ends Michael Shellenberger TEDx CalPoly
https://www.youtube.com/watch From 6:00 mark to 12:00

3. Popular Mechanics – Top automotive engineering failures: the ford Pinto fuel tanks
https://www.popularmechanics.com/cars/a6700/top-automotive
-engineering-failures-ford-pinto-fuel-tanks/

4. YouTube – Nuclear waste James Conca
https://www.youtube.com/watch?v=0JfJEK3R1k0

5. See the supplement "Used Panels vs Used Fuel" at the back of this book.

6. YouTube – Nuclear fuel cask accident testing Sandia Corp. Film 32204
https://www.youtube.com/watch?v=2VMdspuaig0
YouTube – Nuclear fuel transportation packages
https://www.youtube.com/watch?v=3BjaalMR14k

YouTube – Restored Version 720HD - Rocket Powered Train Impact Test of Spent Nuclear Fuel Shipping Cask
    https://www.youtube.com/watch?v=hlextDSoVkQ&t=17s

7. NWMO.ca – Post-closure Safety Assessment of a Used Fuel Repository in Crystalline Rock
    https://www.nwmo.ca/~/media/Site/Reports/2018/01/19/15/54 /NWMO-TR_2017_02_Sixth-Case-Study-Report.ashx
    See frame 80, Section 2.2.2.2 "Hydraulic Parameters"
YouTube – Deep Geologic Repository - OPG's Plan
    https://www.youtube.com/watch?v=7L3H3VILPHQ&t=3s

8. See our supplement "Hydraulic Conductivity" at the back of this book.

9. YouTube – 3D animation nuclear waste disposal
    https://www.youtube.com/watch?v=FOn9tiy51Ok
Deep Isolation.com – technology
    https://www.deepisolation.com/technology/
Deep Isolation.com – Disposal of high-level nuclear waste in deep horizontal drillholes
    https://www.deepisolation.com/wp-content/uploads/2018/12 /DeepIsolationTechnology-White-Paper.pdf
    See Fig. 1

10. CDN science pub – Calculated uranium solubility in groundwater: implications for nuclear fuel waste disposal
    https://cdnsciencepub.com/doi/pdf/10.1139/v82-241
    See "Implications and Conclusions"

11. JM Korhonen – What does research say about the safety of nuclear power?
    https://jmkorhonen.net/2017/03/10/what-does-research-say -about-the-safety-of-nuclear-power/
    See Section 4, "Waste" (two bananas)

12. Springer.com – Neptunium, plutonium and 137Cs sorption by bentonite clays and their speciation in pore waters
    https://link.springer.com/article/10.1007/s10967-006-0356-6
KIT.edu – Sorption studies of actinides/lanthanides onto clay minerals under saline conditions
    https://publikationen.bibliothek.kit.edu/220103001

13. Medium.com – Oklo's natural nuclear reactors

https://medium.com/predict/oklos-natural-nuclear-reactors
-eb2cc3141b48

Mike sez:

Bacteria in Oklo streams ingested a dissolved uranium salt and excreted an insoluble uranium salt. The insoluble salt was gathered as silt by the streams, and when a critical mass was achieved, the U-235 fissioned.

Uranium is a metal, and metals give away their electrons in their outermost orbital. That's why metals used in batteries—they acquire and lose electrons with the greatest of ease.

An "oxidized" metal simply means that the metal has either lost electros or reacted with oxygen to form an oxide molecule, typically as an O_2 (UO_2 = uranium oxide, Fe_2O_3 = iron oxide [rust], etc.)

With six electrons in its outermost orbital, uranium has a whopping six possible oxidation states. This allows it to form a variety of oxides, fluorides, sulfates, etc. Some of these compounds are very soluble, some are "sparingly" soluble, and some are not soluble at all.

NCBI – Interaction of Uranium with Bacterial Cell Surfaces: Inferences from Phosphatase-Mediated Uranium Precipitation

https://www.ncbi.nlm.nih.gov/pmc/articles/PMC4968550/
See introduction

Solubility has a lot to do with the geometry of the molecules in question, and whether or not V-shaped water molecules can gang up and effectively surround an oxide molecule. The degree to which they can is the degree to which a molecule is "soluble."

One soluble form of uranium is a $[UO_2]^{2+}$ dication, with the U in a +6 oxidation state, meaning no electrons in its outermost shell. Oxygen anions usually have a formal charge of -2, so the electron arithmetic is this case is $(+6) + 2(-2) = +2$. It's called a dication (Dye-CAT-eye-on) because it has a net charge of +2. Anions (AN-eye-ons) can have a net negative charge of -1, -2, etc. For example, dianions have a net negative charge of -2.)

Other forms of uranium, such as uranyl nitrate, are soluble. This allows water molecules to surround and encapsulate, or solubilize the pieces.

14.   Wikipedia – Aqueous homogenous reactor
        https://en.wikipedia.org/wiki/Aqueous_homogeneous_reactor
15.   YouTube – Natural nuclear reactor
        https://www.youtube.com/watch?v=pMjXAAxgR-M
        YouTube – How mother nature built a nuclear reactor called Oklo
        https://www.youtube.com/watch?v=zc0lAeFNjDg
        Science Direct – Natural fission reactors in the Franceville
        basin, Gabon: A review of the conditions and results of a
        "critical event" in a geologic system
        https://www.sciencedirect.com/science/article/abs/pii/S001
        6703796002451
16.   Wikipedia – Natural nuclear fission reactor
        https://en.wikipedia.org/wiki/Natural_nuclear_fission_reactor
        See para. 6 in section "Mechanism"
17.   Energy.CA.gov – The Warren-Alquist Act, 2021 edition
        https://www.energy.ca.gov/publications/2020/warren-alquist
        -act-2021-edition
18.   *Ibid.* #15 in Chapter 9
        Ibid. #5 in Chapter 16, second reference
19.   *Ibid.* #2 Chapter 18
20.   Reddit – r/nuclear J Tritten it was clear to us
        https://www.reddit.com/r/nuclear/comments/ynwb6a/j_trittin
        _it_was_clear_to_us_that_we_couldnt_just/
21.   Sierra Club – Reasons to oppose Ohio nuclear bailouts
        https://www.sierraclub.org/sites/www.sierraclub.org/files
        /program/documents/Reasons%20to%20Oppose%20Ohio%20
        Nuclear%20Bailouts.pdf See pg. 2
        Environmental Progress – Sierra Club
        http://environmentalprogress.org/sierra-club
22.   Wikipedia – Friends of the Earth
        https://en.wikipedia.org/wiki/Friends_of_the_Earth_(US)
        See para. 2
        Atomic Insights – Smoking gun: Robert Anderson provided
        initial funds to form Friends of the Earth
        https://atomicinsights.com/smoking-gun-robert-anderson/
23.   Friends of the Earth – New study reveals incompatibility of
        climate safety and gas

https://friendsoftheeearth.eu/press-release/new-study-reveals
-incompatibility-of-climate-safety-and-gas/

24. Atomic Insights – Anti-nuclear strategy April 1991 pdf
    https://atomicinsights.com/wp-content/uploads/Antinuclear-
    strategy-April-1991.pdf

25. Science Daily – Nuclear energy programs do not increase
    likelihood of proliferation, study finds
    https://www.sciencedaily.com/releases/2017/11/171106112256
    .htm
    See Summary, pg. 1
    "Contrary to popular thought, nuclear proliferation is not more
    likely to occur among countries with nuclear energy programs,
    according to research. In a historical analysis of the relationship
    between nuclear energy programs and proliferation from 1954
    to 2000, the study finds that the link between the two has been
    overstated."

26. MIT.edu – Documentation and diagrams of the atomic bomb
    http://people.csail.mit.edu/boris/nuc-bomb.html

27. Oranu.org – The Smyth report - atomic energy for military
    purposes
    https://orau.org/health-physics-museum/files/smyth-report.pdf

28. YouTube – Susan Spotless 1961 PSA
    https://www.youtube.com/watch?v=-BCnGP-ktrQ
    CNN.com – Why you don't throw trash out the window
    https://www.cnn.com/2013/12/01/opinion/greene-lady-bird
    -and-litter
    NY Times – Beneath Her Decorous Demeanor, 'Lady Bird
    Johnson' Was a Political Force
    https://www.nytimes.com/2021/03/16/books/review/lady-bird
    -johnson-hiding-in-plain-sight-julia-sweig.html

CHAPTER NINETEEN: You Can't Get There from Here

1. *Ibid.* #27 Chapter 18
2. New Yorker – Atomic John a truck driver uncovers secrets about
   the first nuclear bombs
   https://www.newyorker.com/magazine/2008/12/15/atomic-john

3. Fissile Materials.org – Report of the Plutonium Disposition Working Group: Analysis of Surplus Weapon-Grade Plutonium Disposition Options

    http://fissilematerials.org/library/doE14a.pdf

    See page B-43, section "Isotopic Degradation", second sentence. PMDA refers to the Plutonium Management and Disposition Agreement between the USA and Russia that was in effect from year 2000 until 2015.

    A "weapons [-grade] plutonium" ratio for Pu-240 /Pu-239 less than 0.10 means that the ratio that must be greater than 10 in order to qualify as weapons-grade. In layman's terms, that's 90% pure Pu-239.

    Pu-239 = 5810 units, Pu-240 = 2840 units. Ratio = 5810 ÷ 2840 = 2.04.

    Since the ratio is less than 10, the material would be classified reactor-grade plutonium, and not weapons-grade.

4. OECD-NEA.org – Spent Nuclear Fuel Assay Data for Isotopic Validation

    https://www.oecd-nea.org/science/wpncs/ADSNF/SOAR_final.pdf

    See page 17 (frame 18), Table 1. Refer to column 3, Content in spent PWR fuel, in units of "g /MTHM" (grams per metric tonne of heavy metal).

5. The U-235 concentration in natural uranium is 0.72%. Thus, there are 7.2 kilograms per tonne. For 25 kilos of U-235: 25 ÷ 7.2 = 3.472, or ~3.5 tonnes.

6. Wikipedia – W88 warhead

    https://en.wikipedia.org/wiki/W88

7. Wikipedia – Nuclear weapon design

    https://en.wikipedia.org/wiki/Nuclear_weapon_design

8. Aqua-calc – weight to volume calculator

    https://www.aqua-calc.com/calculate/weight-to-volume

9. Science and Global Security – North Korean plutonium production pdf

    http://scienceandglobalsecurity.org/archive/sgs05albright.pdf

    See frame 11 (pg. 73 of original doc, pg. 25 of pdf): "When operating at a thermal power of 25 MWt an average of 80 percent of the time, this equation shows that the reactor can produce about 6.6 kilograms of weapons-grade plutonium per year."

Also see frame 16: "For the 25 MWt and 200 MWt reactors, the grade of the plutonium is assumed to be weapons-grade. At a 60 percent capacity factor, the two reactors would produce 5 and 40 kilograms of weapons-grade plutonium, respectively, each year at steady state. The largest reactor is assumed to have an average power of 700 MWt and to produce reactor-grade plutonium. This reactor would produce about 90 kilograms of reactor-grade plutonium per year at steady state."

10. Wikipedia – Chicago Pile-1https://en.wikipedia.org/wiki/Chicago_Pile-1

11. NAP.edu – Monitoring nuclear weapons and nuclear-explosive materials
    https://www.nap.edu/catalog/11265/monitoring-nuclear-weapons-and-nuclear-explosive-materials-an-assessment-of
    See chapters one and five: https://www.nap.edu/read/11265/chapter/1
    https://www.nap.edu/read/11265/chapter/5
    NRC.gov – Nuclear material control and accounting
    https://www.nrc.gov/materials/fuel-cycle-fac/nuclear-mat-ctrl-acctng.html

12. Atomic Heritage.org – Electronics and detonators
    https://www.atomicheritage.org/history/electronics-and-detonators
    PRN Newswire – 800 Nuclear triggers smuggled to Israel, mastermind untouchable - secret FBI files
    https://www.prnewswire.com/news-releases/800-nuclear-triggers-smuggled-to-israel-mastermind-untouchable---secret-fbi-files-143812826.html
    Israel Lobby – FBI Treasury Customs investigation of MILCO - Heli nuclear trigger smuggling to Israeli Ministry of Defense
    https://www.israellobby.org/krytons/

13. CDC.gov – FAQs about dirty bombs
    https://www.cdc.gov/nceh/radiation/emergencies/dirtybombs.htm

14. DHS.gov – Radiological attack dirty bombs and other devices

https://www.dhs.gov/xlibrary/assets/prep_radiological_fact
_sheet.pdf

15. Twitter – That Rad Guy 9am March 2, 2022
    https://twitter.com/ThatRadGuy5/status/1499067341329
    469442

CHAPTER TWENTY: Strange Days, Indeed

1. NY Times – Japan races to build new coal-burning power plants,
   despite the climate risks
   https://www.nytimes.com/2020/02/03/climate/japan-coal
   -fukushima.html

2. Scientific American – Coal ash is more radioactive than nuclear
   waste
   https://www.scientificamerican.com/article/coal-ash-is-more
   -radioactive-than-nuclear-waste/

3. OSTI.gov – Civilian Nuclear Power a report to the president 1962
   https://www.osti.gov/servlets/purl/1212086

4. Wikipedia – Solar power in Japan
   https://en.wikipedia.org/wiki/Solar_power_in_Japan
   Solar capacity: 5000 MW in 2011 increased to 62,000 MW
   in 2019
   Knoema.com – Japan's CO_2 emissions per capita
   https://knoema.com/atlas/Japan/CO2-emissions-per-capita
   A 6.6% decline from 2011 to 2019

5. CNBC.com – Japan just signaled a big shift in its post Fukushima
   future
   https://www.cnbc.com/2022/08/24/japan-just-signaled-a-big
   -shift-in-its-post-fukushima-future.html
   Bloomberg – Threats of blackouts drive japan to embrace
   nuclear again
   https://www.bloomberg.com/news/articles/2022-08-24
   /japan-wants-to-restart-more-reactors-to-avoid-power-shortages

6. Statista – Annual carbon dioxide emissions in Germany from
   1990 to 2021
   https://www.statista.com/statistics/449701/co2-emissions
   -germany/

(Note that from 2011 to 2018, there has only been a 4.6% decrease in CO_2 emissions.)

Clean Energy Wire – Report: Germany suffers more coal-linked deaths than rest of EU

https://www.cleanenergywire.org/news/merkel-says-climate-be-g20-priority-coal-related-deaths/report-germany-suffers-more-coal-linked-deaths-rest-eu

Washington Examiner – German nuclear phaseout is causing 1,100 additional deaths a year: Study

https://www.washingtonexaminer.com/policy/energy/german-nuclear-phaseout-is-causing-1-100-additional-deaths-a-year-study

7. Worldometer – Germany CO_2 emissions

https://www.worldometers.info/co2-emissions/germany-co2-emissions/

Germany's electricity:

9.44 t /person × (1.2 cm / 2.7 cm) = 4.20 t /person.

Electricity segment = 1.2 cm. Total bar height = 2.7 cm.

Worldometer – France CO^2 emissions

https://www.worldometers.info/co2-emissions/france-co2-emissions/

For France's electricity:

5.13 t /person × (0.3 cm / 2.9 cm) = 0.53 t /person.

Light blue electricity segment = 0.3 cm. Total bar height = 2.9 cm.

4.20 t /person Germany ÷ 0.53 t /person France = *7.9X greater emissions in Germany.*

8. World Nuclear.org – Nuclear power in Sweden

https://www.world-nuclear.org/information-library/country-profiles/countries-o-s/sweden.aspx

Balkan Green Energy News – Poland to build three nuclear power plants

https://balkangreenenergynews.com/poland-to-build-three-nuclear-power-plants/

World Nuclear.org – Nuclear power in Czech Republic

https://world-nuclear.org/information-library/country-profiles/countries-a-f/czech-republic.aspx

9. YouTube – I can't put my arms down! A Christmas Story

https://www.youtube.com/watch?v=HW4IZ0Flh3M

10. NuScale Power – NuScale Builds Upon Unparalleled Licensing Progress With Second Standard Design Approval Application Submittal
https://www.nuscalepower.com/en/news/press-releases/2023/nuscale-builds-upon-unparalleled-licensing-progress-with-second-standard-design-approval

11. American Action Forum – Putting nuclear regulatory costs in context
https://www.americanactionforum.org/research/putting-nuclear-regulatory-costs-context/

12. Why Nuclear Power Has Been A Flop – The Gordian Knot of the 21st Century
https://gordianknotbook.com
See Chapter 9

13. Breakthrough.org – The Diablo we know
https://thebreakthrough.org/issues/energy/diablo-canyon-nuclear-power-shutdown-risk
See para. 33
KQED.org – Water board weighs phasing out diablo canyon's cooling system
https://www.kqed.org/news/10349447/water-board-weighs-phasing-out-diablo-canyons-cooling-system
San Luis Obispo.com – Diablo Canyon faces deadline
https://www.sanluisobispo.com/news/local/environment/article39459930.html
Research Gate – A process for evaluating adverse environmental impacts by cooling-water system entrainment at a California power plant
https://www.researchgate.net/publication/6986621_A_Process_for_Evaluating_Adverse_Environmental_Impacts_by_Cooling-Water_System_Entrainment_at_a_California_Power_Plant

14. YouTube – Sharing the Robust Marine Environment at Diablo Canyon Cove and Tidepools
https://www.youtube.com/watch?v=QCf1_oeGGKs13
John Lindsey – The remarkably productive coastline at diablo canyon power plant

https://www.youtube.com/watch?v=j5T5sEOaFB0

15. Power-eng.com – Global electricity consumption to rise 79 percent higher by 2050
https://www.power-eng.com/2019/09/24/global-electricity-consumption-to-rise-79-percent-higher-by-2050-eia-says/

16. WE Forum.org – Renewable energy could power the world by 2050. Here's what that future might look like
https://www.weforum.org/agenda/2020/02/renewable-energy-future-carbon-emissions/

17. See the supplement "Global Electricity Demand in 2050" in the back of this book.

18. Mashable – Air conditioner use will triple by 2050. That's bad news for a warming planet
https://mashable.com/2018/05/15/air-conditioner-use-2050-climate-change/
IAEA.org – The future of cooling
https://www.iea.org/reports/the-future-of-cooling
Science.org – The emergence of heat and humidity too severe for human tolerance
https://www.science.org/doi/10.1126/sciadv.aaw1838
Huffpost – As temps rise, Europe considers a/c
https://www.huffpost.com/entry/europe-air-conditioning-environment_n_5d800030e4b077dcbd62e3f4
National News.com – Working from home could create surge in Gulf's AC bills and emissions
https://www.thenationalnews.com/uae/environment/working-from-home-could-create-surge-in-gulf-s-ac-bills-and-emissions-1.1077001

19. YouTube – Peanut butter & jelly, natural gas & renewables ExxonMobil
https://www.youtube.com/watch?v=6K9f2uy2JzU

20. Tim sez:
In 2018, US natural gas consumption was 30.1 TCF (trillion cubic feet.) That's a 10.8% increase over 2017. Taking into account our exports, US total production was a bit more: 30.6 TCF.
A preliminary report for 2019 production indicates another 10% annual increase, to 33.7 TCF:

EIA.gov – Natural gas data
https://www.eia.gov/dnav/ng/hist/n9070us2A.htm
EIA.gov – Short-term energy output
https://www.eia.gov/outlooks/steo/report/natgas.php
US gas reserves have recently been revised upward to 3,400 TCF. If annual production remained constant (which is unlikely): 3,400 TCF ÷ by 33.7 TCF per year = 101 years until exhaustion. The year 2050 is 30 years from now. So 30 ÷101 = 30% of our natural reserves gone by 2050.

21. Natural Gas Intel.com – LNG Exports Hit Record Volumes, EIA Says
https://www.naturalgasintel.com/u-s-lng-exports-hit-record-volumes-eia-says/

22. EIA.gov – Natural gas explained Natural gas imports and exports
https://www.eia.gov/energyexplained/natural-gas/imports-and-exports.php
See natural gas exports graph

23. Forbes – Trump administration rebrands fossil fuels as molecules of US freedom
https://www.forbes.com/sites/jamesellsmoor/2019/05/30/trump-administration-rebrands-carbon-dioxide-as-molecules-of-u-s-freedom/

24. Mother Jones – Freedom Gas the Trump administration's ridiculous new plan
https://www.motherjones.com/environment/2019/05/freedom-gas-the-trump-administrations-ridiculous-new-plan-to-rebrand-fossil-fuels/

25. The Onion.com – The Onion's guide to renewable energy
https://www.theonion.com/the-onion-s-guide-to-renewable-energy-1849774228

26. Wikipedia – Enron
https://en.wikipedia.org/wiki/Enron
LA Times – How Enron manipulated state's power market
https://www.latimes.com/archives/la-xpm-2002-may-09-fi-scheme9-story.html

27. CNN.com – Deep freeze sends Texas electricity prices soaring 10,000%

https://www.cnn.com/2021/02/16/business/nightcap-texas
-energy-mcdonalds-citibank/index.html

28. Apple podcasts – Decouple with Dr. Chris Keefer
https://podcasts.apple.com/us/podcast/decouple/id1516526694
Ecomodernism.org – An ecomodernist manifesto
http://www.ecomodernism.org/
Note: The term "decouple" evolved through the efforts of a group of forward-thinking environmentalists, now referred to as Ecomodernists, to articulate a new vision of humanity's relationship to the environment in *The Ecomodernist Manifesto*.

29. World in Data – How many people does synthetic fertilizer feed?
https://ourworldindata.org/how-many-people-does-synthetic
-fertilizer-feed

30. Energy.gov – Hydrogen production natural gas reforming
https://www.energy.gov/eere/fuelcells/hydrogen-production
-natural-gas-reforming

31. Gen-4.org – Very-high-temperature reactor (VHTR)
https://www.gen-4.org/gif/jcms/c_42153/very-high-temperature
-reactor-vhtr

32. Oneear.2019 – Impacts of green new deal energy plans on grid stability, costs, jobs, health, and climate in 143 countries – supplemental Information
https://www.cell.com/cms/10.1016/j.oneear.2019.12.003
/attachment/cc629ac5-7712-4a5c-b7f7-a5122f77b6bf/mmc1.pdf
See frame 54 / page 53, Table S10: United States, column 8: 11.58 teragrams /year of H_2 = 11.58 million tonnes /yr. Compare to present H_2 production of 9 tonnes /year.

33. White House.gov – Fact sheet the American Jobs Plan
https://www.whitehouse.gov/briefing-room/statements-releases
/2021/03/31/fact-sheet-the-american-jobs-plan/

CHAPTER TWENTY-ONE: Make Energy, Not War

1. Science Daily – Nuclear energy programs do not increase likelihood of proliferation
https://www.sciencedaily.com/releases/2017/11/171106112256
.htm Summary, pg. 1

"Contrary to popular thought, nuclear proliferation is not more likely to occur among countries with nuclear energy programs, according to research. In a historical analysis of the relationship between nuclear energy programs and proliferation from 1954 to 2000, the study finds that the link between the two has been overstated."

2. Forbes – A nuclear primer what is an atomic bomb?
https://www.forbes.com/sites/jamesconca/2014/09/23/a-nuclear-primer-what-is-an-atomic-bomb/ - 275f997c50c1

3. Forbes – EROI a tool to predict the best energy mix
https://www.forbes.com/sites/jamesconca/2015/02/11/eroi-a-tool-to-predict-the-best-energy-mix/ - 5dc27be0a027
Festkoerper – Energy intensities, EROIs, and energy payback times of electricity generating power plants
https://festkoerper-kernphysik.de/Weissbach_EROI_preprin.pdf

4. Space.com – Meet Au-Spot, the AI robot dog that's training to explore caves on Mars
https://www.space.com/ai-mars-robot-dogs-agu

5. Business Insider – Bill Gates disses tech billionaires spending their free time on space rockets: I don't see how it improves humanity
https://www.businessinsider.com/bill-gates-on-space-exploration-2013-8
New Atlas – Bill Gates's next-gen nuclear plant packs in grid-scale energy storage
https://newatlas.com/energy/natrium-molten-salt-nuclear-reactor-storage/

6. Global Construction Review – World's first nuclear barge fully commissioned in Russia
https://www.globalconstructionreview.com/news/worlds-first-nuclear-barge-fully-commissioned-russ/

7. ThorCon Power Isle
https://thorconpower.com/
Seaborg Power Barge
https://www.seaborg.com/

8. Wikipedia – Megatons to Megawatts program
https://en.wikipedia.org/wiki/Megatons_to_Megawatts_Program

Centrus Energy – Megatons to Megawatts

https://www.centrusenergy.com/who-we-are/history/megatons
-to-megawatts/

The predecessor of Centrus, the United States Enrichment Corporation, as executive agent for the US government, and Joint Stock Company "TENEX" (TENEX), acting for the Russian government, implemented this 20-year, $8 billion program at no cost to taxpayers.

9. Wikipedia – Jevons Paradox

https://en.wikipedia.org/wiki/Jevons_paradox

10. Al Jazeera – Emissions possible: Streaming music swells carbon footprint

https://www.aljazeera.com/ajimpact/carbon-big-foot-climate
-impact-streaming-music-videos-200221220408755.html

11. The Guardian – Tsunami of data could consume one fifth of global electricity by 2025

https://www.theguardian.com/environment/2017/dec/11
/tsunami-of-data-could-consume-fifth-global-electricity-by-2025

12. Independent.co.UK – Bitcoin uses more energy than the whole of Switzerland

https://www.independent.co.uk/life-style/gadgets-and-tech
/news/bitcoin-energy-switzerland-cryptocurrency-mining
-a8989816.html

13. The Story of Stuff

https://www.storyofstuff.org/movies/story-of-stuff/

14. Phys.org – Deadly heat waves will be common in South Asia, even at 1.5 degrees of warming

https://phys.org/news/2021-03-deadly-common-south-asia
-degrees.html

The Guardian – One billion people will live in insufferable heat within 50 years

https://www.theguardian.com/environment/2020/may/05/one
-billion-people-will-live-in-insufferable-heat-within-50-years-study

NY Times – Extreme heat will change us

https://www.nytimes.com/interactive/2022/11/18/world
/middleeast/extreme-heat.html

15. Scientific American – Oceans are warming faster than predicted

https://www.scientificamerican.com/article/oceans-are-warming
-faster-than-predicted/

The Guardian – 'Headed off the charts': world's ocean surface
temperature hits record high

https://www.theguardian.com/environment/2023/apr/08
/headed-off-the-charts-worlds-ocean-surface-temperature-hits
-record-high

NY Times – Oceans are absorbing almost all of the globe's
excess heat

https://www.nytimes.com/interactive/2016/09/12/science/earth
/ocean-warming-climate-change.html

16. NBC Miami – Miami Beach prepares for higher than average
king tide flooding

https://www.nbcmiami.com/news/local/miami-beach-prepares
-for-higher-than-average-king-tide-flooding/2290027/

17. IAEA.org – Achieving net-zero emissions by 2050

https://www.iea.org/reports/world-energy-outlook-2020
/achieving-net-zero-emissions-by-2050

YouTube – Ocean warming and acidification - Dr. Alex
Cannara

https://www.youtube.com/watch?v=T1WzsUeQSdk

Washington Post – The carbon skyscraper: A new way of
picturing rapid, human-caused climate change

https://www.washingtonpost.com/weather/2021/01/12
/carbon-skyscraper-rapid-climate-change/

18. IPCC – Climate change 2021 the physical science basis

https://www.ipcc.ch/report/sixth-assessment-report-working
-group-i/

See B.1 on frame 14 / pg. 14

19. NREL.gov – Blending hydrogen into natural gas pipeline
networks: a review of key issues

https://www.nrel.gov/docs/fy13osti/51995.pdf

20. Wikipedia – Hydrogen storage

https://en.wikipedia.org/wiki/Hydrogen_storage

MDPI.com – Ammonia as Effective Hydrogen Storage: A
Review on Production, Storage and Utilization

https://www.mdpi.com/1996-1073/13/12/3062/htm

21. YouTube – KEPCO E&C's APR1400 Nuclear Power Plant
https://www.youtube.com/watch?v=T5QlVZVcLCw

22. Next Big Future – Carnival of Nuclear Energy 6 – the real world
and arithmetic confounds nuclear skeptics
https://www.nextbigfuture.com/2010/06/carnival-of-nuclear
-energy-6-real-world.html
APR-1400 for $2,333 /kW
Global Construction Review – China approves $10bn plan
to build four nuclear reactors
https://www.globalconstructionreview.com/news/china-approves
-10bn-plan-build-four-nuclear-reacto/
$10.2 billion for four Huolong Ones @ 1100 MW each =
$2300 /kW
Powermag – Hualong One Reactor Now Operating in China
https://www.powermag.com/hualong-one-reactor-now
-operating-in-china/
(Five-year build time for Hualong One was from 2015 to 2020)
Wikipedia – Barakah nuclear power plant
https://en.wikipedia.org/wiki/Barakah_nuclear_power_plant
(South Korea's APR-1400s in the UAE)
The National – UAE's first nuclear plant to begin major
testing on third reactor
https://www.thenational.ae/uae/government/uae-s-first-nuclear
-plant-to-begin-major-testing-on-third-reactor-1.900569
Reuters – Poland sees cost of building 6-9 GW of nuclear
energy at $30 bln
https://www.reuters.com/article/poland-nuclear/poland-sees
-cost-of-building-6-9-gw-of-nuclear-energy-at-30-bln-idUKL8
N2ID2K2 (South Korea proposes building six APRs in Poland
for $3,700 /kW)
The Science Council – The cost of nuclear power
https://www.thesciencecouncil.com/index.php/advisors
/active-advisers/dr-barry-brook/175-the-cost-of-nuclear-power

23. Huffpost – Coal-Addicted Poland Is Going Nuclear With U.S.
Help. It'll Be A Test For Both Nations
https://www.huffpost.com/entry/coal-poland-nuclear_n_636
41f8be4b024c30191f8d6

24. YouTube – ThorCon's Thorium Converter Reactor - Lars Jorgensen in Bali
www.youtube.com/watch?v=oB1IrzDDI9g
See especially 31:00 in

25. Wikipedia – Liberty ship
https://en.wikipedia.org/wiki/Liberty_ship

CHAPTER TWENTY-TWO: We Will Now Begin Boarding

1. YouTube – Aaron Bloom: Transgrid-X 2030 Symposium
https://www.youtube.com/watch?v=5S2e00PqrIk
Aaron Bloom's slides for the above talk:
TransGrid X v10Bloom.pd
https://iastate.app.box.com/s/vfgn9nikl1rz7r8x0vaoauzpm2210t35

2. Energy Matters – Worldwide investment in renewable energy reaches US$ 4 trillion – with little to show for it
http://euanmearns.com/worldwide-investment-in-renewable-energy-reaches-us-4-trillion-with-little-to-show-for-it/

3. NY Times – Indian Point Is Shutting Down. That Means More Fossil Fuel.
https://www.nytimes.com/2021/04/12/nyregion/indian-point-power-plant-closing.html
CaliforniaGlobe.com – Grid Expert: Replacing Diablo Canyon Nuclear Plant with Renewables 'Can't Be Done'
https://californiaglobe.com/articles/grid-expert-replacing-diablo-canyon-nuclear-plant-with-renewables-cant-be-done/

4. Riverkeeper.org
https://www.riverkeeper.org/

5. NYISO.com – Generator deactivation assessment Indian Point energy center
https://www.nyiso.com/documents/20142/1396324/Indian_Point_Generator_Deactivation_Assessment_2017-12-13.pdf/f673a0f8-5620-1d7b-4be2-99aaf781ac5c
See "Assumptions" final sentence

6. Lohud.com – After Indian Point: How does the Hudson Valley keep its lights on?

https://www.lohud.com/story/opinion/2021/05/13/hudson-valley-power-what-do-we-do-after-indian-point/5059068001/
Lohud.com – NY's fossil fuel use soared after Indian Point plant closure. Officials sound the alarm
https://www.lohud.com/story/news/2022/07/22/new-york-fossil-fuels-increase-after-indian-point-nuclear-plant-shutdown/65379172007/

7.   Bloomberg.com – New York $4.5 Billion Plan to Bring Hydropower to Big Apple
https://www.bloomberg.com/news/articles/2022-04-14/new-york-approves-4-5-billion-plan-to-bring-hydropower-to-nyc#xj4y7vzkg

8.   Stanford.edu – Stanford energy and environment experts examine strengths and weaknesses of the Green New Deal
https://news.stanford.edu/2019/03/28/strengths-weaknesses-green-new-deal/
During an interview with two others, Mark Jacobson remarked: "The intent of the GND as originally written is to 'transition off of nuclear and fossil fuels as soon as possible,' so the nuclear folks will try to oppose it as well."
Congress.gov – H.R.2454 - American Clean Energy and Security Act of 2009
https://www.congress.gov/bill/111th-congress/house-bill/2454
(Sec. 199A) Requires the Secretary, by February 1, 2011, to report to Congress on the results of a study on the use of thorium-fueled nuclear reactors for national energy needs, including a response to the International Atomic Energy Agency study entitled, "Thorium fuel cycle – potential benefits and challenges."
(Sec. 242) Requires the Secretary to establish a program to make monetary awards to encourage the owners and operators of new and existing electric energy generation facilities or thermal energy production facilities using fossil or nuclear fuel to use innovative means of recovering thermal energy

9.   John Kerry (climate envoy):
You Tube – John Kerry - From Anti-Nuclear to "Go For It"
https://www.youtube.com/watch?v=f15rSTy7Spg
Senator Sheldon Whitehouse (D – RI):

YouTube – Sen. Whitehouse - Environment and Public Works
Committee on the American Nuclear Infrastructure Act
https://www.youtube.com/watch?v=BjzNHpexmmk
Senator Corey Booker (D – NJ):
YouTube – Sen. Cory Booker on why he supports clean nuclear
energy
https://www.youtube.com/watch?v=JXD-jvlmwbY
Rep. Elaine Luria (D – VA) (retired US Navy reactor operator):
Daily Press.com – Luria calls for national effort on advanced
nuclear technology
https://www.dailypress.com/government/local/dp-nws-luria
-nukes-20190619-story.html

10. A wide spectrum of people are pro-nuclear, even die-hard
"greenies"
Generation Atomic
https://www.generationatomic.org/
Mothers for Nuclear
https://www.mothersfornuclear.org/
US News – Environmentalists warm to nuclear amid climate
change threat
https://www.usnews.com/news/politics/articles/2018-11-15
/environmentalists-warm-to-nuclear-amid-climate-change-threat
Medium.com – For The First Time, World Learns Truth
About Risk Of Nuclear
https://medium.com/generation-atomic/for-the-first-time-world
-learns-truth-about-risk-of-nuclear-6b7e97d435df
Climate and Capitalism – Socialist arguments for nuclear power
https://climateandcapitalism.com/2011/06/14/socialist
-arguments-for-nuclear-power/
NEI.org – Bipartisan nuclear energy legislation passes out of house
https://www.nei.org/news/2018/bipartisan-nuclear-energy
-bill-passes-out-of-house
BBC.com – Extinction Rebellion: Nuclear power 'only option'
says former spokeswoman
https://www.bbc.com/news/uk-54103163
Facebook – Progressives for nuclear power
https://www.facebook.com/nuclearprogress/

Planning Beyond Capitalism.org – The conspiracy against nuclear energy: how big oil built the ecology movement to demonize nuclear energy competition

https://socialistplanningbeyondcapitalism.org/the-conspiracy-against-nuclear-energy-how-big-oil-built-the-ecology-movement-to-demonize-nuclear-energy-competition/

Facebook – People's Fission: Progressives for Nuclear Energy

https://www.facebook.com/groups/2336208663269193

WIN US.org – Women in nuclear

https://www.winus.org/

Environmental Progress

https://environmentalprogress.org/

Climate Coalition

http://climatecoalition.org/

4th Generation

https://4thgeneration.energy/

11.   *Ibid.* #7 of Chapter 3, second reference

12.   *Ibid.* #8 of Chapter 3

13.   The Guardian – Rare earth mining in China: the bleak social and environmental costs

https://www.theguardian.com/sustainable-business/rare-earth-mining-china-social-environmental-costs

Volkswagen AG.com – What you should know about the contentious issue of lithium

https://www.volkswagenag.com/en/news/stories/2020/03/lithium-mining-what-you-should-know-about-the-contentious-issue.html

14.   See the supplement "Global Primary Energy in 2050" at the back of this book.

YouTube – Alex Cannara - Ocean Acidification

https://www.youtube.com/watch?v=wtQxF_3BSxQ&t=21s

15.   Statista – Number of container ships in the global merchant fleet from 2011 to 2022

https://www.statista.com/statistics/198227/forecast-for-global-number-of-containerships-from-2011/

Pier Next Barcelona – The new wave of container ships: Bigger? More sustainable?

https://piernext.portdebarcelona.cat/en/logistics/the-new-wave
-of-container-ships/

16. Michael Klare – Rethinking our relationship to the natural
world after Covid-19
https://www.thenation.com/article/environment/coronavirus
-nature-humans/
From the section "Heeding Mother Nature's Warning":
"What humanity may need to do is institute a new policy of
"peaceful coexistence" with Mother Nature. This approach would
legitimize the continued presence of large numbers of humans
on the planet but require that they respect certain limits in their
interactions with its ecosphere."
City AM.com – Former Extinction Rebellion activist now
embraces nuclear power
https://www.cityam.com/a-message-from-a-former-extinction
-rebellion-activist-fellow-environmentalists-join-me-in-embracing
-nuclear-power/

Supplements

SIEVERTS AND GRAYS
(Chapter 2, endnote 7)

First, please review:

Wikipedia – Absorbed dose

> https://en.wikipedia.org/wiki/Absorbed_dose

(See section: Stochastic risk—Conversion to Equivalent Dose

Figure: Ionizing radiation—protection dose quantities in SI units)

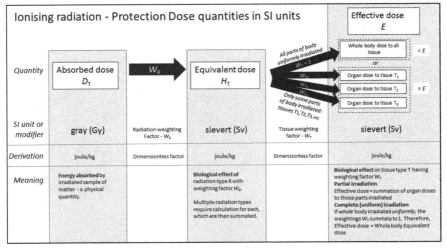

Fig. S-1: Equivalency Of Sieverts And Grays
"Ionizing radiation – protection dose quantities in SI Units"
Credit: By Doug Sim, own work, CA-BY-SA-3.0

Of the three common modes of radioactive decay (alpha, beta, and gamma), alpha has a greater biological effect than the other two. That is, alpha's effectiveness is greater than beta or gamma, when all three modes are normalized to the same unit of radiative energy. That measurement unit is usually "mega-electron volt," or MeV.

Alpha's effect factor is usually regarded as 20 times greater than the other two modes. For example, 1 milliGray of alpha energy, if absorbed by a living cell, is regarded as producing 20 milliSieverts of "effective" energy. This means an alpha-emitter that has been swallowed or inhaled delivers 20 times as great an "effective dose" as an equal-energy beta- or gamma-emitter.

However, alpha particles cannot penetrate animal skin. They cannot even move very far through air, and lose all their kinetic energy within a few centimeters of travel. So for externally-sourced radiation, alpha particles are biologically irrelevant.

Therefore, with *external* emitters more than a few centimeters away, there is no difference between absorbed dose, measured in Grays, and effective dose, measured in Sieverts.

URANIUM MINING AND PROCESSING

Supplement

PART ONE
(Chapter Four, endnote 20)

Determining the Global Waste-to-Ore Ratio of Dirt-Mined Uranium

NERD NOTE: The term "ore concentration" can be confusing. Broadly speaking, ore is mineral-bearing rock. So it is natural to assume that ore concentration (sometimes called ore grade) refers to the amount of ore rock found among the waste rock.

That would be incorrect. Ore concentration refers to how much of the desired mineral (in this case, uranium) is contained within the mineral-bearing ore that gets dug up along with the waste rock.

A mine's ore concentration is often called its "ore grade," and is usually expressed in weight-percent (wt%) units. For example, an 0.05% ore grade would be written as 0.05 wt%. That is, 0.05% of the ore's weight is due to the desired mineral within the ore.

Getting the facts and figures on some of these mines is like pulling teeth. Even WISE-Uranium.org, the nuclear-skeptic watchdog website we link to below, hasn't been able to dig up all the info on every mine, though they've been hard at it for years. So we did what we could to compile some reasonably accurate estimates.

Working with the numbers we could gather from WISE-Uranium. org, World-Nuclear.org, and the mines' own websites and reports, we estimate an global average waste-to-ore ratio for dirt-mined uranium of *3.9:1*. That is, for every tonne of uranium-bearing ore, about 3.9 tonnes of waste rock will be exhumed as well.

To be clear, the amount of uranium extracted from the ore depends upon the ore grade. But whatever the ore grade happens to be, about 3.9 tonnes of waste rock, on average, will accompany each tonne of ore extracted ("dirt-mined") from the earth. On average, ore grades at existing uranium mines, whether by dirt mining or in-situ leaching (ISL), average just 0.078%.

Source: World Nuclear.org – Uranium mining overview https://www.world-nuclear.org/information-library/nuclear-fuel-cycle/mining-of-uranium/uranium-mining-overview.aspx

Country	Mine Name	Year	Tonnes U/yr	Ore Grade (wt%)	Mined Ore (Megatonnes)
Kazakhstan	Inkai 1, 2, 3	2021	3449	0.038	7.09
	Kharasan 1, 2	2021	2809	0.108 (av)	2.6
	Akdala (Betpak)	2021	1579	0.057	4.07
	Inkai 4	2021	1509	0.043	3.51
	Karatau (Budenov)	2021	2561	0.063	3.91
	Akbastou (Budenov)	2021	1545	0.091	1.7
	S. Moinkum / Kanzhugan	2021	1493	0.038	3.93
	N. Moinkum / Tortkuduk	2021	2840	0.118	2.41
	C. Mynkuduk (Kendala)	2021	1579	0.038	4.16
	W. Mynkuduk	2021	805	0.036	2.24
	Irkol	2021	962	0.042	2.29
	N. and S. Kararmurun	2021	800	0.070 (av)	1.14
	Zarechnoye	2021	655	0.064	1.02
Australia	Olympic Dam	2021	1922	0.056	5.47
	Four Mile	2021	2241	0.320 (av)	0.56
	Ranger	2020	1574	0.165	0.95
Canada	Cigar Lake	2021	4693	4.732	0.17
	McArthur River	2017	7303	2.417	0.30
	Rabbit Lake	2015	1912	0.526	0.36
Namibia	Husab	2021	3309	0.0518	6.39
	Rossing	2021	2444	0.033	7.41
	Langer Heinrich	2017	1294	0.029	4.46
Niger	SOMAIR (Arlit)	2021	1996	0.140	1.43
	Akouta (Cominair)	2018	1294	0.270	0.24
Russia	Krasnokomensk	2015	1997	0.2	0.97
			53,891 t		72.21 Mt
Global Average Ore Grade: (total U tonnes ÷ total mined ore)				0.078%	

Fig. S-2: World's Top Twenty-Five Uranium Mines—80% of Global Uranium Production
Credit: By the authors

Our table is a list of the world's top twenty-five uranium mines, which together produce about 80% of world uranium. The first 13 mines are in Kazakhstan. All of them are ISL mines, as is the Four Mile mine in Australia.

The remaining eleven mines are dirt mines. The first one on the list, Australia's Olympic Dam, is a mixed-metal mine, with uranium as a by-product of copper production. This makes it difficult to determine which waste belongs to which mineral, so we excluded Olympic Dam from our calculations. That leaves the last ten mines on our list, some of them underground mines but mostly open-pit digs. We could only get solid information on five of them, so we went with it. (As we said, this is an estimate.)

SOURCES:
World Nuclear.org – Country profiles
   https://world-nuclear.org/information-library/country-profiles.aspx

WISE-uranium.org – Uranium mining industry info
   https://wise-uranium.org/indexu.html#UMMSTAT

These are the five dirt mines we considered:

CIGAR LAKE
World-Nuclear.org – Uranium mining overview
   https://www.world-nuclear.org/information-library/nuclear-fuel-cycle/mining-of-uranium/uranium-mining-overview.aspx
   This unique Canadian mine has an unheard of waste-to-ore ratio of just 1.26:1. See frame 15, Table 1-2, Note 6 of the link below: "Mineral reserves have been estimated with an average allowance of 26% dilution . . ."
Cameco – Cigar Lake operation technical report (2016)
   https://s3-us-west-2.amazonaws.com/assets-us-west-2/technical-report/cameco-2016-cigar-lake-technical-report.pdf

in 2021, Cigar Lake produced 4,693 t of uranium, with an ore grade of 4.73 wt% from Inferred Resources. This came from about 0.1 Mt of ore: (4693 ÷ 0.0473) = 99,200 t of ore, or about 0.1 Mt (megatonnes).

(NOTE: In mining surveys, resource values are categorized as either Measured, Indicated, or Inferred. We used Inferred Values, which incorporate all available field data.)

Cigar Lake's mining method is unique as well: First, a section of the ore body is frozen solid. Then, a tunnel is dug below the ore body, and a high-pressure water jet blasts out pieces of the ore body above the tunnel. Chunks of ore drop down into the tunnel below, where they are processed into a slurry and pumped to the surface. And notice the backfilled cavities—they're putting the waste rock back in the hole, as they proceed. When mining is done right, displaced earth becomes replaced earth.

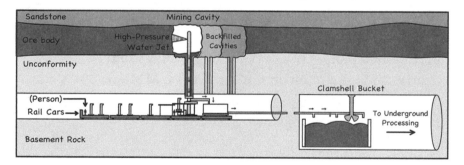

Fig. S-3: Cigar Lake Mining Method (Freezing and Slurry Extraction)
Source: https://www.cameco.com/businesses/mining-methods

HUSAB MINE

The open-pit Husab mine in Namibia is estimated to have a waste-to-ore ratio of 7:1. See section "Uranium Production":
Mining Technology.com – Husab uranium project

https://www.mining-technology.com/projects/husab-uranium -project-namibia-swakop/

The mine produced 3,309 t of uranium in 2021:
World Nuclear.org – Uranium mining overview (see table)

https://www.world-nuclear.org/information-library/nuclear-fuel-cycle/mining-of-uranium/uranium-mining-overview.aspx

The IAEA describes the ore grade as 518 ppm (see para.1):

Inis.IAEA.org – Preliminary study on uranium ore grade control techniques for the Husab mine Namibia

https://inis.iaea.org/collection/NCLCollectionStore/_Public/49/097/49097422.pdf

Mining one tonne of this ore will also bring up 6.39 megatonnes of waste rock:

(3,309 t ÷ 0.0518 wt%) = 6.388 Mt

ROSSING MINE

The open-pit Rossing Mine in Namibia is estimated to have a waste-to-ore ratio of 2.1:1. See "Mining Operations" section:

Rossing.com – Our operations

https://www.rossing.com/mining_processing.htm

World-Nuclear.org says Rossing produced 2,444 t of uranium in 2021. See the table titled "The Largest Producing Uranium Mines in 2021":

World Nuclear.org – Uranium mining overview (see table)

https://www.world-nuclear.org/information-library/nuclear-fuel-cycle/mining-of-uranium/uranium-mining-overview.aspx

WISE-Uranium.org says Rossing has a wt% of 0.033:

WISE-uranium.org – Uranium mine ownership – Namibia (scroll for Rossing)

https://www.wise-uranium.org/uona.html#LANGERH

So, (2,444 t ÷ 0.033 wt%) = 7.41 MT of ore.

SOMAIR MINE

In 2021, this open-pit mine in Niger had a waste-to-ore ratio of 7:1. See "Open Pit Mining" on frame 3:

Goviex.com – Goviez updates Madaouela project pre-feasibility study (2021)

https://goviex.com/site/assets/files/4156/2021-02-18_gxu_nr_pr_madaoeula_updated_pfs_final.pdf

SOMAIR has a Inferred ore grade of 0.14%:
WISE-uranium.org – Uranium mine ownership - Africa (scroll for Somair)
https://www.wise-uranium.org/uoafr.html
In 2021, the mine produced 1,996 t of uranium:
World nuclear.org – Niger
https://world-nuclear.org/information-library/country-profiles/countries-g-n/niger.aspx
So, (1,996 t U ÷ 0.0014) = 1.43 Mt of ore.

LANGER HEINRICH MINE

This open-pit mine in Namibia has a waste-to-ore ratio of 1.48:1. Finding this number took us a bit of digging. To follow the bread-crumb trail, go to:
WISE-uranium.org – Uranium mine ownership - Namibia (scroll for Langer Heinrich mine)
https://wise-uranium.org/uona.html#LANGERH
At the bottom of the Langer Heinrich section, click on "Calculate Mine Feasibility." Scroll down to the hot pink "Parameter Entry" section. In the blue "Process Parameters" box, note the 1.48 waste / ore ratio for this mine.

(NOTE: Select some of the other mines we discuss. You will see that even this dedicated nuclear watchdog website hasn't tracked down all the digits.)

In 2017, the Langer Heinrich mine produced 1,294 t of uranium from 4.46 megatonnes of ore, with an ore concentration of 0.029%. (See "Langer Heinrich" section in above link.)

(1,294 t U ÷ 0.029 wt%) = 1,294 t U.
And, (1,294 ÷ 0.000 29) = 4.46 Mt ore.

DETERMINING THE WEIGHTED AVERAGE
(WASTE-TO-ORE RATIO)

The combined total tonnage of ore extraction from these five top uranium dirt mines (Cigar Lake, Husab, Rossing, Somair, and Langer Heinrich) is as follows:

0.010 Mt + 6.39 Mt + 7.41 Mt + 1.43 Mt + 4.46 Mt = *19.79 Mt* (total megatonnes of ore).

Now we have to "weight" each mine's contribution to the whole. The weighting factors are derived as follows (see the far right column of our table in Fig. S-1 above for the ore megatonne values):

Cigar: 0.10 Mt ÷ 19.79 Mt = 0.0051 weighting factor

Husab: 6.39 Mt ÷ 19.79 Mt = 0.323 weighting factor

Rossing: 7.41 Mt ÷ 19.79 Mt = 0.374 weighting factor

Somair: 1.43 Mt ÷ 19.79 Mt = 0.072 weighting factor

Langer-Heinrich: 4.46 Mt ÷ 19.79 Mt = 0.225 weighting factor

Therefore, the global weighted average waste-to-ore ratio for dirt-mined uranium is about:

(0.0051 x 1.26) + (0.323 x 7) + (0. 374 x 2.1) + (0.072 x 7) + (0.225 x 1.48) = *3.9:1*.

We will be using this number in the uranium section of Part Two of this supplement.

PART TWO
(Chapter Four, endnote 22)

Determining the Global Average Mining Waste for All Mined Uranium (Dirt Mining & In-Situ Leaching)

We derived our numbers on material throughput from two authoritative sources. First is the 2021 World Nuclear Association (WNA) compilation of material throughput estimates by the World Bank, the US Department of Energy, the International Energy Agency, and others:

1.) World Nuclear.org – Mining requirements for electricity generation
https://www.world-nuclear.org/information-library/energy-and-the-environment/mineral-requirements-for-electricity-generation.aspx

The above compilation gives reliable finished material figures, but our analysis also considers the volume of mining waste ("displaced earth") from whence those finished materials come. For this, we used the RMR methodology recently adopted (April 2022) by the USGS (United States Geological Survey) and introduced in this paper:

2.) Pubs.acs.org – Rock-to-metal ratio: a foundational metric for understanding mining wastes
https://pubs.acs.org/doi/pdf/10.1021/acs.est.1c07875

The RMR of a finished material is its "Rock-to-Metal Ratio." That is, how much rock (dirt, sand, etc.) must be dug up (displaced) to obtain a certain volume of finished material. For finished materials like concrete and glass, the ratio would more properly be termed a rock-to-rock ratio, but the concept is the same.

The link below is the supporting data for the RMR paper above. See frame/page S-5 of the link for a table of the twenty-five principal energy-industry minerals analyzed in the study. Their RMRs are listed in the far right column. Unlike fuel, which is consumed as a power plant operates, structural materials are amortized over the lifespan of the system:

Pubs.acs.org – Supporting information for rock-to-metal ratio pdf
    https://pubs.acs.org/doi/suppl/10.1021/acs.est.1c07875/suppl_file/
es1c07875_si_001.pdf

The compilation by the World Nuclear Association (source link #1 above) incorporates various studies that use a 60-year lifespan for nuclear and 30-year lifespans for solar and wind. The numbers are in general agreement, from one study to the other, but they do vary somewhat. Rather than choosing one study over the other, we used the middle-of-the-road numbers of 30-year lifespans for wind and solar and a 60-year lifespan for nuclear, with capacity factors of 85% for nuclear, 35% for wind, and 25% for solar.

Beyond the data available in these sources, we also accounted for the fact that several of the throughput materials they examine are recycled—steel, copper, aluminum, and more. Since our particular focus is on the total mining footprints of nuclear, wind, and solar that underlie their finished material throughput per terawatt-hour of production, we factored in a recycling discount wherever we could. This is reflected in our Throughput Table (Fig. S-4 below).

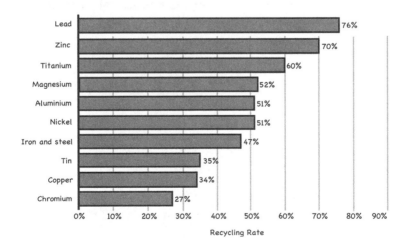

Fig. S-4: Recycled Metal Rates in US (2021)
Sources: https://www.statista.com/statistics/251345/
percentage-of-recycled-metals-in-the-us-by-metal/
https://galvanizeit.org/hot-dip-galvanizing/what-is-zinc/zinc-recycling

Recycled material in a modern industrial economy is no small matter: In the US, for example, about 70% of zinc is recycled, 47% of steel, and about 34% of copper. That's a lot of mining that doesn't have to be done to keep the lights on.

In nuclear power, the reprocessing (recycling) of used fuel for a second run in a light-water reactor would cut the volume of mining waste nearly in half. Fast-neutron reactors (see Chapters 5 and 6) will ideally get about *thirty times* as much energy from the same load of fresh fuel "burned" in an LWR. This alone would have a tremendous effect on uranium mining.

ROCK AND METAL

Rock and metal are the two basic categories of material under consideration. The mining and refining of metal is addressed by the RMR concept described above. The mining and refining of cement, concrete, glass, and silicon can be similarly assessed by their "rock-to-rock" ratios.

> **NERD NOTE:** In this supplement, the material expressed in tonnes (t) refers to the material throughput needed to produce one terawatt-hour of energy production (TWh).

Some notes and examples before we get to our throughput table and bar graph:

CEMENT

Wind and nuclear use concrete, while the footings for ground-mount solar racks are typically cement, which lacks the aggregates like rock and gravel that turn cement into concrete. The WNA study doesn't cover cement; here is the source we used:
Wikipedia – Concrete industry in the US
https://en.wikipedia.org/wiki/Cement_industry_in_the _United_States

CONCRETE

Similarly, the WNA paper discusses finished concrete, but does not discuss the raw material behind the finished product. Neither does the RMR method used by the USGS (the second source link listed above), which of course is a rock-to-metal ratio and not a rock-to-concrete ratio. So we explored this ourselves. For concrete's raw-to-finished ratio, we used:

Wikipedia – Concrete (see "cement" section)

> https://en.wikipedia.org/wiki/Concrete#Cement

See the section "Mix Ratios." We selected the 1:2:4 mix ratio for "cement: sand: aggregate." If the sand and aggregate (gravel / crushed rock) inputs are assumed to have no mining waste, the overall discarded material for concrete is just the discarded material from its cement production. That's 1.6 parts for cement from above, minus 1 part of finished cement = 0.6 parts discarded.

A 1:2:4 mix ratio totals 7 parts by mass. Those seven parts, plus the 0.6 parts discarded, totals 7.6. So, 7.6 ÷ 7 = about 1.1 RTR for concrete.

GLASS

To determine glass throughput of a solar farm, the average capacity factor of photovoltaic (PV) solar must be taken into account, as well as PV panel lifespan. The CF (capacity factor) for solar is typically in the 23%–25% range:

EIA.gov – Electricity overview

> https://www.eia.gov/electricity/monthly/epm_table_grapher.php

If we assume a generous 25% capacity factor, a 100-MW solar farm will produce 25 MW of average power. Over the course of a calendar year (8,760 hours) this totals about 219,000 MWh. With a 30-year solar panel lifespan, the lifetime output for this 100-MW solar farm will be 6.57 terawatt-hours (219,000 MWh x 30).

We used the SunPower E20-327 panel for our solar glass calculations: Solar Design Tool.com – Sunpower SPR-327NE-WHT-D

http://www.solardesigntool.com/components/module-panel-solar/
Sunpower/1501/SPR-327NE-WHT-D/specification-data-sheet.html

https://us.sunpower.com/sites/default/files/media-library/data-sheets/ds-e20-series-327-residential-solar-panels.pdf

The industry-standard SunPower E20-327 panel produces 301.4 Wdc under PTC conditions. We got this number using the solar design tool in the first link above. PTC means "Practical Test Conditions," a laboratory standard that simulates outdoor conditions. Assuming a standard 85% dc-to-ac conversion efficiency, we get 0.85 x 301.4 W = 256 Wac PTC per panel, meaning 256 watts of alternating-current electric power under practical test conditions.

At these performance specs, a 100-megawatt solar farm would need 390,625 panels. (100 MWac ÷ 256 Wac /panel = 390,625 panels)

An E20-327 panel has a surface area of 1.629 668 m² (see SunPower link above.) So the farm's total panel area is 390,625 panels × 1.629 668 m² /panel = 636,589 square meters.

The glass protecting a panel's solar cells is typically 3.2 mm thick: Sinovoltaics.com

https://sinovoltaics.com/

So the solar farm's overall glass volume will be 0.0032 m × 636,589 square meters = 2,037 cubic meters.

The volume density of glass is 2,520 kilograms per cubic meter (kg /m³). Search for "tempered glass" in this link: Omni calculator.com – construction glass weight

https://www.omnicalculator.com/construction/glass-weight

Therefore, the total amount of glass in a 100-MW solar farm will be 2,520 kg /m³ × 2,037 m³ = 5,133,000 kg, or 5,133 tonnes.

Suppose the farm has an average capacity factor of 25%, and the panels have a 30-year service life. The farm's total lifetime energy production will therefore be 6.57 TWh: (100 MW × 0.25) × (8760 h /yr × 30 yr) = 6.57 TWh.

5,133 tonnes of glass, amortized over the farm's energy production lifespan comes to: 5,133 t ÷ 6.57 TWh = 781 t /TWh. This is the finished material glass throughput for PV solar technology.

Glass is made from about 75% silica sand (silicon dioxide, or SiO_2), along with soda ash and limestone. The contents are easily gathered with little to no waste rock:

Lenntech.com – What is glass and how is it produced

https://www.lenntech.com/glass.htm

SILICON

"Solar cells" are the black silicon wafers beneath the glass cover of the solar panel that produce electricity from sunlight. There are 96 solar cells in the E20-327 panel.

The USGS says the RMR (Rock-to-Metal Ratio) for silicon is 3:1 See frame 5 Table S-4 in the link below. The RMRs are listed in the far right column:

Pubs.acs.org – Supporting information for rock-to-metal ratio pdf

https://pubs.acs.org/doi/suppl/10.1021/acs.est.1c07875/suppl_file/es1c07875_si_001.pdf

Although silicon is not a metal, it is classified as a "metalloid" because it has some metal-like properties. Since it's mined like a metal, the RMR formula can be applied to determine the amount of displaced rock required to make photovoltaic cells.

According to the WNA compilation, the silicon throughput for one TWh of solar energy is 57 tonnes. For your convenience, here's the WNA source link again. See Table 3 in the section "Mineral Requirements for Clean Electricity" (below the 7th paragraph):

World Nuclear.org – Mineral requirements for electricity generation

https://www.world-nuclear.org/information-library/energy-and-the-environment/mineral-requirements-for-electricity-generation.aspx

The RMR of silicon is 3:1. Therefore, the displaced earth that must be mined to produce one TWh of solar energy is 171 tonnes (57 × 3).

COPPER

The USGS says the RMR for copper is 513:1, meaning that for every 513 tonnes of displaced earth (rock) at a copper mine, we can expect to refine one tonne of fresh copper.

According to the WNA, copper throughput for solar is 68 tonnes per TWh. If it was all freshly mined, it would be recovered and refined from 34,900 t of displaced rock (68 X 513).

34% of the copper used in the US is recycled. Therefore, the actual throughput is 44.9 tonnes of fresh copper (0.66 × 68 t /TWh) derived from 23,030 t of rock (44.9 × 513).

The solar Industry's rule-of-thumb is 5 tonnes of copper per 1 MW of PV solar capacity. So for a 100-MW installed capacity solar farm, Cu = 500 metric tonnes (5.5 US tons):

Copper.org – Renewables
https://www.copper.org/environment/sustainable-energy/renewables

The throughput would be 500 t ÷ 6.57 TWh lifetime = 76 t / TWh. This industry rule-of-thumb agrees pretty well with 68 t /TWh from the WNA.

ZINC

Zinc is used as a protective coating for outdoor metal, such as the steel found in renewable energy equipment. The RMR for zinc is 71:1, and the recycling rate in the US is a healthy 70%:

Galvanize It.org – Zinc recycling
https://galvanizeit.org/hot-dip-galvanizing/what-is-zinc/zinc-recycling

CHROMIUM

Generous amounts of chromium are used in the nuclear industry to strengthen steel and protect it from corrosion. The RMR for chromium is 18:1 and the recycling rate is 27%.

URANIUM

One of our favorite metals. We covered most of this in parts One and Two of this mining supplement, but to review:

This is an unusual metal, because when uranium is fully utilized it ceases to be uranium. "Recycling" uranium fuel actually means fissioning some or all of whatever wasn't fissioned the first time around.

The global average waste-to-ore ratio for dirt-mined uranium is about *3.9:1*. In-situ leaching (ISL) has no waste rock and minimal tailings. About 57% of the world's uranium is now mined this way.

Uranium ore concentrations (wt%) can vary wildly: from Cigar Lake in Canada, with a whopping 4.73 wt% (uranium content by weight), down to less than 0.1 wt% for other uranium dirt mines, such as the large open-pit mines in Namibia and Niger.

After running the numbers, we estimated a global average ore concentration of *0.078%* (see our uranium mine table, Fig. S-2 above). Our survey covered more than 80% of the world's uranium production. We assume the same waste ratio and ore concentration applies to the remaining 20%.

By applying the global in-situ discount of 57% to the mines we surveyed, we get a weighted average global rock-to-uranium ratio of 3,000:1 and a weighted global average rock-to-fuel-pellet ratio of 21,700:1 (see the calculations below.)

At the present time, and unless things change (see Chapters 5 and 6), the uranium we use in today's reactors will be subject to the numbers and parameters we've discussed in this mining supplement.

To see how these ratios apply to the material throughput for one terawatt-hour of nuclear energy produced by a light-water reactor using fresh fuel pellets, let's explore the Nuclear Fuel Material Balance Calculator at the WISE Uranium Project, a uranium mining watchdog group:

WISE-Uranium.org – Nuclear fuel material balance calculator

https://www.wise-uranium.org/nfcm.html

USING THE NUCLEAR FUEL CALCULATOR

Relax—this is easy and fun. In the above link, at the bottom row of the calculator's flow-chart, in the Nuclear Power Plant section above the Calculate button, enter 1000 in the left-hand Electricity Production box. This sets the calculator at the material throughput for 1,000 gigawatt-hours of electricity (GWh$_e$), equal to one terawatt-hour (1 TWh).

Below the Calculate button is a Process Parameters section. First, click on the Default button to set normal parameters encountered in U mining. Then in the top row, lower the Waste/Ore ratio to our estimated global average of 3.9. Next to that, set our estimated global Ore Grade of 0.078 wt% of uranium.

Hit the Calculate button, and go to the top of the flow chart to see the results. For the parameters we entered, it shows that one terawatt-hour of energy from dirt-mined uranium requires digging up 132,098 tonnes of waste rock in the process of digging up 33,871 t of ore, which contains 26.4 t of uranium. (Note: That's 26.4 tonnes, not 26,400 tonnes.)

Once the uranium is leached from the ore, this depleted ore becomes 33,844 t of additional waste rock (33,871 − 26.4 = 33,844). So the total waste material is:

165,942 t (132,098 + 33,844) = 165,942 tonnes.

This would be correct for dirt-mined uranium. But with a 57% in-situ discount (remember, only 43% of global uranium is obtained by digging up rock), the waste material is reduced to:

0.43 × 165,942 = *71,355 tonnes*. This is the raw material through-put per TWh of energy produced for "mined uranium," a term which includes both dirt mining and ISL.

The 26.4 tonnes of uranium we extracted from this raw material is converted and enriched further down the flow chart, and finally fabricated into 3.29 tonnes of fuel pellets. This is the amount of fuel required by conventional light-water power reactors to produce one TWh of electricity.

(Notice that the final 2.93 tonnes of uranium coming out of the Enrichment Plant then comes out of the Fuel Fabrication Plant as 3.29 t of uranium oxide (UO_2) fuel pellets. The difference in weight is all the oxygen content—two O atoms for every U atom.)

In the uranium calculator, notice that the finished uranium tonnage from the Conversion Plant is 23.5 tonnes. This is the total amount of natural uranium metal (not uranium oxide) produced from all the mining, processing, and conversion. Therefore, the global average rock-to-uranium ratio is about 3,000:1 (71,355 t ÷ 23.5 t = 3,036).

Note: This applies only to existing Light-Water Reactors (LWRs) running on standard Low-Enriched Uranium (LEU) fuel.

Enriching and oxidizing the uranium gives a rock-to-fuel pellet ratio of about 21,700:1. We derived this number from the uranium calculator's Fuel Fabrication Plant output. Notice that the mass of uranium oxide (UO_2) is 3.29 tonnes per terawatt-hour: (71,355 ÷ 3.29) = 21,688. We rounded this up to 21,700.

SUMMARY

Now that we have the tedious stuff out of the way, here is our throughput table:

Mineral	Tech	Through-put t/TWh	Newly-mined portion	Newly-mined Throughput t/TWh	Rock-Metal-Ratio RMR	Displaced earth t/TWh
Steel WNA	Solar	940		498		4,480
	Wind	1,450	53%	768	9 USGS	6,910
	Nuclear	130		68.9		620
Aluminum WNA	Solar	287.5		141		987
	Wind	17.4	49%	8.5	7 USGS	59
	Nuclear	0.3		0.1		0.7
Copper WNA	Solar	68.0		44.9		23,030
	Wind	39.1	66%	25.8	513 USGS	13,240
	Nuclear	2.5		1.7		872
Zinc WNA/IEA	Wind	61.0 [a]	30% [d]	18.3	71 USGS	1,300
Chromium WNA/IEA	Wind	4.4 [b]	73%	3.2	18 USGS	57.6
	Nuclear	4.9 [c]		3.6		64.8
Non-Metals					Rock-to-Rock	
Glass DoE	Solar	2,700	NA	2,700	NA	NA
Silicon DoE	Solar	57	NA	57	NA	NA
Concrete WNA	Solar	1,220		1,220		1,340
	Wind	4,470	NA	4,470	1.1 authors	4,920
	Nuclear	1,060		1,060		1240
Cement DoE	Solar	3,700	NA	3,700	1.6 authors	5,920

Fig. S-5: Material Throughput Table Of Wind, Solar, And Nuclear With Recycling
Credit: By the authors

Notes for the above table:

[a] The zinc for constructing energy technologies is stated in units of tonnes per MW of power capacity. In this case, how much zinc is used to build one MW of wind. We use the WNA/IEA estimate of 5.6 t /MW. To convert this 5.6 t /MW to t /TWh units, we multiply by a conversion factor that expresses the lifetime energy per MW of that technology, based on its assumed capacity factor and lifespan.

With a 35% CF and a lifespan of 30 years, the conversion factor is 10.9 MW /TWh: (1 MW x 35% x 8760 h /y x 30 y = 0.920 TWh /MW over 30-year lifespan. Then, 1 ÷ 0.920 gives us a factor of 10.9 MW /TWh.) The zinc throughput for Wind is:

5.6 t /MW x 10.9 MW /TWh = 61.0 t /TWh.

[b] Chromium for constructing energy technologies is also stated in units of tonnes /MW of power capacity. Chromium use for wind is 0.4 t /MW. As derived in [a] above, Wind's conversion factor is 10.9. The chromium throughput for Wind is:

0.4 t /MW x 10.9 MW /TWh = 4.4 t /TWh.

[c] Chromium usage for nuclear is 2.2 t /MW, stated in units of tonnes per MW of power capacity. With an 85% CF for nuclear and a lifespan of 60 years, the conversion factor is 2.24 MW /TWh: (1 MW x 85% x 8760 h /y x 60 y) = 0.447 TWh /MW, over lifetime operation. And (1 ÷ 0. 447) = 2.24 MW /TWh.) Therefore, the chromium throughput for Nuclear is:

2.2 t /MW x 2.24 MW /TWh = 4.928 t /TWh.

[d] For zinc recycling in the US, see:
Galvanize It.org – Zinc recycling

https://galvanizeit.org/hot-dip-galvanizing/what-is-zinc/zinc-recycling

For your convenience, here's our Material Throughput graph once again:

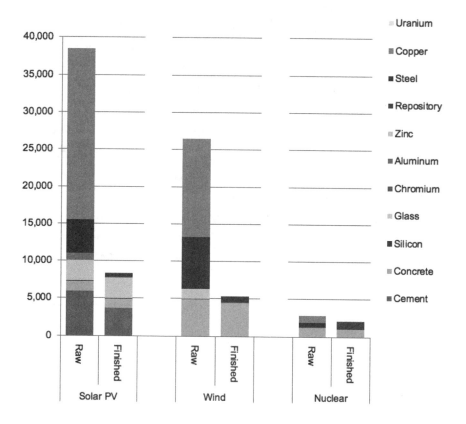

Fig. S-6: Raw and Finished Material Throughput of
Wind, Solar, and Nuclear (no fuel mining)
Credit: By the authors

	A	B	C	D	E	F	G	H	I	J
1										
2		Solar PV			Wind			Nuclear		
3		Raw	Finished		Raw	Finished		Raw	Finished	
4	Cement	5,920	3,700		-	-		-	-	
5	Concrete	1,340	1,220		4,920	4,470		1,240	1,060	
6	Silicon	57	57		-	-		-	-	
7	Glass	2,700	2,700		-	-		-	-	
8	Chromium	-	-		58	3		65	4	
9	Steel	4,480	498		6,910	768		620	69	
10	Copper	23,030	45		13,240	26		872	2	
11	Zinc	-	-		1,300	18		-	-	
12	Aluminum	987	141		59	9		1	0	
13	Repository	-	-		-	-		-	1,000	
14	Uranium	-	-		-	-			3	
15	White			50,000			50,000			50,000
16										
17	Total	38,514	8,361		26,487	5,294		2,798	2,138	
18										

Fig. S-7: Authors' Spreadsheet

413

RESERVES OF DEPLETED URANIUM
(Chapter Five, end note 8)

DU Energy reserves:

World Nuclear.org – Uranium and depleted uranium

http://www.world-nuclear.org/information-library/nuclear-fuel-cycle/uranium-resources/uranium-and-depleted-uranium.aspx

See section **Depleted Uranium**, 2nd para. DU world stockpiles in storage = 1.6 megatonnes, or 1,600,000 tonnes. The US portion is about 750,000 tonnes.

See para. 1:

NRC.gov – Background information on depleted uranium

https://www.nrc.gov/waste/llw-disposal/llw-pa/uw-streams/bg-info-du.html

Assume 1 tonne of U-238 in a fast-neutron reactor yields 1 GWe-year.

Therefore:

One GWe-yr per 1 tonne DU × 750,000 tonnes DU = 750,000 GWe-yr of total energy from USA's DU stockpile.

One GWe-yr × 8760 hrs /yr = 8.76 TWh per 1 GWe-yr.

So 750,000 GWe-yr of total energy × 8.76 TWh /1 GWe-yr = 6.57E6 TWh of total energy from USA's DU stockpile. (6.57 million TWh)

See:

EIA.gov – What is US electricity generation by source

https://www.eia.gov/tools/faqs/faq.php

In first table, US electricity in 2021 was 4108 TWh /yr.

Therefore: 6.57E6 TWh of US future DU energy divided by 4108 TWh /yr =

1,599 years of US electricity supplied by our depleted uranium.

In other words, we could operate our US electric grid with DU for about 1,600 years at present consumption rate.

Mark Jacobson *et al* call for converting US society to all-electric status by year 2050. Under this plan, our total societal energy consumption would become 1179 GW average, throughout the year: Stanford.edu – Mark Jacobson zero air pollution 2021

https://web.stanford.edu/group/efmh/jacobson/Articles/I/21-USStates-PDFs/21-USStatesPaper.pdf

US total annual electric load would then become 1179 GW × 8760 h /yr = 10,330 TWh /yr. Which would be greater than our 2021 load of 4108 TWh by a factor of 2.51. (10,330 TWh ÷ 4108 TWh = 2.51)

Divide 1599 years by 2.51 factor to obtain **637 years.** This would be the duration of all-electric societal energy supply obtainable from US depleted uranium reserves.

DIETARY POTASSIUM VS FUKU WATER
(Chapter 14, endnote 25)

We want to give a big shout-out to DJ LeClear (MHP, Master's in Health Physics) and Philip Hult (M. Phil. Nuclear Engineering, University of Cambridge) for their excellent work on the arcane calculations behind Generation Atomic's Tritium Calculator.
You guys rock!

PART ONE
Radiative Intensity of Dietary Potassium:
0.453 µSv (microSieverts) per day

Initially valued as fertilizer, "potash" or potassium carbonate (K_2CO_3) is a valuable soil additive, originally made from boiling down the ashes of burnt wood and leaves. The ancient Greeks called it *kalium*, and modern Germanic languages do the same. This is why the element potassium is symbolized by the letter K.

Like zinc, copper, and selenium, potassium is an absolutely essential mineral for our muscles and nerves. Serious health consequences can occur if our K+ levels stray even a little from proper amounts.

Potassium, including its naturally-occurring radioisotope K-40, is essential to pretty much all lifeforms. With a proper diet, potassium levels remain relatively constant throughout life. Since natural potassium has a K-40 concentration of 120 ppm (see second link below), one atom out of every 8,300 potassium atoms in the human body is radioactive.

These are the calculations on dietary potassium:

1) A 70-kg person with proper diet has about 140 grams of bodily potassium.
NIH.gov – Potassium Fact Sheet for Health Professionals
https://ods.od.nih.gov/factsheets/Potassium-HealthProfessional/
Para. 2

2) The K-40 content of Earth's potassium is about 120 ppm.
Wikipedia – Potassium-40
    https://en.wikipedia.org/wiki/Potassium-40
Therefore a person has: 140 g x 120 ppm = 16.8E-3 grams of K-40.

3) K-40 has a specific radioactivity of 263,000 Becquerels per gram -
Bq /g.
HPSchapters.org – Human Health Fact Sheet
    http://hpschapters.org/northcarolina/NSDS/potassium.pdf
See Table: Radioactive Properties of Potassium-40
Specific radioactivity is 0.000 007 1 Curie /gram (Ci /g). The Becquerel
Conversion Factor from Curies to Becquerels is 37E9 Bq /Ci.
So, K-40 has 0.000 007 1 Ci /g x 37E9 Bq /Ci = 263,000 Bq /g

4) Therefore, a person has 16.8E-3 g of K-40 x 263,000 Bq /g = 4,418
emission events per second, internally (i.e., within their body).

5) K-40 emission produces a beta-particle with energy of 0.52 MeV.
Ibid. Third link above (the potassium pdf).

6) So, 4,400 events /sec x 0.52MeV /event = 2,290 MeV /s.

7) The conversion from MeV to joules is 6.24E12 MeV /J. So, 2,290
MeV /s ÷ 6.24E12 MeV /J = 367E-12 J /s.

8) Dividing by 70 kg body mass gives 367E-12 J /s ÷ 70 kg = 5.24E-
12 Gray /second, which is equivalent to 5.24E-12 Sievert /sec with
the radiation being beta type.

9) There are 86,400 seconds in one day, so the daily radiation dose
from our internal K-40 is 5.24E-12 Sievert /sec x 86,400 s /day =
453E-9 Sv /day. Or **0.453 µSv /day.**

PART TWO
Radiative Intensity of Fuku Water
(as of July 2023):
1.192 µSv per liter

In 2014, the segregated (old) water in the early-use storage tanks at Fukushima had a volume of 460,000 cubic meters, or 460 million liters. The water's total radio-intensity was 830 trillion Becquerels, or 830 TBq. (Radio-intensity refers to radioactivity per unit mass.) Therefore, the radio-intensity of Fukushima's tritiated water at that time was 1.8 MBq per liter (830 Tbq ÷ 460M liters = 1.80 MBq per liter.

Fukushima.JAEA.go.jp – Status of Contaminated water (Ishikawa 2014)

https://fukushima.jaea.go.jp/fukushima/result/pdf/pdf1410/4a-1_Ishizawa.pdf
(Pgs. 13 and 19)

In July of 2023, after 9.3 years of radio-decay, radioactivity has decreased by 41% to 490 TBq (59% remaining) down from 830 TBq. This total is derived by running the numbers through a scientific calculator (see our "Deep Dive Into Tritiated Water" below).
Current radio-intensity = 490 TBq ÷ 460,000 m3 = 1.06 MBq /L
1.06 MBq /L x 0.0057 MeV x 1 L swallowed = 6.04E3 MeV /sec internal
6.04E3 MeV /sec x 1 J /6.24E12 MeV (conversion factor) = 968E-12 J /s, immediately upon swallowing.
Divided by 70 kg person = 13.8E-6 µSv /second, immediately from **1 liter** swallowed.
Multiply by 86,400 seconds /day = **1.192 µSv.**
With potassium's 0.453 µSv /day, and with 1.192 µSv /day from Fuku water, one liter of Fuku water gives more radiation than K-40 by a factor of 2.6. (1.192 ÷ 0.453 = 2.63)

So to equal their K-40, a person would have to drink *in one sitting* (1 ÷ 2.6) = **0.38 liter.**

Of course, this is an imperfect analogy, since drinking 0.38 liters in one sitting would give you a spike of radiation (neither lethal nor dangerous) which would taper off to zero in about three months' time. If you wanted to match the steady, everyday radiation you receive from your dietary potassium, you would need to drink a much smaller daily maintenance dose instead. This would gradually build up an internal inventory, similar to many pharmaceuticals, eventually establishing a steady-state daily dose, similar to what we get from our dietary potassium.

PART THREE
"Maintenance Dose" of Tritiated Tank Water
About 25 ml per day

Determining a daily maintenance dose is tricky. For one thing, potassium levels are scrupulously maintained by the body, while water levels are not. The potassium atom (including the radioactive K-40 variety) becomes an integral part of bodily tissue, and any ingested excess is excreted. (This is why comparing different radiation exposures to eating X amount of bananas has limited utility.)

The body handles water molecules much differently than it does potassium atoms. A water molecule's half-life residence time in the body is roughly 10 days, and can vary from 2 days to about 17 days. But on average, after 10 half-lives, or about 100 days, virtually all ingested water molecules will have been replaced by more recently-ingested water.

During this time, the tritium in the water (if any) is decaying much more rapidly than K-40 potassium: While tritium's half-life is 12.3 years, the half-life of potassium-40 is 1.25 billion years. Further complicating matters is that while most water molecules stay intact during their residence in the body, some small amount (2% - 5%) of ingested water is metabolized (broken down) into its constituent hydrogen and oxygen atoms, which then become bound to our cells, similar to most potassium atoms.

With a 12.3-year half-life, tritium atoms undergo some small amount of radiative decay, but that is negligible compared to the more rapid biologic half-life (residence time) of water in humans. Slightly complicating matters is that while most water molecules stay intact during their residence in the body, some small portion (2% - 5%) of ingested water is metabolized (broken down) into its constituent hydrogen and oxygen atoms, which then become bound to our cells.

Since tritiated water has one and sometimes two tritium hydrogen isotopes, the concern is that these faintly radioactive atoms will become "organically bound" to our bodily tissue, and keep radiating us from within for years on end. Concerns over Organically Bound Tritium (OBT) have thus been raised, equating it to other bio-accumulations, such as the mercury in the oceanic food chain.

Such concerns over OBT are unfounded. Experiments have shown that while fish raised in artificially-tritated seawater do indeed have elevated levels of tritium, but their tritium levels drop when the fish are transferred to clean seawater, whereupon they quickly achieve and maintain a "tritium equilibrium" with their environment. TEPCO.co.jp – Fukushima Daiichi Nuclear Power Station Status of Progress of the Marine Organisms Rearing Test

https://www.tepco.co.jp/en/hd/decommission/information/news-release/reference/pdf/2023/reference_20230525_01-e.pdf

Unlike mercury, which will bio-accumulate in fish to much higher concentrations than the surrounding seawater, tritium levels in fish stay in balance with the surrounding aquatic environment. In short: The water a fish lives in would itself have to be highly tritiated for the fish to have high levels of tritium.

So, with different rates of decay, different residence times in the body, and different degrees of bio-accumulation, comparing dietary potassium to "dietary tritium" is tricky. DJ and Philip, our intrepid brainiacs, had to employ the differential equations used in pharmacology to determine a maintenance dose of Fuku tank water (prior to seawater dilution) that would match our daily dose of radiation from dietary potassium.

We won't drag you through the math, but suffice it to say that the maintenance dose comes to right around 25 ml, or a generous half-shot per day. (A shot, sometimes called a jigger, is 1.5 fluid oz, or 44.36 ml.)

This shows us how un-dangerous the Fukushima tank water already is. And just to be sure, they're diluting more than 5,500 times with seawater before releasing it into the ocean.

Behold the power (and expense) of Nuclear Fear.

PART FOUR
A Deep Dive into Tritiated Water
by
Tim Maloney

In March of 2014 the overall volume of stored water at Fukushima was 460,000 m³. The water's total radioactivity was 830 trillion becquerels (830E12 Bq.) See:

Fukushima.JAEA.go.jp – Ishizawa
    https://fukushima.jaea.go.jp/fukushima/result/pdf/pdf1410/4a-1_Ishizawa.pdf
    See page 15, frame 8, Volume of Water Stored. Also page 19, frame 10, Tritium Status.
    Therefore initial volumetric Radioactivity in 2014 was 830 trillion Bq divided by 460,000 m³ equals 1.80E9 Bq / m³ or 1.80E6 Bq /liter.
    In July 2023, original Fukushima water has experienced a decline in its radioactivity to 59.1% of its initial value, to 1.06E6 Bq /liter.

This is the radioactivity value used by Generation Atomic's on-line Calculator for determining effective doses in humans from ingesting such water.

To understand the current volumetric Radioactivity of FUKU water in 2023, consider this:

1)
Tritiated water's molecular mass: 20 atomic mass units (amu) per water molecule (equivalent to 20 grams per mole). This comes from:

one oxygen atom = 16 (8 protons + 8 neutrons)
one normal hydrogen atom = 1 (one proton only)
one tritium atom = 3 (one proton + 2 neutrons)

Total for one molecule = 16 + 1 + 3 = 20 amu

2)
Tritium's half-life (symbol $t_{1/2}$, or HL) = 12.3 years
Half-life is the amount of time that must elapse for one half of the original atoms or molecules to return to their normal stable (non-radioactive) state.

In basic time units of seconds, tritium's half-life is:
HL = 12.3 yr ×365 days × 24 hr × 3600 s = 388E6 s = 388 million seconds

3)
a) The exponential decay formula is $Q(t) = Q_0 \times$ **e** raised to the power (- *lam* × *t*)
Where $Q(t)$ symbolizes the quantity of atoms (or molecules) remaining at a particular time moment, symbolized *t*.

b) Q_0 is the original quantity of atoms, at starting time, $t = 0$.

c) Letter e (sometimes Greek letter epsilon – ε) is the numeric value of the base of the natural logarithm system, approximately 2.72. But we don't need to know that value because in the era of hand-held scientific calculators, the number e is accessed by the calculator's key labeled "e^x", meaning "e raised to the x power." It is often the 2nd function of the "ln x" key – the natural logarithm key.

d) (*lam* is short for the Greek letter *lambda: λ*) It is called the "decay constant" for a collection of radioactive atoms.
lam is related to half-life *HL* by the formula *lam* = 0.693 ÷ *HL*. Therefore from

2) above, tritium's decay constant is *lam* = 0.693 ÷ 388E6 s = 1.79E-9.

e) Radioactivity R of a particular quantity of atoms (or molecules) is equal to the rate at which those atoms are disappearing or changing. That is, R measured in decay events per second, called becquerels, unit symbol Bq, specifies the number of atoms per second that are changing from active (tritium) to inactive (plain hydrogen.)

f) The rate at which atoms disappear or change into stable atoms is always equal to the number of atoms that are present at that moment, multiplied by that atom's decay constant *lam*. So for tritium (tritiated water), the rate of disappearance is $R = Q(t) \times 1.79E-9$, from **D)** above. In other words, tritium's radioactivity equals the absolute number of tritium atoms that are present at that moment, multiplied by 1.79E-9.

g) Avogadro's number (N_A) is 6.02E23, or 602,000,000,000,000,0 00,000,000.
This is the number of molecules present in 20 grams of tritiated water. That is true because the mass of a tritiated water molecule is 20 amu, from **1)** above.

(Alternatively, the mass of one mole of tritiated water equals 20 grams per mole. Called molar mass.)

Therefore, the molecular count (number of molecules per gram) for tritiated water equals 6.02E23 ÷ 20 g = 30.1E21 molecules per gram.

In March 2014, Fukushima's stored water had radioactivity of 830E12 Bq.

R(0) = 830E12 Bq, at t = 0.

From F) above, with R (t) = Q (t) ×1.79E-9

Q (t) = R (t) ÷ 1.79E-9, so:

Q (0) = R (0) ÷ 1.79E-9

= 830E12 Bq ÷ 1.79E-9

= 464E21 atoms of tritium existed at t = 0 in March, 2014.

In July of 2023 , that tritiated water had been decaying (disappearing) for 9.33 years, or 294 million seconds (294 E6 s).

So from **A)** above:

Q(t) = Q(0) × e raised to the power (- *lam* × t)

= 464E21 molecules × e raised to the power (- 1.79E-9 x 294E6 seconds)

= 464E21 molecules ×e raised to the power (- 526E-3)

{Use a scientific calculator to raise e to this power.}

= 464E21 molecules x 591E-3

= 274E21molecules of tritiated water still present in July 2023.

This represents **59.1%** of the original 464E21 molecules. (274E21 ÷ 464E21 = 0.5906 or 59.1%) So the radioactivity of FUKU water has also declined to 59.1% of its initial value of 1.80 M Bq /liter.

59.1% x 1.80 M Bq /L = **1.06 M Bq /L** becomes the basis for the Calculator's function.

IODINE-129 FROM FUKUSHIMA
(Chapter 15, endnote 9)

Iodine-129, a radioactive isotope emitted from Fukushima, is often used as a "tracer" in scientific research. A mass spectrometer is a laboratory instrument that can detect and identify differing masses of molecule fragments, as well as individual atoms or isotopes, such as iodine-129 versus, say, iodine-125.

A British Columbia study in 2015 detected an iodine-129 concentration of 211,000,000 atoms/liter in groundwater. In the Abstract of the paper linked in (a) below, this number is stated as 211×10^6 atoms. This seems to be a large number. But that is deceptive because the mass and consequent radioactivity of that number of atoms is a tiny fraction of the natural radioactivity of a human body.

An adult human body has radioactivity of about 4,500 becquerels (Bq), arising from our bodily potassium derived from diet. See (b) below. The radioactivity of 211 million atoms of iodine-129 is only 0.000 000 295 becquerel. See (c) below. That is, 295E-9 Bq per liter of groundwater.

Therefore, a person would have to drink *more than 15 billion liters* of such water to equal our normal internal radioactivity derived from our dietary potassium.

(4,500 Bq divided by 0.000 000 295 Bq /liter = 15,300,000,000 liters.)

That's over 6,000 Olympic swimming pools. See note (d) below.

NOTES:

(a)

AGUpubs – Atmospheric transport of iodine-129 from Fukushima
https://agupubs.onlinelibrary.wiley.com/doi/full/10.1002/2015WR0
17325

"The atmospheric transport of iodine-129 from Fukushima to British Columbia, Canada and its deposition and transport into groundwater" (see Abstract, Line 10.)

(b) The body of a 155-pound human contains about 140 grams of dietary potassium in internal organs and body fat. Of that, about 120 parts per million is the radioactive isotope potassium-40, symbol K-40. So 140 grams ×120 ppm = 0.0168 gram of K-40 resides in the adult body.

The radioactivity of K-40 is 265,000 Bq /gram. Refer to footnote (e) below.

So an adult human has radioactivity of 0.0168 gram × 265,000 Bq /gram = 4,450 Bq.

(c) One gram of iodine-129 consists of 4.67E21 atoms.

Avogadro's number N_A = 6.02E23, divided by 129 (atomic mass of I-129) gives 4.67E21 atoms /gram. (6.02E23 ÷ 129 = 4.67E21).

The BC Canada study stated 211E6 atoms /liter, so 211E6 divided by 4.67E21 atoms /gram = 45.2E-15 gram /liter, called 45.2 femtogram of iodine per liter of groundwater.

The radioactivity of iodine-129 is 6,540,000 becquerels /gram. See (e) below. (6.54E6 Bq /gram) So one liter of groundwater, with 211E6 atoms, has radioactivity of 45.2E-15 gram /liter × 6.54E6 Bq /gram = 295E-9 Bq /liter. (0.000 000 295 Bq /liter).

(d) An Olympic swimming pool is 50 meters in length by 25 meters wide by 2 meters deep = 2,500 cubic meters. Multiply by 1,000 liters /m³ gives 2.5E6 liters per pool. 15.3 billion liters divided by 2.5E6 liters /pool = 6,120 Olympic swimming pools.

(e) The number of atoms in one gram of any isotope is equal to Avogadro's number, N_A, divided by the atomic mass of that isotope, namely its isotope identification number. For I-129, atomic mass = 129 amu (atomic mass units, or its number of protons plus neutrons.)

Avogadro's number (N_A) equals 6.02E23, the factor by which 1 gram is greater than 1 amu. That is, 602,000,000,000,000,000,000,000 times greater.

So the number of I-129 atoms per gram = 6.02E23 divided by 129 = 4.67E21 atoms /gram.

For any radioactive material, its radioactivity (measured in decay events per second, or becquerels) is equal to the number of atoms that are present, divided by the average lifetime of those atoms. For 1 gram of I-129, the radioactivity (R) in becquerels (Bq) equals 4.67E21 atoms divided by their average lifetime.

The average lifetime for any exponentially decaying atom is mathematically called the "time constant," symbolized by the Greek letter "tau," which looks like an English "t" with a curved top. Average lifetime *tau* is related to atomic half-life (*HL*) by the formula:

tau = *HL* divided by natural logarithm of 2

tau = *HL* divided by 0.693

The half-life of iodine-129 is 15.7 million years. That is equivalent to 495 trillion seconds, or 495E12 seconds. Therefore, its average lifetime *tau* = 495E12 seconds divided by 0.693 = 714E12 seconds, or 714 trillion seconds.

So the radioactivity of 1 gram of I-129 equals 4.67E21 atoms divided by 714E12 seconds = 6.54E6 decay events per second. Therefore R = 6.54E6 Bq /gram for iodine-129.

SAVANNAH THORIUM
(Chapter 17, footnote 11)

A certain beach ridge near Savannah, Georgia is made of sand containing 127 parts per million of thorium. If they wanted to, they could, in theory, power their city with this alone. Let's explore the science:

See page 29, Table 2, sample No. 81

Jstor.org – Heavy minerals in the beach sands of Florida
    https://scholarship.rice.edu/bitstream/handle/1911/14135/6307170.
PDF?sequence=1

The Ford F-150 has a maximum load of 2320 pounds, or 1053 kilograms. See:

Kimber Creek Ford.com – F-150 towing and payload https://www.kimbercreekford.comblog/2017-ford-f-150-towing-and -payload-capacities/

So a pickup truckload may contain about 0.13 kilogram of thorium: (1053 kg /load × 127 parts /1E6 = 0.13 kg /load).

With full burnup, a thorium-fueled MSR will consume about 1000 kilograms of thorium for 1 gigawatt-year of electric energy production. **See (a) below.** One GW-yr is equivalent to 1e9 W × 8760 h /yr = 8.76e9 kWh.

(NOTE: Thorium-232 is fertile, but not fissile. So it is not strictly a fuel, but a feedstock. It is converted inside the MSR to fissile uranium-233, which is the actual fuel that burns.)

So a pickup truckload containing 0.13 kg of thorium could produce 1.1 million kWh per truckload:

(0.13 kg × 8.76E9 kWh /1000 kg = 1.1E6 kWh).

Savannah's 2021 population is 147,000. The US average electric power consumption is 1.4 kW per capita. **See (b) below.** So Savannah consumes about 4.9 million kWh per day:

(1.4 kW /person × 147,000 persons × 24 hrs = 4.9 million kWh /day).

Therefore, the thorium in one truckload of beach sand will satisfy Savannah for this amount of time:

1.1 million kWh per load ÷ 4.9 million kWh /day =
0.22 days, or about 5-1/2 hours.

(a) Course hero.com – Nuclear energy: Fuel, waste, and economics
https://www.coursehero.com/file/10350101/L17-Nuclear-waste/
See the table labeled "Fuel consumption and waste generation
from various electricity generation sources for 1GWe-year."
That table shows 30 tons (27 tonnes) of fissile fuel is needed to
produce 1 GWe-yr. It assumes a standard 3.5% burnup of uranium
during the functional life of a fuel rod in a Gen 3 LWR. So:
27 tonnes = 27,000 kg fuel /GWe-yr.
With Gen 4 reactors achieving 100% burnup of fuel, rather than
just 3.5%, their energy production will increase by a factor of about
28X. (100% ÷ 3.5% = 28.5)
Therefore, 28 GWe-yr can be produced from that same 27,000
kilograms of fissile fuel in a Gen 4 fast-neutron reactor, rather than
just 1 GWe-yr. Fuel consumption normalized to 1 GWe-yr becomes
27,000 kilograms ÷ 28 GWe-yr = 960 kilograms per GWe-yr. This
is rounded to 1000 kg /GWe-yr.
Alternatively, use the WISE-uranium.org Nuclear fuel calculator:
http://www.wise-uranium.org/nfcm.html
Enter the numeral 1 in the lower right-most box, meaning 1 GWae
("a" for "annum," or year, and "e" for electricity). Observe 25.4 tonnes
uranium fuel input, two steps higher in the flowchart. This is similar
to 27 tonnes of uranium from the table referenced above

(b) See EIA.gov – What is US electricity generation by energy source
https://www.eia.gov/tools/faqs/zfaq.php
See first sentence. US electricity in 2018 = 4178 TWh /yr, or:
(4178E12 Wh)
4178E12 Wh /yr ÷ 8760 h per year = 477E9 Wavg, for entire US.
US population in 2018 = 327.2 million persons. 477E9 Wavg ÷
327.2E6 persons = 1.46E3 Wavg /person. Or about 1.4 kW per capita.

FUKU FISH
(chapter 17, footnote 22)

See Wikipedia – Cesium-137
https://en.wikipedia.org/wiki/Caesium-137
See 3.22 terabecquerels per gram, or 3.22E6 Bq /microgram. Each decay event emits energy of 1.17 MeV.

So 3.22E6 Bq /microgram ×1.17 MeV /event = 3.77E6 MeV /sec per microgram of Cesium.

Mass amount of Cs-137 in one cubic meter (1000 liters) of seawater: 4 Bq /m³ increase in total radioactivity ÷ 3.22E6 Bq /microgram = 1.2E-6 micrograms Cs /1000 liters seawater.

Cesium's radiative energy in 1000 liters of seawater = 4.5 MeV /s.

1.2E-6 microgram Cs /1000 liters × 3.77E6 MeV /s per microgram of Cs =

4.5 MeV /s, due to Cs-137 alone.

The conversion factor from MeV units to basic SI energy units is 6.24E12 MeV per 1 joule. So 4.5 MeV /s ÷ 6.24E12 MeV /J = 720E-15 J /s per cubic meter.

With a single fish filtering 50 cubic meters through its gills, 720E-15 J /s × 50 m³ = 36E-12 J /s per fish. In that scenario, each fish contains an amount of cesium that emits 36E-12 J /s.

Eaten by a 70-kilo human, 36E-12 J /s ÷ 70 kg = 0.51E-12 Sv /s, or 0.51E-9 mSv /second. On a daily basis, 0.51E-9 mSv /s × 3600 s / hr × 24 hr = 0.000 044 mSv /day.

Eating such fish every day throughout the year gives 0.000 044 mSv /day × 365 days – 0.016 mSv /ycar. That's thc intcrnal radiation dose caused by eating Fuku fish every day. This is *less than 2%* of our normal background dose.

ENERGY DENSITY OF LEU
(Chapter 17, endnote 26)

See Euro Nuclear.org – fuel comparison
  https://www.euronuclear.org/glossary/fuel-comparison/
Enlarge the graphic, either on-screen or by printing. On the graphic, project up from the 1 kg mark of natural uranium. It contains the energy equivalent of about 14,000 kilograms of coal, which is sufficient to produce about 45,000 kWh of electric energy, as shown on the top line. (Coal's electric energy density is 45,000 kWh ÷ 14,000 kg = about 3 kWh per kg.)

One kilogram of natural uranium contains only 0.007 kilogram of the fissile isotope U-235. Therefore, the coal-equivalent of pure U-235 is 14,000 kg divided by 0.007, which is 2,000,000 kg. In other words, the energy content of U-235 is 2,000,000 times greater than for the same mass of coal.

45,000 kWh from 1 kg of natural uranium is enough to supply one American person for more than 3 years. This includes personal use as well as per-capita share of all societal consumption—commercial, industrial, and governmental.

Regarding Light Enriched Uranium, LEU, refer to:
World nuclear.org – Fuel cycle overview
  https://www.world-nuclear.org/information-library/nuclear-fuel-cycle/introduction/nuclear-fuel-cycle-overview.aspx
See section "Material balance in the nuclear fuel cycle." In the 5th row of the table, 24.3 tonnes of 4.5% enriched uranium are needed to fuel a 1000-MWe reactor for one year. [NOTE: The World Nuclear Association now uses 4.5% for its calculations, rather than the traditional 3.5%.]

Assuming that 1000 MWe refers to average power (1100 MW-peak capacity operating at 91% capacity factor), the reactor's annual energy production is 1000 MW × 8760 hours /yr = 8.76 billion kWh.

So the energy density of 4.5% LEU equals 8.76E9 kWh ÷ (24.3 tonnes LEU × 1000 kg) = 360,000 kWh per kilogram of LEU fuel.

Therefore, LEU is greater than coal's energy density by:

360,000 kWh /kg ÷ 3 kWh /kg = 120,000 times greater.

We're not done yet! Check out the following short supplement on natural gas:

ENERGY DENSITY OF NATURAL GAS
(Chapter 17, footnote 26)

World nuclear.org – Fuel cycle overview

https://www.world-nuclear.org/information-library/nuclear-fuel-cycle/introduction/nuclear-fuel-cycle-overview.aspx

Refer to the section "Power Generation and Burn-Up," para.5: Natural gas usage of 8.5 million cubic meters applies to a combined cycle gas turbine facility (CCGT).

So volumetric natural gas energy volume density is 44 million kWh ÷ 8.5 million m³ = approximately 5 kWh /cubic meter.

In our supplement "Energy Density of LEU" we show that LEU fuel energy density is equal to 360,000 kWh per kilogram.

Therefore, to match one kilogram of 4.5%-enriched LEU, the required volume of natural gas would be: 360,000 kWh ÷ 5 kWh / m³ = 72,000 cubic meters (2,500,000 cubic feet.) [NOTE: The World Nuclear Association now uses 4.5% for its calculations, rather than the traditional 3.5%.]

Converting from kilograms to pounds of LEU: 2,500,000 ft³ / kg LEU ÷ 2.2 pounds /kg LEU = 1,100,000 ft³ of natural gas. This will match the energy in 1 pound of 4.5% LEU.

At standard temperature and pressure, the mass density of natural gas is about 0.7 kg /m³. So its electric energy density is 5 kWh /m3 ÷ 0.7 kg /m3 = approximately 7 kWh per kilogram of natural gas.

Comparing to LEU, we find that 360,000 kWh /kg ÷ 7 kWh /kg = a factor of difference of about 50,000. In other words, 4.5% LEU produces about 50,000 times more energy than the same mass of natural gas.

NOTES ON THE ABOVE: All of the above applies to the low-enriched uranium (LEU) fissioned in a light-water reactor (LWR). Reprocessing this spent nuclear fuel (SNF) would access the remaining 95% of fission energy in the LEU, which is mostly from the uranium-238 filler, and partly from the unfissioned plutonium-239 that some of the U-235 fuel transmuted into.

This is because only about 5% of the potential fission energy of the fuel is actually released when LEU is burned in a Generation-III LWR. If the "spent" nuclear fuel from these reactors were then reprocessed to fuel a Generation-IV fast-neutron reactor, the SNF could further produce:

95% ÷ 5% = 19 times as much energy.

With 3.5% LEU, it would be 30x as much energy. We mention 30x in the book because nearly all of our historical stockpiles of spent fuel were enriched at 3.5%, not the more recent 4.5%.

50,000 times more energy × 19 = nearly 1,000,000 times more energy, compared to the same mass of natural gas.

USED PANELS VS. USED FUEL
(Chapter 18, endnote 5)

The issue we examine here is the amount of waste material from a PV solar farm that has produced the same lifetime energy as the Connecticut Yankee NPP (nuclear power plant). In other words:

How much waste for how much energy?

The second issue we examine is the amount of land required for such a solar farm, versus the land required for Connecticut Yankee. That is:

How much land to produce how much energy?

The third issue we examine is the comparable amount of land and solar waste if the Connecticut Yankee NPP had a fast-neutron reactor that fully fissioned its fuel.

The light-water reactor at Connecticut Yankee operated for just 28 years before its premature shutdown in 1996. The plant, and its "dry cask" used fuel storage facility, are described at:
Gordian Knot Book.com – Why nuclear energy has been a flop

http://gordianknotbook.com/wp-content/uploads/2022/05/gordian_wQ.pdf

See pages 235 and 236 for a description of the plant's dry cask storage arrangement (we use the same photo in Chapter 18). The storage facility consists of a 228 ft × 70 ft (69.5 m x 21.3 m) concrete pad storing 43 steel & concrete casks. The casks are 12 ft (3.6 m) in diameter and 20 ft (6 m) in height.

Of those 43 casks, **40 casks** contain the used fuel rods that enabled the 600-MW reactor at Connecticut Yankee to produce 110 billion kilowatt-hours of electric energy. The reactor's internal components, along with various irradiated tools and instruments, are stored in the last three casks.

Originally licensed to operate for 40 years, the reactor was prematurely decommissioned after just 28 years of service. So we will examine this subject in two steps:

1. We will first compare the 28-year collection of used fuel depicted in the photo to a tight pile of used solar panels stored on the same-sized pad, from a solar farm that produced the same amount of energy.

2. We will then extrapolate our results to compare the used fuel from a 40-year operating life of the same light-water reactor, stored on a larger (longer) pad, with the used panels needed to make that same energy stored on a matching extended pad.

Mark Jacobson's 100% Wind, Water, and Solar "Roadmap" for the US, which we examine in our book *Roadmap to Nowhere*, proposes SunPower's E20 series solar panel for all utility scale and rooftop PV installations. (While the panel isn't called out by name, the specs are an exact match to the Sunpower E20-327.) See page 40 of the US Roadmap, frame 53, Table S16, the third paragraph of the notes below the table:
Stanford.edu – 2021 Jacobson US Roadmap
    https://web.stanford.edu/group/efmh/jacobson/Articles/I/21-USStates-PDFs/21-USStatesPaper.pdf
The individual panel surface area specified in the notes is 1.629 668 m^2, matching the SunPower model E20-327. This is the SunPower panel's Spec Sheet:
Sunpower.com – E20 series 327 residential panel specs
    https://us.SunPower.com/sites/default/files/media-library/datasheets/ds-e20-series-327-residential-solar-panels.pdf
See page 2 of the spec sheet, bottom right: Panel dimensions are 1558 mm length and 1046 mm width. Converting to meters and multiplying gives 1.629 668 m^2, the exact surface area specified

in the authors' notes below Table S16 in the US Roadmap (see above link.)

The notes below Table S16 also anticipate a substantial improvement in photovoltaic solar-to-electric conversion efficiency by midcentury, rising to an optimistic 23.9% from the Roadmap's present estimate of 20.5%. This would be an improvement by a factor of 1.19 times, or 19%: [23.9% ÷ 20.5% = 1.19]

If such an improvement becomes possible by midcentury in a mass-produced panel, the SunPower E20's 327 W rated dc power would rise to 389 Wdc. [327 W × 1.19 = 389 Wdc] This is the justification the Roadmap authors use to round up the panel's nameplate capacity to 390 W (see para. 3 of their notes below Table S16).

However, if the US Roadmap actually does get built over the next few decades, it will of course be with existing or similar panel technology, rather than the rosy numbers the Roadmap authors project for midcentury. Therefore, we will first calculate the volume of solar panel waste based on existing PV technology. We will then calculate the difference in waste, should panel performance improve by 19% during the buildout.

The SunPower spec sheet linked above explains that the model E20-327 produces 327 watts of dc power under laboratory Standard Test Conditions, or STC. (See footnote 11 on page 2.)

However, under actual operating conditions, in a solar field or on a rooftop, denoted as PTC (PV USA Test Conditions), the output power of the E20-327 panel decreases to 301.4 Wdc:
Solar Design Tool.com – SunPower SPR-E20-327 spec sheet
http://www.solardesigntool.com/components/module-panel-solar/SunPower/2494/SPR-E20-327/specification-data-sheet.html

People often take PTC to mean Practical Test Conditions. The term implies a derating factor of 0.922, which is considered by many people to be an industry standard: (301.4 W ÷ 327W = 0.922)

For calculation purposes, and to be more than fair, we assume a dc-to-ac conversion efficiency of 85%, though some authorities believe this value is too optimistic. See:

US.sunpower.com – Understanding solar system ratings

https://us.SunPower.com/sites/default/files/media-library/white-papers/wp-understanding-solar-system-ratings.pdf

Here's where things get a little mathy:

The ac output power produced by an E20-327 panel will be 85% of 301.4 Wdc = 256 Wac, under actual outdoor conditions. Therefore, the E20 panel's ac power density is:

256 Wac ÷ 1.629 668 m². = 157 Wac /m².

Ac power density is the most useful metric for grid-scale electric power discussions.

For discussion purposes, we also accepted the Roadmap's optimistic midrange life expectancy of 48.5 years for PV panels. (See p. 41, frame 54, Table S17, row Utility-Scale PV of the Jacobson link above.)

What follows are the calculations for the main issue at hand:

"How Much Waste for how much energy?"

We will examine two scenarios here. The first one is:

1. Premature plant closure after 28 years

To compare PV panel waste to Connecticut Yankee's used fuel, we will imagine a large PV solar farm operating for 48.5 years (the full lifecycle of the panels) and producing the same amount of electric energy (110 billion kWh) as Connecticut Yankee produced in 28 years.

To produce 110 billion kWh over 48.5 years, the annual energy production of the solar farm must be 110E9 kWh ÷ 48.5 yr = 2.27E9 kWh /yr.

For this amount of annual energy output, the average power output of our solar farm must be 2.27E9 kWh ÷ 8760 hours /yr = 259E6 watts, average. That is, an average output of 259 MWac.

By concentrating US solar farms in sunny locales, the annual average capacity factor for US solar has risen to about 25%:

24.6% CF for 2021:

EIA.gov – Electric power monthly table 6.07.B

https://www.eia.gov/electricity/monthly/epm_table_grapher.php

24.7% CF for 2014 – 2017:

Statista.com – Solar photovoltaic capacity factors in the United States 2014 – 2017

https://www.statista.com/statistics/1019796/solar-pv-capacity-factors-us-by-state/

With a CF of 25%, our farm's maximum generating capacity must be four times greater than this amount, or 259 MWac ÷ 0.25 = 1,036 MWp ac. This would be called a one-gigawatt solar farm. With an installed power capacity of 1.036 GW, it would be one of the ten largest solar installations in the world.

With a power density of 157 Wac /m^2 (see above), the total panel surface area would be 1,036 MWac peak ÷ 157 Wac /m^2 = 6.60 million square meters of panel area.

From the SunPower Spec Sheet linked above, on page 2 at bottom right, an E20-327 panel has a frame thickness of 46 mm, or 0.046 meter. With the panels packed tightly together and stacked, the overall waste volume from this farm, after 48.5 years of production, would be 6.60 million m^2 × 0.046 m = **about 303,000 cubic meters of waste**.

The 70 ft × 228 ft storage pad at the Connecticut nuclear facility has area of 1,480 m^2 (21.3 m × 69.5 m = 1,480 m^2). To store the worn-out panels on the same-sized pad, we would have a stack of junk over 200 meters high: 303,000 m^3 ÷ 1480 m^2 = 204 meters, or **670 feet high** on the same sized storage pad. That is, the pile of panel waste would be **34x taller** than the fuel storage casks (204 ÷ 6 = 34).

With the Roadmap's optimistic efficiency improvement of 19% by midcentury, the solar panel waste pile would shrink to "only" 562 feet high. [670 ft ÷ 1.19 = 562 ft.], or "just" 28x as high.

Regarding the secondary issue:

"How Much Land for how much energy?"

Let's examine this question in light of the plant's premature closure at 28 years.

The outsized land requirements for a grid-scale solar farm can be a delicate issue when seeking public approval and permitting. In the Roadmap link above, refer to p. 51, frame 64, Table S22. In the row labeled "Solar PV Plant", the area given for solar density is 12,220 m^2 /MWdc. (Note: This is expressed in dc units.) This is more commonly expressed as 0.012 220 km^2 /MWdc (Also note: 1 km^2 = 1 million m^2)

As we mentioned, the SunPower E20-327 panel used in the Roadmap has an in-field ac power density of 157 Wac /m^2. To know the panel's land density in terms of more meaningful ac units (since the electricity will be converted to ac before use), its dc land density of 0.0122 km^2 /MWdc must be multiplied by the ratio of its rated dc power output to its actual installed ac peak power output.

According to the SunPower link above, the panel's STC-rated dc output is 327 Wdc, and the panel's area is 1.629 668 m^2. Therefore, the SunPower E20-327 panel's dc power density is 327 Wdc ÷ 1.629 668 m^2 = 200.7 Wdc /m^2. So its ratio of dc power density to ac power density is 200.7 Wdc ÷157 Wac = 1.28.

The Roadmap Table S22 cites a land density value of 0.0122 km2 /MWdc. Multiplying by the 1.28 ratio gives 0.0156 km2 /MWac for utility PV solar's land density, expressed in ac units.

Therefore, this one-gigawatt solar farm would require a land area of 1,036 MWp ac × 0. 0156 km^2 /MWac = 16.2 km^2 of land. Or: 16.2 km^2 ÷ 2.59 km^2 per square mile = 6.3 square miles, or **about 4,000 acres**. Compare this to Connecticut Yankee's land area of approximately 10 acres.

Conclusions: **Solar needs 400X the land, and would produce a waste pile, on the same-sized storage pad, about 34 times the height of the dry casks.**

2. Licensed 40-year Nuclear Operation

The Connecticut Yankee NPP had a 600-MW electric generating capacity, and operated at an average capacity factor CF of 74.8% over 28 years, producing 110 billion kWh.

CF = actual electric energy divided by potential electric generation. So, CF = 110 billion kWh ÷ (600 MW X 8760 hours /year X 28 years) = 110E9 kWh ÷ 147E9 kWh = 0.748, or 74.8%

If it had operated for its entire 40-year design lifetime at that same 74.8% capacity factor, it would have produced 600 MW X 0.748 CF X 8760 hr /yr X 40 yr = 157 billion kWh. That is, it would have produced a greater amount of electric energy by a factor of 1.43. (157B kWh ÷ 110B kWh = 1.43) The plant's operating life would have increased by the same factor: 40 yr ÷ 28 yr = **1.43**.

For a larger solar farm, also lasting 48.5 years, to match that greater amount of energy production with the same Sunpower model E20-327 PV panel, the second solar farm's annual energy production would have to be 157E9 kWh ÷ 48.5 yr = 3.23E9 kWh /year. In that case, the farm's average annual power delivery would be 3.23E9 kWh ÷ 8760 hours /yr = 369E6 watts, average. That is, an average output of 369 MWac.

Assuming a 25% solar capacity factor as before, this larger farm's peak ac generating capacity must be 369 MW ÷ 0.25 = 1,470 MWac. This is almost the size of the 1,547 MW Tengger Desert Park in Ningxia, China, the world's 5th largest solar installation.

With ac power density of 157 Wac /square meter, the solar site's total panel area would have to be 1,470 MWac ÷ 157 Wac /m^2 = 9.36 million square meters of panel area. This is greater than the former farm's 6.60 million m^2 by the same factor of 1.43. (9.36 M ÷ 6.60 M = about 1.43).

With the E20-327 panel's frame thickness of 46 mm or 0.046 meter, the overall panel volume would be 9.36 million m² X 0.046 m = **431,000 cubic meters**. This becomes waste volume at the end of 48.5 years.

Lengthening the concrete pad by a factor of 1.43 to accommodate the greater number of fuel storage casks would give 1.43 X 1480 m² = 2,116 m² of pad area. This increases the number of used fuel casks from 40 to 57. With the same three reactor casks for reactor parts, we get a total of **60 casks.**

Keeping the pad the same width, our pad would extend from its original 228 ft × 70 ft (69.5 m x 21.3 m) to **326 ft x 70 ft** (99.38 m x 21.3 m). The solar waste pile would be longer as well, while its height would be the same, at 204 meters. (431,000 m³ ÷ 2116 m² = 204 meters).

With Jacobson's optimistic solar-to-electric conversion efficiency improvement of 19%, the solar waste pile shrinks to 204 m ÷ 1.19 = 171 meters, or 562 feet high.

As for our secondary issue:

"How Much Land for how much energy?"

The 1,470 MW solar farm would need to be enlarged by 1.43 times. Referring to the land density value of 0.0156 km2 /MWac shown in the former calculation, gives 1,470 MWac X 0.0156 km² / MWac = 22.93 km². This is greater by a factor of about 1.43. (22.93 km² ÷ 16.2 km² = 1.42)

Therefore land requirement would be 22.93 km² ÷ 2.59 km² per square mile = 8.9 square miles, or about **5,700** acres. Compare this to Connecticut Yankee's 10 acres.

Finally:

"What if Connecticut Yankee was a fast reactor?"

The answers are simple: Since a light-water reactor only burns 3% of its fuel, a fast reactor can theoretically consume all of its fuel. However, since no process is perfect, we estimate a conservative 80% burnup with a fast reactor, or about 27x the efficiency of an LWR. This reflects the 2,000:1 EROI of a fast reactor compared to the 75:1 EROI of an LWR (2,000 ÷ 75 = 26.6). Matching fast-reactor production with solar would mean 27x the land and 27x the waste. This is depicted in the second drawing in Fig. 68 of Chapter 18.

The land calculation would then be 27 x 5,700 acres = about 154,000 acres (240 square miles), compared to Connecticut Yankee's 10 acres.

The solar waste volume would be 27 x 431,000 cubic meters = about **11,600,000** cubic meters. Another way of looking at it, 11,600,000 cubic meters is **the volume of about 12 Empire State Buildings—compared to just 60 dry storage casks on a concrete pad.**

This is what energy density is all about.

HYDRAULIC CONDUCTIVITY
(Chapter 18, endnote 8)

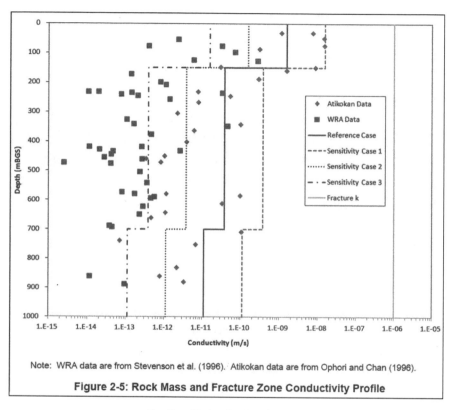

Note: WRA data are from Stevenson et al. (1996). Atikokan data are from Ophori and Chan (1996).

Figure 2-5: Rock Mass and Fracture Zone Conductivity Profile

Fig. S-9: Hydraulic Conductivity

The original color graph is here:
NWMO.ca – Postclosure safety assessment used fuel repository crystalline rock
https://www.nwmo.ca/~/media/Site/Reports/2018/01/19/15/54/ NWMO-TR_2017_02_Sixth-Case-Study-Report.ashx (frame 81)

The red squares and the blue diamonds in the original graph denote rock samples taken at different depths below Canada's Whiteshell Research Area (red squares of WRA data) and below their Atitokan research area (blue diamonds), respectively. We'll focus on the solid vertical Reference Case line, and the blue diamond to the left of the line at 730 meters. The water migration time from a depth of 730 meters up to the surface can be calculated using the solid Reference Case line.

The x-axis shows rates of conductivity (water migration) in fractions of a meter per second, with the speed of travel increasing to the right of the graph. The x-axis intersection conductivity rate of 1E-11 per second for the Reference Case at depths below 700 meters translates to 0.000 000 000 01 meters per second. With 31.55E6 seconds in a year (31,550,000 seconds), this amounts to one meter of travel every 3,170 years:

1.E-11 × 31.55E6 = 315.5E-6 meters of hydraulic conductivity per year. So 1 ÷ 315.5E-6 meters /yr = 3,170 yr /m.

At this rate, it would take 91,500 years for any material released from a failed cask buried at 730 meters below the Atikokan site to rise 30 meters to a depth of 700 meters (30 m × 3,170 = 95,100 yr).

From 700 meters to 150 meters, conductivity rates speed up (note how the Reference Case line jogs to the right at 700 meters.) Repeat the calculation for the reference line segment from 700 m up to 150 m depth, a 550-meter rise. The projected conductivity rate of this segment is approximately 3.5E-11* m /sec, which is 1.1E-3 m /yr. Divide 550 m by 1.1E-3 m /yr to get 500,000 years.

Repeat again for the final 150 meters at the top of the violet path. By projection to the x-axis, conductivity rate = 2E-9 m /s, or 63E-3 m /yr. So 150 m ÷ 63E-3 m /yr = 2,400 years.

Add the three segments. 95,000 + 5000,000 + 2,400 = about 598,000 years total migration time for water-borne nuclides to reach the surface.

---

* The graph is a logarithmic scale, which could be seen better if it were wider. Then we could see whether the projection line intersected the axis closer to the "3" position or to the "4" position. With this scaling we can only estimate the intersection to be 3.5.

GLOBAL ELECTRICITY DEMAND IN 2050
(Chapter 20, endnote 17)

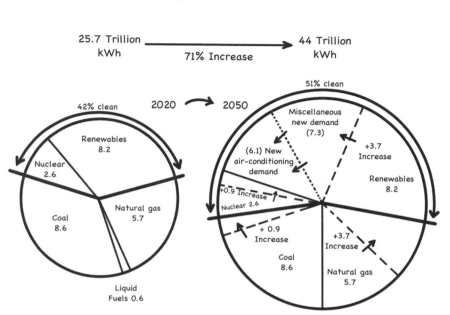

Fig. S-10 Projected Generation to 2050
Source: https://www.weforum.org/agenda/2020/02/
renewable-energy-future-carbon-emissions/

In the World Economic Forum paper (above link), see section "Projected electricity generation worldwide to 2050." (Suggestion: Print the bar graph in landscape view, with the image occupying the entire printable area.)

Measure the y-axis distance from 0 to 50. Get 143 millimeters. The y-axis scale factor is 50 : 143 = 350 terawatt-hours per millimeter, 350 TWh /mm.

For year 2020, measure the entire bar height as 73 mm. This shows global electric energy consumption as 73 mm × 350 TWh / mm = 25,600 TWh /year.

Repeat for 2050, get 125 mm, showing global electric energy consumption projected to be 125 × 350 = 43,800 TWh in year 2050.

So global electricity is projected to increase by 18,200 TWh /yr. [43,800 − 25,600 = 18, 200] This is a 71% increase. [18,200 TWh ÷ 25,600 TWh = 0.71] Note that this supports the low end of the "70% to 80%" that we mentioned in Chapter 19.

Repeat both measurements for the green bar segments, representing renewable sources. For 2020, green = 22.5 mm, or 7900 TWh from RE. [22.5 × 350 = 7900 TWh /yr] For 2050, green = 61.5 mm, or 21,500 TWh from RE. [61.5 × 350 = 21,500 TWh /yr]

So RE electricity is projected to increase by 13,600 TWh /yr. [21,500 − 7900 = 13,600 This is a 172% increase. [13,600 TWh ÷ 7900 TWh = 1.72] This is better thought of as an increase by a factor of 2.72.

The renewables increase can be expressed as a portion of the world's total electric increase. RE increase ÷ total electric increase = 13,600 TWh ÷ 18,200 TWh = 0.75. That is to say, renewable sources are projected to provide 75% of the world's total electric *increase* by year 2050. That's three-fourths.

28% fossil fuel increase:

Repeat the measurements described above for the "liquids" and the "coal" bar segments, representing combined fossil fuels, for years 2020 and 2050.

Get fossil fuel electricity in 2020 of 14,900 TWh, and in 2050 of 19,100 TWh. This is a factor of 1.28, showing a 28% increase. (29,100 ÷ 14,900 = 1.28)

GLOBAL PRIMARY ENERGY IN 2050
(Chap. 22, endnote 14)

The US Energy Information Administration—EIA—projection of global primary energy demand for year 2050 is about 962 exajoules—962E18 J. See (A) below. This is about a 43% increase over the world's 2020 demand. See (B) below. It would be possible to meet this entire midcentury demand with a global fleet of 30,000 large nuclear reactors. See (C) below.

(A) EIA.gov – EIA projects nearly 50% increase in world energy usage by 2050
   https://www.eia.gov/todayinenergy/detail.php?id=41433
(Suggestion: Print the graph Global primary energy consumption by region 2020–2050)
Measure the vertical axis height to obtain 51 mm. The scale factor is 1000 Q / 51 mm = 19.6 Q /mm.
1 Quad (quadrillion BTUs) is equal to 1055E15 joules.
2050 bar height is 46.5 mm. PRI NRG = 46.5 mm × 19.6 Q /mm × 1055E15 J/ Q = 962E18 J.
So 962 EJ in 2050

(B) *Ibid.* https://www.eia.gov/todayinenergy/detail.php
2020 bar height is 32.5 mm. PRI NRG = 32.5 mm × 19.6 Q /mm × 1055E15 J/ Q = 672E18 J.
So 672 EJ in 2020. 672 exajoules.
962 EJ ÷ 672 EJ = 1.43. That is, EIA projects a 43% increase in global PRI NRG for year 2050.

(C) A large light-water reactor (LWR) may have electric power capacity of 1100 MW. It is capable of operating with annual capacity factor of 91%. Its average power output is 1100 MW × 91% = 1000 MWavg.

Its electric energy production will be 1000E6 W × 8760 hours /year
= 8.76E12 watt-hours /year.
8.76E12 watt-hours /yr × 1 J /sec / W × 3600 sec /hour =
31.54E15 J /year
So about 32 petajoules per year, per reactor.
For year 2050: 962E18 J ÷ 32E15 J per reactor = **30,060 1-GW reactors**

Afterword

Dear Reader,

WE'RE CONFIDENT THAT THIS BOOK has provided you with an excellent foundation for understanding our nuclear planet, from the dawn of time until now. Knowledge is power, and with power comes responsibility. So even though you are armed with all this new information, resist the temptation to fire at will. Conversations about nuclear energy can be sensitive, and we need to meet people where they're at before we can ask them to come along with us.

In service to that goal, we've reinvented the ALARA principle. This book teaches us that ALARA is the radiation safety principle of keeping exposures "As Low As Reasonably Achievable." Generation Atomic has taken some artistic license to give it a new meaning:

- **A**sk a question to understand their viewpoint

- **L**isten to their response carefully

- **A**cknowledge their concern respectfully

- **R**eframe the conversation with stories, facts, or metaphors

- **A**sk a follow-up question to check their understanding

We know, it's quite a stretch from the traditional ALARA definition. But desperate times call for desperate acronym reappropriation! When having sensitive conversations about nuclear power, this new ALARA provides a handy mnemonic for maintaining an open, respectful dialogue instead of, well, exposing the other person to excess "zoomies" in the form of heated arguments and zingers.

Hopefully, the radiation safety folks won't revoke our creative license. ;) After all, communication and education are important safety measures, too!

The nuclear story remains unfinished, but you can help write the next chapter through advocacy and education. Generation Atomic is here to help. If you feel the call, we encourage you to join the effort. Stay curious, stay radiant. With an empowered community, we can write an incredible next chapter in nuclear's unfolding story.

Eric G. Meyer
Executive Director

Index